CURIOSITY

Philip Ball is a freelance writer and a consultant editor for *Nature*, where he previously worked as an editor for physical sciences. He writes regularly in the scientific and popular media, and his many books on scientific subjects include *Critical Mass: How One Thing Leads to Another*, which won the 2005 Aventis Prize for Science Books. His latest books include *The Music Instinct, Universe of Stone* and, most recently, *Unnatural: The Heretical Idea of Making People.* Philip obtained a PhD in physics from the University of Bristol.

ALSO BY PHILIP BALL

Designing the Molecular World:
Chemistry at the Frontier
Made to Measure: New Materials for the 21st Century
H₂0: A Biography of Water
The Self-made Tapestry: Pattern Formation in Nature
Bright Earth: The Invention of Colour
Stories of the Invisible: A Guided Tour of Molecules
The Ingredients: A Guided Tour of the Elements
Critical Mass: How One Thing Leads to Another
Elegant Solutions: Ten Beautiful Experiments in Chemistry
The Devil's Doctor: Paracelsus and the World of
Renaissance Magic and Science
Nature's Patterns: A Tapestry in Three Parts
Universe of Stone: Chartres Cathedral and the
Triumph of the Medieval Mind
The Sun and Moon Corrupted
The Music Instinct: How Music Works and
Why We Can't Do Without It
Unnatural: The Heretical Idea of Making People

PHILIP BALL

Curiosity

How Science Became Interested in
Everything

VINTAGE BOOKS
London

Published by Vintage 2013

2 4 6 8 10 9 7 5 3 1

First published in Great Britain in 2012 by
The Bodley Head

Vintage
Random House, 20 Vauxhall Bridge Road,
London SW1V 2SA

www.vintage-books.co.uk

Addresses for companies within The Random House Group Limited
can be found at: www.randomhouse.co.uk/offices.htm

The Random House Group Limited Reg. No. 954009

A CIP catalogue record for this book
is available from the British Library

ISBN 9780099554271

The Random House Group Limited supports the Forest
Stewardship Council® (FSC®), the leading international
forest-certification organisation. Our books carrying the FSC
label are printed on FSC®-certified paper. FSC is the only
forest-certification scheme supported by the leading
environmental organisations, including Greenpeace.
Our paper procurement policy can be found at:
www.randomhouse.co.uk/environment

Typeset in Dante MT by Palimpsest Book Production Limited,
Falkirk, Scotland

Printed and bound by CPI Group (UK) Ltd, Croydon, CR0 4YY

Contents

Preface

While I was talking about this book with the literary scholar Mary Baine Campbell, we considered the idea that curiosity could become pathological. Isn't there something problematic about responding to an analysis of, let's say, domestic accounting during the Thirty Years War or the detailed manoeuvres of a gastric enzyme not with glazed eyes but with a breathless 'But that's so interesting!'? Recalibration of one's curiosity threshold is a necessary preparation for most PhD students, but in the wider world mightn't there be something ill-disciplined, even improper, about a voracious curiosity that permits nothing to be too trivial or obscure?

It was a sobering thought, I suspect for the both of us, as we acknowledged what seemed to me a guilty complicity. Was there after all something in the old accusation that it is weak-willed to succumb to the wiles of curiosity? But the problem of our times – and also its great good fortune – is that temptation is everywhere. Not only is it now acceptable to be curious – and this book is largely about how that came to be the case – but it is easier than ever, because of the knee-trembling quantity of information we have at our fingertips. We no longer have to seek out this stuff in dusty vaults and ancient libraries; it sits waiting for us at our desk, humming gently, perhaps even in scanned, gorgeously browning facsimile. More, we carry it everywhere in our bags and pockets. Yes, of course this is all just so much data, amorphous and indeed meaningless unless we have some notion of how to select and organize and filter. And yes, of course it is in some ways a mere side effect of, or accompaniment to, new opportunities for turning our back on curiosity altogether and immersing ourselves in the empty immediacy of a virtual now, of chatter and gossip and a numbing profusion of 'choice'.

But is all this, too, really so new? It has always been a complaint levelled at curiosity that it is the enemy of productivity, an unwelcome distraction from our daily duties. Meanwhile, the Enlightenment's mockers of curiosity were, as we'll see, often not utilitarian Gradgrinds but gossipy, solipsistic wits and libertines. And a surfeit of information has always given cause for grumbling. Alexander Pope felt that the printing press, 'a scourge for the sins of the learned', would lead to 'a deluge of Authors [that] covered the land'.

The relationship between access to information and curiosity about it has, I think, yet to be well explored. But it is clear that the first 'professors of curiosity' who flourished in the century of Pope's birth had to work tremendously hard to get their knowledge, and curiosity was, before profit or fame or reputation, their most significant motivation. This situation has been rightly celebrated, but rarely has it been examined or explained. Mary is one of the scholars who have begun that exploration, and as such, is one of those to whom I owe an immense debt for this book. All the more so because she kindly agreed to read the manuscript and offered insightful and important views on it. For similar acts of generosity I offer sincere thanks to Brian Ford, Michael Hunter, Neil Kenny and Catherine Wilson.

It has been my great pleasure to have published several books under the editorship of Will Sulkin at Bodley Head, who I (and many others) shall miss greatly after his retirement in 2012. Will's enthusiasm, erudition and passion about writing and ideas has been a vital source of encouragement. I am consoled to know that I will still enjoy the thoughtful and diligent editorial support of Jörg Hensgen, and of his colleagues Kay Peddle and Hannah Ross. David Milner has done another splendid job of copy-editing the text. And my good fortune to have Clare Alexander as my agent is one of those things that I always find a little astonishing. As ever, the greatest consolation, support and inspiration comes from my family, within which these days I can delight in watching curiosity bloom in its purest form.

Philip Ball
January 2012

1 Old Questions

To whatever object the eye first turns, the same is a wonder and full of wonder, if only we examine it for a little.

Giovanni Dondi (*c*.1382)

The important thing is not to stop questioning . . . Never lose a holy curiosity.

Albert Einstein (1955)

'The Large Hadron Collider is a discovery machine. Its research program has the potential to change our view of the universe profoundly, continuing a tradition of human curiosity that's as old as mankind itself.' This is Robert Aymar, former director general of CERN, the European centre for particle physics in Geneva, explaining why the collider has been constructed.

The LHC is the world's most powerful particle accelerator. It uses electromagnetic fields to accelerate protons to 99.999999 per cent of the speed of light, so fast that they travel around the entire 27-km circumference of the tunnels in less than a ten-thousandth of a second. Then the protons are smashed into one another in collisions energetic enough to recreate the conditions in the first instants of the Big Bang from which the universe began. The hope is that this will spawn particles never before seen, which will help us to understand some deep questions about the nature of matter, such as why certain types of particle have mass.

At a cost of $6 billion and twenty-five years in the planning, the LHC is as big as Big Science gets. Why go to all this effort and expense?

Aymar appeals to the role of human curiosity. He implies that this is just the latest development in an unbroken history of curiosity about nature that stretches back to our own origins. It is, he says, an extension of what we have always done.

Perhaps it is therefore no surprise that, as the LHC's inaugural run in late 2008 approached, the media became fixated on ludicrous fears that the experiment would destroy the world, if not the universe. For tradition teaches us that curiosity – especially curiosity about the Creation – cannot be indulged with impunity. Even if this latest threat of apocalypse was more a public plaything than a genuine cause for dread, it showed that we have still not quite made our peace with curiosity.

But there's more to the LHC than a desire for knowledge. It seeks justification in practical spin-off benefits. 'We are constantly being told that we live in a competitive world in which innovation is the main driver towards growth and prosperity', says Aymar:

> History teaches us that big jumps in human innovation come about mainly as a basic result of pure curiosity. [Michael] Faraday's experiments on electricity, for example, were driven by curiosity but eventually brought us electric light. No amount of R&D on the candle could ever have done that.

The underlying assumption here is that voiced by Stephen Hawking in support of the LHC: 'modern society is based on advances in pure science that were not foreseen to lead to practical applications'. Leaving aside the fact that this presents a distorted view of the symbiosis (in fact it is an intimate merging) between science and technology, it is striking how the narrative that Aymar insists on here about curiosity, science and technology contrasts with the defence of curiosity offered by the French philosopher Michel Foucault:

> Curiosity is a vice that has been stigmatized in turn by Christianity, by philosophy, and even by a certain conception of science. Curiosity, futility. The word, however, pleases me. To me it suggests something altogether different: it evokes 'concern'; it evokes the care one takes for what exists and could exist; a readiness to find strange and singular what surrounds us; a certain readiness to break up our familiarities and to regard otherwise the same things; a fervor to grasp what is

happening and what passes; a casualness in regard to the traditional hierarchies of the important and the essential.

Foucault seems to wish to be enchanted and beguiled by curiosity, to be awakened to wonder, to feel a hunger for experiences strange and new that will break down old ideas and distinctions. Here curiosity is a radical force. In science, on the other hand, curiosity is more often enlisted in the name of taming the world – it is a compulsion to *understand*. The curiosity (if that's what it is) motivating the Large Hadron Collider is likely to lead to *new* hierarchies in our conception of matter and space, while this and other 'curiosity-driven' research is advocated as a source of unforeseen practical bonuses. This is very much the sober view espoused by Francis Bacon in the early seventeenth century: curiosity as an engine of knowledge and power.

Why has curiosity come to stand for these rather different agendas? Can we reconcile them? Is either borne out by history? Those are some of the questions I seek to examine in this book.

The turning point in Western attitudes to curiosity occurred in the seventeenth century, which began with an essentially medieval outlook and ended looking like the first draft of the modern age. This change can be seen in dramatic fashion simply by charting the use of the word 'curiosity' (and its cognates) in the European literature of the period, as historian Neil Kenny has done. The frequency of usage varies little from the mid-sixteenth century until 1650, when it takes off suddenly, peaking in 1700 but remaining high thereafter.

The transformations in thought, particularly in the natural sciences, that characterize the span of nigh on a hundred years between the

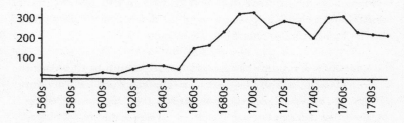

The number of uses of 'curious', 'curiosity' and cognate words in books of the 1500s to the 1700s rises dramatically in the mid-seventeenth century. Note that this graph does not allow for variations in the total number of books published each year.

death of Elizabeth I (1603) and the coronation of Queen Anne (1702) have often been called the Scientific Revolution. Its stories are familiar: Galileo validating Copernicus's sun-centred universe; Isaac Newton explaining the motions of the celestial bodies with his theory of gravity and outlining the basic laws of all mechanical motion; the Anglo-Irish scientist Robert Boyle tolling the death knell of alchemy; the endlessly inventive Robert Hooke exploring the microscopic world and the Dutch cloth merchant Antony van Leeuwenhoek discovering microbes wriggling therein. The conventional narrative identifies the *scientific method* itself as the key innovation of the age: a logical system for investigating and interpreting all of nature.

This cosy tale tends to imply that natural philosophers simply got better at asking and answering questions, forsaking tautological or mystical reasoning in favour of explanations that invoked cause-and-effect mechanisms, amenable to measurement and testing. There is some truth in that, but it will not get us very far in understanding what these proto-scientists thought and why. Least of all does it justify the conventional narrative in which science merely expands to crowd out superstition. The now well-known interest of Newton and Boyle in alchemy is merely one manifestation of the true origins of the new philosophy in a mode of thought that arose largely outside the formal university system. To the new philosophers, the natural world was replete with secrets that they must hunt down diligently with an experimental approach that was closely allied to the tradition of natural magic. This 'hunt' was to be engaged by international, sometimes occult fraternities of virtuoso-scientists, themselves a construct of utopian visions of which Francis Bacon's *New Atlantis* (1624), the foundational text of the Royal Society of London, was the most influential.

Underpinning all this was a profound change in the nature of the questions one might ask. Nothing was too mean or trivial to be neglected, for as Boyle said, it was all God's work and therefore worthy of attention. A glance at Boyle's own notebooks reveals the dizzying consequence. His lists of 'things to be remembered' suggest that he would, if he could, have made an exhaustive inventory of all that existed or occurred under the sun: 'Remember', he wrote,

the use of a Coach
the eyes of Puppys newly whelpt
the Feathers, Claws and Beaks of birds yet in the shell
the Gunpowder whole and ground
Insects and other Creatures that lye as it were dead in the Winter
Moses's Serpent and the Transmuted water
that Beauty do's not make the Parts, but result from them as do also
    Health, Harmony, Symmetry
that Internal Forms may be but lasting Dispositions to be wrought upon
    by External Objects
the seal'd weather glass &c. and the consequences of such thing
Monsters, and the longings and frights of teeming women
the unskilfull Restitution in Springs made by hammering &c.
to breake a Glass buble in a Barometre.

Popular accounts of the Scientific Revolution rarely stop to think how odd this is. While there was a long, if controversial, history of asking questions about nature and human activity, such enquiries had tended to limit themselves to what was obviously useful, or important, or universal: why plants grow, why winds blow, why we get sick, how the stars and planets progress across the sky. But suddenly, the slightest blemish seen on the surface of a distant planet might spark earnest and learned debate, or the question why fleas can jump so high, or why concentric coloured rings could be seen in mineral flakes under the microscope. The early meetings of the Royal Society embraced a phantasmagoria of phenomena and inventions – some evidently valuable, such as watches for helping chart longitude, others sounding like superstitious or fantastical rumour, such as monstrous births and weird lights in the sky.

And this glimpse of Boyle's restless mind hints at the problem of such eclecticism: how does one make sense of it all? If you can ask anything, then there is no end to the questions. How do you organize all the observations? How do you decide which phenomena are important and which are frivolous? Is anything truly frivolous? But then the task of science is hopeless, because you must always suspect that the next question will challenge your current theory.

Because my survey of the roles of curiosity in science is located almost wholly in the seventeenth century, it might seem unlikely that it will have much to say about the particle physics of the twenty-first.

On the contrary, I contend that we can only truly understand what today's scientists – what people like Aymar – say and believe about curiosity if we examine this critical period in which it first came to be explicitly claimed for the purposes of science. It was in the seventeenth century that science first emerged as a modernizing force and altered both our conception of the world and our ability to manipulate it. Pronouncements like those used to justify the LHC are very much predicated on a narrative that roots itself in the conventional, triumphal picture of this 'Scientific Revolution'.

Historians of science tend now to look askance at the bald assertion of a Scientific Revolution. Or rather, they typically adopt the view eloquently expressed by Steven Shapin in his 1996 book on the subject, which begins 'There was no such thing as the Scientific Revolution, and this is a book about it.' Which is to say, perhaps, that the traditional accounts give us the right facts but connect them together in a warped manner. I agree with Shapin that something changed profoundly in natural philosophy in the seventeenth century, but that it prejudices an examination of this transformation to imagine that we know what we are talking about even in using the words 'scientific' and 'revolution'. I argue that one better way to understand this critical period is to look at changes in the meanings and the values it attached to the notion of curiosity. In homage to one of the most perceptive historians of this era, I suggest that there is no such thing as what Robert Aymar calls a 'tradition of human curiosity that's as old as mankind itself' – and this is a book about it.

This singular passion

For the seventeenth-century English philosopher Thomas Hobbes, curiosity was one of the defining characteristics of humankind (and as such, a good thing):

> Desire, to know why, and how [is called] CURIOSITY; such as in no living creature but *Man*; so that Man is distinguished, not onely by his Reason; but also by this singular Passion from the other *Animals*.

It was curiosity, said Hobbes, that motivates 'the continuall and indefatigable generation of knowledge'. It was 'a more than ordinary curiosity' about a particular optical phenomenon that made Isaac Newton determined to discover 'from whence it might proceed' – to search for the principles behind it. Unlike carnal passion, said Hobbes, curiosity was not expended with 'short vehemence' but was inexhaustible – as his one-time mentor Francis Bacon said, 'of knowledge there is no satiety'.

But curiosity does not mean and has never meant just a single thing. Even if we accept the modern definition of 'eagerness to know or learn something',* there are many ways to be curious. One can flit in gadfly manner from one question to another, acquiring little bits of knowledge without ever allowing them to cohere and mature into a real understanding of the world's mechanisms. One can store up snippets of information like a miser, never putting them to good use. One can pose questions idly or flippantly, with no plan for coherent enquiry into nature. One can be curious about matters that really are none of one's business, such as the sexual habits of one's neighbours. But one can also seek knowledge with serious and considered intent – and may then do so either in the manner of Isaiah Berlin's fox who would know many little things, or as the hedgehog who knows a single thing profoundly. One can be curious obsessively, or passionately, or soberly, or with clinical detachment.

But this is only to scratch the surface of what the word could connote in earlier times. 'Curiosity', says Neil Kenny, 'was understood in so many ways that it had no ineliminable core that always characterized it.' People could be curious, but so could objects: it was an attribute as well as a state of mind. If we call something curious as Alice did ('curiouser and curiouser'), we generally mean to allude to some quality of strangeness in it. This sense is implied in the cult of cabinets of curiosities (explored in Chapter 3), where the 'curios' may be objects that are unusual and intricate but which offer little purchase for the enquiring mind that wants to understand and explain the world. To call an object curious could mean that it was rare, exotic, elegant, beautiful, collectable, valuable, small, hidden, useless, expensive – but conversely, in certain contexts, common, useful or cheap. At any rate, the curious object was one towards which curiosity might properly be directed: to call it 'curious'

---

* When I shall use the word, it is generally in this provisional sense unless the meaning is explicitly refined or modified by the context.

was not simply to label it as singular, odd or worthy, but to say 'Look at this – and look closely.'

Small wonder, then, that it is often impossible to say whether a particular writer is for or against curiosity. At least three of the key figures in this story – Francis Bacon, René Descartes and Galileo – use 'curiosity' to mean different things at different times. Yet when writers, philosophers and moralists of all ages have pronounced on curiosity, they have typically had only one or some of these many meanings in mind, and their judgements both good and bad can therefore hardly be said to speak to the full range of what it means to be curious. Lorraine Daston and Katharine Park, two of the foremost 'historians of curiosity' (a small but perceptive bunch), say of the curiosity praised by Hobbes and that condemned by medieval theologians that, while they share some kinship, they are 'not of the same emotional species'.

'Curious' derives ultimately from the Latin *cura*, meaning care, and until at least the seventeenth century a 'curious' person could simply refer to one who undertook investigations with diligence and caution.* When Robert Hooke said of the blue fly under his microscope that 'the hinder part of its body is cover'd with a most curious blue shining armour', he meant that it appeared to be carefully crafted. From *cura* also comes the curator, the person who looks after obligations or objects with care, whose modern incarnation as a collator and administrator of collections in a museum or gallery stems directly from the tradition of collecting that spawned the cabinets of curiosities.

Incurious ancients

Francis Bacon described Pliny's encyclopaedic work *Natural History* as 'fraught with much fabulous matter, a great part not only untried, but notoriously untrue, to the great derogation of the credit of natural history with the grave and sober kind of wits'. While he might equally have been describing almost any of the encyclopaedias of the natural world produced between the age of Imperial Rome and the Renaissance, he has a point. Pliny, a Roman administrator of the first century AD, had a credulous, sprawling, magpie-like infatuation with the most peculiar and unlikely of stories. Cut open a hairy phalangium spider

---

* It is telling, however, that for the Romans, *curiosi* were spies or secret informers.

and take out the two little grubs in its belly, put them in a pouch made from the leather of a red deer, tie this to the arm of a woman before sunrise, and she will then be incapable of conceiving a child. Rubbing mouse dung on a bald head will replenish its hair. A wayfaring man with a rod made of myrtle will never tire. The bodies of drowned men always float face upwards, but those of women downwards, 'as if Nature had provided to save their honesty and cover their shame'. And so on. To Pliny we owe the notion that elephants 'cannot abide a mouse' and that ostriches 'hide' by burying their heads when frightened. (Less well known and perhaps more contentious is his assertion that the cornered male beaver will bite off its own testicles.)

This is now part of the immense charm of *Natural History*, every page offering wonders so bizarre and whimsical that their invention would vex the imagination. This tradition was still thriving in the semi-fantastical bestiaries and accounts of faraway marvels during the late Renaissance, such as *La minera del mondo* (*Riches of the World*) by the Italian Giovanni Maria Bonardo, which attests that 'On the top of Mount Palombra there is a wonderful fountain and those that drink of its waters will never feel pain of any sort for as long as they live and will also preserve their youthful appearance for ever.' It was no coincidence that one of the most popular English translations of Pliny's books, by the scholar Philemon Holland, was published in 1601 during the Elizabethan infatuation with curiosities of nature. Holland's version was probably read by Shakespeare, who seems to allude to some of its strange claims in *Othello* (1604). John Donne, a connoisseur of occult philosophies, refers to the elephant and the mouse in *Progress of the Soul* (1601).

But Pliny intended that his miscellany of marvels should be seen as a serious guide to all that was known to humankind. '[I] take it upon me', he wrote, 'to speak of everything, and to gather as it were a complete body of arts and sciences (which the Greeks call *enkyklapaideios*) that are either altogether unknown or become doubtful, through the overmuch curiosity of fine wits.' Other men may have written some of these things before, Pliny admitted, but they had done so obscurely or at tedious length. His was an unashamedly populist project: to present all knowledge in a convenient, easily digestible form. To gather the material, he claimed to have read 2,000 books by a hundred different writers, and from them to have extracted '20,000 things, all worthy of regard and consideration'.

From the existence of such compendia, one might be inclined to infer an ancient tradition of curiosity. But it was not exactly that; in some ways it was quite the opposite. Pliny's compatriots were notoriously indifferent to anything resembling a scientific study of the world. They were content to take instruction from the ancient Greeks, who, the Romans believed, had already discovered everything (even if much of it was now lost). 'There was nothing left untried or unattempted by [the ancients],' wrote Pliny, 'nothing kept secret, nothing which they wished to be of no benefit to posterity.' The medieval encyclopaedias that are so obviously indebted to Pliny – the bestiaries, lapidaries, herbaria – never sought to explain or understand what they reported. They were a strange mixture of sensationalist display and down-to-earth guidebook. And since in the Middle Ages truth was often expressed and perceived in symbolic rather than literal terms, these collections might also supply a palette of moral metaphors. It was for this very reason that a need for descriptive accuracy was not strongly felt.

For the Greeks, curiosity was not even a clearly articulated concept. To the extent that it was acknowledged at all, it stands in contrast to its mercurial sibling, wonder. Aristotle believed that all humans naturally desire knowledge, but he felt that curiosity (*periergia*) had little role to play in philosophy. It was a kind of aimless, witless tendency to pry into things that didn't concern us. Wonder (*thauma*) was far more significant, the true root of enquiry: 'It is owing to their wonder', he wrote, 'that men both now begin and at first began to philosophize.'* Daston and Park argue that until the seventeenth century, wonder was esteemed while curiosity was reviled.

It was after all the ultimate cause of the ills of the world, unleashed from the jar by the meddlesome Pandora. 'It was not her cunning or wiliness that prompted her to open the jar', says the classicist Willem Jacob Verdenius in his commentary on Hesiod's version of the myth, 'but her curiosity.' In his *Moralia*, Plutarch considers curiosity the vice of those given to snooping and prying into the affairs of others: the kind of busybody known in Greek as a *polypragmon*.† It is true that

---

* The aphorism that philosophy, science or knowledge 'begins and ends in wonder' has been attributed to many others since, including Samuel Coleridge and Alfred North Whitehead.

† Both *periergia* and *polypragmosyne* were usually translated into Latin as *curiositas*.

Plutarch recommends that the *polypragmon* might cure himself by directing his attentions instead to questions about nature – why does the moon wax and wane? Why are fruits of different shapes? – and insists that 'these truly are the secrets of nature, neither is she offended and displeased with those who can find them out'. But the overwhelming sense in the classical world is that the curious person is a meddler and a nuisance or hazard to society.

Against curiosity

In early Christianity it was worse than that. Now curiosity was not merely frowned upon but condemned as a sin. 'We want no curious disputation after possessing Christ Jesus,' wrote the second-century Christian apologist Tertullian, 'no inquisition after enjoying the Gospel.' The Bible told us all we needed – and should expect – to know.

In Christian Scripture, the dangers of curiosity were apparent from the outset. It might seem surprising, given the medieval hierarchy of nature in which humankind is unambiguously at the apex, that Adam and Eve were the last living beings to be created in Genesis – they were made only after God had filled the seas with fish and the skies with birds. But the Roman writer Lactantius in the third century AD had an explanation for this: being last on the scene, Adam would not see how it had all been done. (His descendants are evidently now striving to evade that divine precaution by running the tape of Creation once more.)

That some knowledge was forbidden to humankind is of course central to the Christian Creation myth: this is the basis of the Fall. 'When you eat of it your eyes will be opened and you will be like God', the serpent tells Eve of the fruit on the tree of knowledge. The transgressive aspect of curiosity is an insistent theme in Christian theology. Time and again the student of the Bible is warned to respect the limits of enquiry and to be wary of too much learning. 'The secret things belong to the Lord our God', proclaims Deuteronomy. Solomon (if it was he who wrote Ecclesiastes) cautions that:

> with much wisdom comes much sorrow;
> the more knowledge, the more grief.

And then again, he says:

> Do not pry into things too hard for you
> Or examine what is beyond your reach . . .
> What the Lord keeps secret is no concern of yours;
> Do not busy yourself with matters that are beyond you.

Or, as the King James version has it:

> Be not curious in unnecessary matters:
> For more things are shewed unto thee than men understand.

St Paul was considered to have echoed this sentiment in the admonition 'Seek not to know high things.' The fact that he did not actually
write this at all speaks volumes in itself, suggesting that the mistranslation fitted with prevailing prejudice. In the fourth-century Vulgate
Bible, the phrase is rendered as *noli altum sapere, sed time*, for which a
fair translation is 'do not become proud, but stand in awe'. It was a
rebuke against false claims to moral wisdom; but *sapere*, to be wise,
became interpreted as something closer to *scire*, secular knowledge
– the root of 'science'. 'Seek not to know high things' is then the
result, as it appears in a late fifteenth-century translation of the Bible
into Italian, by which time Paul's words had become irretrievably
associated with a condemnation of curiosity. 'Do not take pride in
the arts or sciences,' wrote Thomas à Kempis in the fifteenth century,
'rather, fear what has been told to you.' The British monk Pelagius
disputed the idea that St Paul intended to discourage learning – but
who would listen to Pelagius, a notorious heretic? The sixteenth-
century scholar Erasmus of Rotterdam, as ever speaking calm and
erudite wisdom into a world that did not care for it, argued that 'the
words do not condemn learning, but attempt to free us from pride
in our worldly excesses'. But distrust of curiosity and of the desire for
worldly knowledge ran too deep in medieval Christian thought for
these objections to be heeded.

No one, after all, was likely to question the greatest authority
among the early Christian Fathers, St Augustine of Hippo, who
proclaimed in his *Confessions* that curiosity is a 'disease', one of the
vices or lusts at the root of all sin. 'It is in divine language called the

lust of the eyes', he wrote. 'From the same motive, men proceed to investigate the workings of nature, which is beyond our ken – things which it does no good to know and which men only want to know for the sake of knowing.' He claimed that curiosity is apt to pervert, to foster an interest in 'mangled corpses, magical effects and marvellous spectacles'. And it leaves us prey to pride:

> Nor dost [God] draw near, but to the contrite in heart, nor art found by the proud, no, not though by curious skill they could number the stars and the sand,* and measure the starry heavens, and track the courses of the planets.

Thus the astronomer, viewing an eclipse, is wont to assert grandly that he understands it, rather than to submit in awe to this demonstration of divine power.

This aversion to curiosity as an impulse to know more than is good for you did not originate in the Christian world – Socrates is attributed as saying 'We should not concern ourselves with things above' – but Christianity established a robust moral basis for it. Augustine's injunctions were repeated by the twelfth-century Cistercian theologian St Bernard of Clairvaux, for whom curiosity was 'the beginning of all sin':

> Seek not what is too high for you, peer not into what is too mighty . . . Stay in your own place lest you fall if you walk in great and wonderful things above you.

According to St Bernard, Lucifer 'fell from truth by curiosity when he turned his attention to something he coveted unlawfully and had the presumption to believe he could gain'. As a result, he says:

> The Seraphim set a limit to impudent and imprudent curiosity. No longer may you, Satan, investigate the mysteries of heaven or uncover the secrets of the Church on earth.

---

* In *The Sand Reckoner*, Archimedes set out to estimate how many grains of sand it would take to fill the universe. This apparently pointless exercise in fact requires Archimedes to estimate the size of the universe and to develop a new notation for recording very large numbers.

Not all curiosity was on so grand a scale; it could also be petty seeking after trivia and things not worth knowing – a 'passion for knowing unnecessary things' as William of Auvergne, Bishop of Paris, put it in the thirteenth century. This is how Thomas Aquinas – who was no enemy of the desire for knowledge – expressed his aversion to curiosity, casting it in the classical mould as being associated with a certain mental and moral inertia and idleness. But this was scarcely less deplorable. The central problem with curiosity was that it was thought to be motivated by excessive pride. The accumulation of pointless learning ran the risk not that one would become another Lucifer but that one would primp and preen rather than bow one's head before the Lord. 'O curiosity! O vanity!', cried the late twelfth-century theologian Alexander Neckam. 'O vain curiosity! O curious vanity!'

The imperative of pious humility was what commended wonder to Augustine at the same time as it indicted curiosity. There was nothing frivolous or hedonistic about wonder. It instilled awe, reminding us of our powerlessness and insignificance before the glory of God. That is why wonder in the face of nature's splendour was seen as the educated response, and a willingness to believe in marvels and prodigies was not only praiseworthy but virtually a religious duty. Curiosity, like scepticism, was a sign that you lacked devotion and faith.

Thinking for yourself

All this can too easily feed the stereotype of an austere, anti-intellectual Middle Ages ruled by proscriptive priests. That is only part of the story. The increasing availability of Latin translations (via the Arabic) of the works of the ancient Greeks, particularly Aristotle, from the twelfth century brought with it a genuine interest in nature – not as a Platonic allegory but as an entity worthy of study for its own sake. Aristotle was no experimentalist, but nonetheless he took an interest in the particulars of the world, the distinctions between species of animal, plant and mineral, and he was a careful observer. The rise of Aristotelianism in the thirteenth century was accompanied by a greater realism in the visual and plastic arts: plants and animals are less stylized and more recognizable as particular species.

Yet Aristotle's advocacy of compilations ('histories') of natural things and phenomena did not imply, neither did it engender, a perceived need to explain and understand them except as particulars that illustrated general rules. It was these generalities that were the quarry of the philosopher. Anyone could see that nature was full of variety, but these 'accidents' were of no account in themselves. The aim was not to explain all of natural history, but to rub away its bumps and blemishes until only the broad outlines remained.

By the thirteenth century, Aristotle's thinking and methods had come to dominate natural philosophy as part of the intellectual movement known as scholasticism, which held sway in the cathedral schools and in the universities that began to flourish in Paris, Montpellier, Oxford, Bologna and other great cities of Europe. There has been a tendency until recent decades to contrast both the 'Scientific Revolution' and Renaissance humanism with a preceding era of alleged stagnation and dogmatism in which timid, hidebound scholastics spent their time amassing tedious arguments pro and con by scouring the works of the ancient authorities. This image, which mirrors the exaggeration of scholars from the sixteenth and seventeenth centuries who were keen to emphasize the novelty of their own ideas, does scant justice to the variety and often the vibrancy of medieval thought. An era that produced philosopher-theologians as dextrous and diligent as the great Dominicans Thomas Aquinas in Oxford and Albertus Magnus at Cologne, or Roger Bacon, Robert Grosseteste, William of Ockham, John Duns Scotus and Jean Buridan, scarcely deserves to be dismissed as an intellectual wasteland.

And yet . . . it remains true that much of what passed for learning in the Middle Ages amounted to rearranging old knowledge (much of it spurious) rather than adding to it anew. New ideas were often greeted with scepticism, for why should anyone trust them when they had not been through the rigorous filter of the ages? Originality of thought was a sign of unhealthy pride, and the pedantic, twisting paths of logic evident in some medieval works can hardly be mistaken for curiosity. The natural philosopher was expected to conform to the tenets of Aristotelianism – an orthodoxy that, after Aquinas had 'Christianized' it, became almost as unassailable as the Scriptures. Natural phenomena were 'explained' by making them fit with some permutation of Aristotelian first principles, an a priori approach known as deductive reasoning. This deployment of indisputable reason and logic was deemed to be the only way to achieve knowledge of the world as sound as the principles of geometry. It not only prescribed a particular method but also defined the scope of permissible (or at least, worthwhile) questions.

The very methods of scholastic Aristotelianism helped to make it impervious to serious challenge. The university academics would consider all questions about the world in isolation, listing arguments for and against a particular explanation of a phenomenon before offering their own interpretation. There was no attempt to find connections between different phenomena, so that contradictions and inconsistencies in Aristotle's teleological epistemology were barely noticed, let alone resolved. This atomization of knowledge meant that the overarching framework was never itself exposed to doubt. And if experience itself conflicted with Aristotle, so much the worse for experience: deviations from 'the norm' were by definition of little consequence.

Even some of the most innovative thinkers found it expedient to detach learning from curiosity, which they regarded *pace* Aristotle as an aimless wish to pry into trivial matters, distinct from true devotion to learning (*studiositas*). 'No wrongful curiosity can attend intellectual knowledge', Aquinas insisted, while Albertus Magnus wrote that:

> Curiosity is the investigation of matters which have nothing to do with the thing being investigated or which have no significance for us; prudence, on the other hand, relates only to those investigations that pertain to the thing or to us.

It was curiosity, Albertus said, that led to an inappropriate fixation on details and particulars rather than the true objective of identifying Aristotelian generalities. When he wrote about plants and animals he tended to describe what is 'typical', and only at the end to list specific features. It is, he admitted condescendingly, 'pleasureable for the student to know the nature of things and useful to the life and preservation of the cities' – but that is hardly the concern of the philosopher. When he lists particular species of plant in his *De vegetabilibus*, he makes it clear that he is merely 'satisfying the curiosity of students rather than philosophy, for there can be no philosophy of particulars'.

By this means, scholastic learning could be maintained on a higher intellectual plane than the knowledge of the untutored folk. Craftspeople, labourers and farmers generally knew far more than philosophers about plants, animals and minerals, but that didn't count, because they knew only the secondary, superficial details. The philosopher did not need to explain why the world is as we find it, but rather, to extract from it (and most importantly, from what the ancients had said about it) universal truths which they would pass on to students.

Rebel angels

A few individuals defied these constraints on what one should know and ask. Inevitably, they suffered accusations of heresy, blasphemy and witchcraft as a consequence. Even becoming a pope (Sylvester II) did not save the formidable tenth-century scholar Gerbert of Aurillac, an authority on astronomy and mathematics, from rumours of having studied magic under the Arabs and of having made a Faustian pact with the Devil. The pugilistic Norman scholar William of Conches poured scorn on those who denounced his penchant for asking awkward questions: 'Ignorant themselves of the forces of nature and wanting to have company in their ignorance, they don't want people to look into anything; they want us to believe like peasants and not to ask the reason behind things.' Like his spiritual heirs in the seventeenth century, he defended the impulse to enquire about nature on the grounds that it was our Christian duty to understand all we can about the world God has made. He believed that the Lord had fashioned it with reason, making of it a system that operated by intelligible laws. To those who argued that it was not only hubristic to seek the laws but heretical to imply that God himself was confined by them, William answered:

> One will say that it conflicts with divine power to say that man is made thus. To this I respond: on the contrary, it magnifies it, since we attribute it to Him to have given things such a nature, and thanks to this nature, to have created thus the human body . . . Certainly God can do everything, but what is important is that he did such and such a thing. Certainly God could make a calf out of the trunk of a tree, as country bumpkins might say, but did he ever do so?

One of the most curious men of the Middle Ages is now unjustly remembered primarily as a mere translator of ancient books. Adelard of Bath travelled far from south-west England in search of knowledge, studying in Tours, Laon and Chartres in France and passing into Sicily around 1116 before journeying into the lands of the Saracens to devote himself to the study of Arabic wisdom. Here he visited Antioch and Tarsus, and among the works of the Greeks that he found there in Arabic translation and translated into Latin was the pre-eminent ancient treatise on geometry, Euclid's *Elements*. He is also considered the likely translator of the *Mappae clavicula* (*The Little Key*), a treatise on the preparation of pigments and other (al)chemical materials that derives from Greek sources and which is now one of the best windows on the chemical technology of the classical world. Several other translations have been attributed to Adelard in error, but it seems clear that he played a major part in bringing Arabic knowledge of geometry, astronomy and mathematics to the West. 'It is worth while to visit learned men of different nations', he wrote,

> and to remember whatever you find is most excellent in each case. For what the schools of Gaul do not know, those beyond the Alps reveal; what you do not learn among the Latins, well-informed Greece will teach you.

If the details of Adelard's life are sketchy, the image of the man that leaps out from the pages of his own original works is nonetheless extraordinarily vivid: calm, wry, sceptical and deeply intrigued by the natural world, he could be a medieval Erasmus. He railed against the conservative tendency of his contemporaries to dismiss any original thinking because it lacked the imprimatur of ancient authorities. This was the reason why so many works from this period were written under the pseudonyms of revered Greeks and Arabs: 'Thus when I have a new idea, if I wish to publish it, I attribute it to someone else and declare: "It is so-and-so who said it, not I."'

But Adelard did publish some works under his own name, the most notable of them being an elegy to the study of philosophy, *De eodem et diverso* (*On the Same and Different*), and an ode to the delight and value of asking questions about the world, *Quaestiones naturales*. (Neither work is precisely dated, but both were written in the early twelfth century.) The former contains one of those uncanny premonitions about future science and technology that seem a speciality of the curious medieval mind. Anticipating the telescope and the microscope, Adelard wrote:

> The senses are reliable neither in respect to the greatest nor the smallest objects. Who has ever comprehended the space of the sky with the sense of sight? . . . Who has ever distinguished minute atoms with the eye?

And *Quaestiones naturales* presents a discourse on a delightful miscellany of issues in natural history, in the form of an imagined dialogue between the narrator and his inquisitive yet somewhat naïve nephew after the author has returned from the Arab lands.* The nephew serves as a foil with which Adelard can prick the stubbornness of the scholastics for whom ancient authority trumps reason. When the nephew asks about animals, he replies, 'It is difficult for me to discuss animals with you. For I learned from my Arabian masters under the leading of reason; you, however, captivated by the appearance of authority, follow your halter.' But the nephew is also haltered by dumb wonder, which obstructs rational thought:

> I know that the darkness that holds you, shrouds and leads into error all who are unsure about the order of things. For the soul, imbued with wonder [*admiratione*] and unfamiliarity, when it considers from afar, with horror, the effects of things without [considering their] causes, has never shaken off its confusion. Look more closely, consider the circumstances, propose causes, and you will not wonder at the effects.

The list of 'natural questions' examined here reveals both an awakening curiosity at the dawn of the Gothic age and the difficulty of harnessing it in the absence of any programme for a systematic enquiry into nature. Here are some of them:

> —when one tree is grafted upon another, why is all the fruit that of the grafted portion?
> —why do some animals ruminate?

* This format of a dialogue involving *quaestiones* was a commonly used vehicle for exploring issues that ventured beyond Aristotle, and prefigures the same device used by Galileo to challenge Aristotle's earth-centred cosmology in his incendiary *Dialogue Concerning the Two Chief World Systems* (1632) (see Chapter 7).

—why do some animals lack a stomach?
—why do some animals drink but do not urinate?
—why is the sea salty?
—why do men grow bald in front?
—why don't humans have horns?
—why do some animals see better at night?
—why can we see objects in the light while standing in the dark, but not vice versa?
—why are the fingers of unequal length?
—why don't babies walk as soon as they are born?
—why do we fear dead bodies?

There was no theoretical basis for answering such questions, unless one was content with the blithe tautologies of the ancients. But for Adelard, there was no harm in asking.

Experiments and secrets

Adelard's frank curiosity and the populist inclusiveness of Pliny's encyclopaedic tradition contrast with a quite different practice of recording and conveying knowledge in the Middle Ages. It came to be seen as something to be hoarded, or at best shared only among a privileged few. It became *secret* knowledge, and in consequence acquired an air of mystery and danger, even of heresy. One of the most popular encyclopaedias of the High Middle Ages was the *Secreta secretorum* (*Secrets of Secrets*), falsely attributed to Aristotle but in fact an Arabic work probably dating from the tenth century and based on older sources. Translated into Latin in the twelfth century, it was an eclectic mix of politics and ethics, alchemy, astrology and medicine. According to the thirteenth-century Franciscan friar Roger Bacon in Oxford, whoever reads and understands the *Secreta* will find 'the greatest natural secrets to which man or human invention can attain in this life'.

Bacon was himself rumoured to know such secrets; some said he was a wizard who dabbled in diabolical activities. Today Bacon has a reputation as a pioneering experimentalist; his readiness to find out about the world through experiment has even earned him the rather meaningless epithet 'the first scientist'. There seems no doubt that

Bacon did use technical materials and equipment in his studies – he was especially interested in optics and the nature of light, and probably conducted alchemical studies too. But most of his 'experiments' were done in his mind: they were descriptions, in the Aristotelian deductive mode, of what *would* happen in such and such a circumstance, rather than empirical investigations of what actually *did* happen.

Bacon perceived concrete reasons for keeping some knowledge secret. He experimented with gunpowder, and is sometimes attributed with introducing it to the West. He advocated the use of what we would now call scientific principles for the development of war engines and other military technologies in defence of Christendom. He appealed to Pope Clement IV to support science for that reason, arguing (rather speciously) that 'by the paths of knowledge Aristotle was able to hand over the world to Alexander'. Knowledge was power for this Bacon too.

But could secret knowledge really be science? For Aristotle, *scientia* was the demonstration of the causes of things. Yet 'secrets' were typically phenomena of a sort that could not be demonstrated to follow from Aristotelian first principles. They were not knowable in the way Aristotle's science was; their causes were hidden, *occult*. Magnetism was the archetypal example: wonderful and undeniable, but at the same time mysterious and inexplicable. Such things were prodigies, one-offs, unpredictable and opaque to the light of reason. That was why in antiquity they had not been deemed an important part of knowledge – not because their empirical validity was in any doubt, but because the phenomena seemed self-contained and hermetic. They were a matter of experience pure and simple – but experience as such was widely distrusted both by Platonists, who considered it superficial, and by Aristotelians, who considered it contingent and untidy. Whether or not Roger Bacon deserves much credit as a hands-on experimentalist, he proposed to alter the rules of the game significantly in suggesting that empiricism was a valid form of knowledge – that there are things worth knowing that cannot be deduced from scratch.

All the same, these two sources of knowledge tended to be held at arm's length: there was a *via rationis* and a *via experimentalis*, the former being generally regarded as superior. The thirteenth-century French physician Bernard de Gordon summed up the prevailing scholastic prejudice with a patronizing remark: 'Because the young greatly enjoy *experimenta*, let me give some here.' Finding things out, rather than working them out, was not beyond the pale, but it was indulged like a harmless hobby.

The medieval view of curiosity as a lust and a sin was by no means eclipsed by the early seventeenth century, and we will see that it was a charge against which Francis Bacon had to defend his new experimental philosophy. In his popular emblem book *Iconologia* (1593) showing classical personifications of the human qualities, the Italian author Cesare Ripa depicted curiosity as a wild, dishevelled woman, driving home the message in the caption: 'Curiosity is the unbridled desire of those who seek to know more than they should.' The Antwerp-born Jesuit priest Martín del Rio, an enthusiast of witch-hunts, wrote in his condemnation of the magic arts *Disquisitionum magicarum* (1599–1600) that the *mala curiositas*, the plague of curiosity, is an affliction among those who try 'to get to know that which they should not know'. By that time, however, moralists like del Rio had plenty of targets – for curiosity had become a European obsession.

'Curiosity' as depicted in Cesare Ripa's *Iconologia* (1593).

2 The Academies of Secrets

It must certainly be allowed, that nature has kept us at a great
distance from all her secrets, and has afforded us only the knowledge of a few superficial qualities of objects; while she conceals
from us those powers and principles, on which the influence of
these objects entirely depends.

David Hume, *An Enquiry Concerning Human Understanding* (1748)

Where secrecy or mystery begins, vice or roguery is not far off.

Samuel Johnson

Leonardo da Vinci was good at acquiring projects but not at completing
them. He was especially enchanted by the character of water, particularly
as manifested in the forms and furies of its flow, a subject that appealed
equally to his artistic, scientific and engineering sensibilities. He studied
these things in great detail, and intended to write a book called *The
Nature of Water*, but it is not clear if he ever did so. All we have to go
on is a sketchy treatise known as the Codex Leicester, which bears the
title 'On the Nature, Weight and Movement of Water'. Here Leonardo
provides an outline of his grand scheme, and it becomes evident soon
enough why it was probably never anything more than a dream.

The work was to comprise fifteen separate 'books', as follows:

Book 1   Of the nature of water
Book 2   Of the sea
Book 3   Of subterranean rivers
Book 4   Of rivers

Book 5   Of the nature of the depths
Book 6   Of the obstacles
Book 7   Of gravels
Book 8   Of the surface of water
Book 9   Of the things that move on it
Book 10   Of the repairing of rivers
Book 11   Of conduits
Book 12   Of canals
Book 13   Of machines turned by water
Book 14   Of raising water
Book 15   Of the things which are consumed by water

These are not all natural or obvious categories of enquiry, but rather, a ragbag of issues that occurred to Leonardo as hydraulic engineer and as natural philosopher. The list could go on indefinitely: what about the freezing of water, or boiling? Waterfalls? Rain and snow and mountain mist?

But it soon becomes clear in the Codex Leicester that the hapless Leonardo was already overwhelmed by such questions. He seeks a taxonomy of vortices and waves. He veers off into speculations about the biblical Deluge. He lists sixty-four descriptive terms for water in motion, from 'rotating' and 'repercussing' to 'submerging' and 'surging'. He attempts to delineate what distinguishes a sea from a lake, a river from a torrent, a well from a pool. And just look at the list of questions he accumulates about the nature of flow through rivers:

Of the different rates of speed of currents from the surface of the water
to the bottom.
Of the different cross slants between the surface and the bottom.
Of the different currents on the surface of the waters.
Of the different currents on the bed of the rivers.
Of the different depths of the rivers.
Of the different shapes of the hills covered by the waters.
Of the different shapes of the hills uncovered by the waters.
Where the water is swift at the bottom and not above.
Where the water is slow at the bottom and swift above.
Where it is slow below and above and swift in the middle.

Where the water in the rivers stretches itself out and where it contracts.
Where it bends and where it straightens itself.
Where it penetrates evenly in the expanses of rivers and where unevenly.
Where it is low in the middle and high at the sides.
Where it is high in the middle and low at the sides.
Where the current goes straight in the middle of the stream.
Where the current winds, throwing itself on different sides.
Of the different slants in the descents of the water.

These are all matters that demand meticulous observation, and often recourse to controlled and methodical experiment. Leonardo did investigate flow down artificial channels and past barriers, sketching the forms of the ripples and eddies with marvellous grace. These records look superficially like modern experimental science: the investigator creates a particular set of circumstances and watches how they constrain the play of nature, carefully noting the outcomes in a lab book. And indeed, not only do some of the flow patterns correspond to those now recognized in the science of fluid mechanics, but Leonardo even introduced some of the experimental techniques still in use today, such as the addition of 'tracers' like floating seeds or dyes to make the flow forms more evident.

But it is a mistake (often made) to turn Leonardo into a modern scientist. For one thing, his experimental programme is incontinent: an outpouring of questions, apparently listed simply in the order in which they popped into his head. This is merely a ledger of phenomena, uncontained by any fundamental hypothesis about how nature behaves. It is well-nigh impossible to imagine emerging from this programme a unified picture of how fluid flow occurs, like the one that scientists possess today.* Rather, if Leonardo had been at liberty

* The flow of liquids can be expressed mathematically by the so-called Navier-Stokes equation, devised in the late nineteenth century, which is essentially an expression of Isaac Newton's second law of motion applied to the liquid everywhere at once. As such, it is a 'theory' of sorts, but of limited utility, since the equations cannot be solved to map out the motions of the fluid except in some very particular circumstances. The problem, then, is not in *understanding* what is happening, but in *predicting* it. While Leonardo considered mechanics the 'paradise of mathematics', the art historian Ernst Gombrich is right to point out that 'a quick glance into the most elementary text-book of fluid mechanics will convince you that this branch is the hell of mathematics'.

Leonardo's sketches of water flow show astonishing attention to detail.

to investigate each of his questions, he would have gathered together a roster of particulars, none of them obviously deducible from the others: a body of facts, not an explicatory framework.

This is not because Leonardo had an ill-disciplined mind – at least, it was not simply because of that. Leonardo could not draw up a coherent research agenda because there was no philosophical tradition of doing so. Worse, the prevailing (scholastic) tradition was to take precisely the route that Leonardo chose, to divide and subdivide, making ever finer distinctions between the categories of things and questions. What distinguishes Leonardo was not the method he used to approach the nature of water flow, but the fact that he considered these things worth studying in the first place: not that he was obsessive in drawing up his lists, but in the object of his obsession.

Leonardo was also set apart from the scholastics by having Neoplatonic sympathies. Thinking about the behaviour of the seas and skies and the circulation of water between them, he was constantly aware of the relationship that Renaissance champions of this ancient tradition perceived between the macrocosm and the microcosm: it was more than metaphor when he called rivers the 'blood of the earth'. This belief in an inner unity of the diverse forms and effects of nature encouraged his use of analogy: light becomes akin to rippling water, and water to hair and smoke.

Leonardo's Neoplatonism explains why he was not in the end quite the faithful recorder of nature that he is commonly made out to be. His flow forms are idealized, exaggerated so that they resemble more closely the patterns of wavy and braided hair – connections all but invisible to ignorant eyes, yet which reveal to the adept the deep structure of the world. To art historian Martin Kemp, these sketches are 'an intricate synthesis of observations and theoretical constructions, with neither separate from the other'. Leonardo's fascination with hidden forms is not exactly that of the scientist as we would recognize it, but that of the philosopher who believes that nature is inherently creative and that the artist only mimics her inventiveness. Painting, said Leonardo, is 'a subtle *inventione* which with philosophy and subtle speculation considers the nature of all forms'. Particular phenomena – specific kinds of flow, for example – were to be understood not as explicit realizations of some under-lying mathematical process so much as capricious variations on it.

Leonardo believed that the artist can hope to make a convincing depiction of nature only by penetrating beneath the caprice – his aim was not to render as skilfully and accurately as possible what he thinks he sees, but to grasp and illustrate the forces at play. As Kemp says, 'Each painting is, in a sense, a proof of Leonardo's understanding.' Recognizing that few artists will have the patience (or perhaps the aptitude) for such dedicated observation, Leonardo concedes that 'At this point . . . the opponent says that he does not want so much *scienza*, that practice is enough for him in order to draw the things in nature.' But that would leave one skating over the surface of the world, bewitched by arbitrary invention and ephemera: 'The answer to this is that there is nothing that deceives us more easily than our confidence in our judgement, divorced from reasoning.'

Hermetically sealed

Abjuring both the arid logic of the scholastics and a slavishly literal empiricism devoid of interpretation, Leonardo thus seems to unite a belief in experiment with the quest for underlying causes that characterizes science today. On the one hand, 'those who fall in love with practice without science are like a sailor who enters a ship without helm or compass, and who never can be certain whither he is going'; on the other hand, 'although nature begins with the cause and ends with the experience, we must follow the opposite course, namely, begin with the experience, and by means of it investigate the cause'. His experimental programme, if we may call it that, even included the need for replication:

> Before you base a law on this case test it two or three times and see whether the tests produce the same effects . . . This experiment should be made many times so that no accident may occur to hinder or falsify this proof, for the experiment may be false whether it deceived the investigator or no.

He seems 'modern' also in his conviction that nature is law-bound and thus predictable and reliable: 'Nature does not break her law; nature is constrained by the logical necessity of her law which is inherent in her.' Furthermore, today's physicists would applaud his notion that these natural laws are as simple as they could be:

> O marvellous necessity, thou with supreme reason constrainest all effects to be the direct result of their causes, and by a supreme and irrevocable law every natural action obeys thee by the shortest possible process . . . Necessity is the mistress and guide of nature.

Equally will they celebrate his suggestion that these necessary laws are mathematical:

> There is no certainty where one can neither apply any of the mathematical sciences nor any of those which are connected to the mathematical sciences . . . Whoever condemns the supreme certainty of mathematics feeds on confusion, and can never silence the contradictions of the sophistical sciences, which lead to an eternal quackery.

But again, Leonardo doesn't always mean what we might like to imagine. Take his reliance on 'experiment'. In the Renaissance this word was often more or less synonymous with 'experience': not the kind of carefully planned, often highly constrained and artificial procedure used by scientists to explore a particular phenomenon in isolation from others and to test a hypothesis about how it happens, but rather, a simple, raw observation of nature. From the seventeenth century, the scientific experiment came increasingly to be an abstraction from and a manipulation of what happens 'naturally', precisely because the natural situation typically involves so many complicating factors and influences that interpretation becomes difficult. This widening gap between nature and the laboratory was to prove controversial.

And Leonardo does not propose to use experiment as we do today, for testing hypotheses. Rather, it supplies a means to identify what it is that reason must explain. 'First', he declares, 'I shall test by experiment before I proceed further, because my intention is to consult experience first and then with reasoning show why such experience is bound to operate in such a way.' This could easily provide a prescription for the tautologies and Just So stories that hamper the natural philosophy of antiquity: although the theory must accord with what we experience, it can nonetheless be formulated with armchair logic and reasoning.

What is more, Leonardo's belief in mathematical order and simplicity in nature did not necessarily impute simplicity to experience. In the Renaissance, 'laws' of nature did not have the binding character that they are awarded today. In the Aristotelian view formulated by Thomas Aquinas, a law describes what is normally the case. But nature, like an artisan, could make mistakes and aberrations: monsters, freaks, wondrous phenomena might occur, not just – as many believed – because they were divine portents, but also because nature had wrought in error or whimsy. Leonardo's position here is somewhat ambiguous. As we've seen, he felt that nature enjoys latitude for invention and variety, although he does not quite specify whether this is simply because the laws permit it, or because they can be relaxed, or because there are many more laws than we imagine: 'Nature is full of infinite causes that have never occurred in experience', he wrote.

Most of all, the Neoplatonism that informed Leonardo's faith in an orderly cosmos is, despite its immense and under-acknowledged influence on the development of early science, quite different from the motivation for this notion today. In *Timaeus* Plato asserted that the universe was made by the Creator through a process of geometrical division according to laws of simple proportion, rather like cutting and folding a strip of paper. As a result, mathematical proportion and harmony is woven into the very fabric of the world. Several of the founding fathers of the Christian Church, including Augustine and Boethius, were sympathetic to this idea, which flourished during the eleventh and twelfth centuries before Platonism was eclipsed by a reliance on Aristotle. The Neoplatonism of the Renaissance mixed this geometrical cosmology with the mystical tradition known as Hermeticism, which originated in Hellenistic Greece and promoted the study of natural magic and alchemy. The Florentine philosopher Marsilio Ficino was one of the principal architects of this synthesis. At the request of the city's prince Cosimo de' Medici he founded the Platonic Academy in a villa at Careggi in the hills outside the city, and in 1462 Ficino translated for Cosimo the newly discovered works of Hellenistic Gnosticism known as the Corpus Hermeticum, as well as a new version of *Timaeus* and the writings of Plato's disciple Plotinus, the founder of classical Neoplatonism.

There's no harm in attaching to Ficino the cliché exhausted on Leonardo – that he was a Renaissance man – if we turn it on its head in meaning precisely that he was a product of his times. While it was not the case that every educated person of that age had multivalent talents – Erasmus never painted altarpieces, Raphael never played the lute – nonetheless the notion that a 'complete man' was competent in several of the roles we regard as distinct today ensured that the philosopher would have little difficulty in becoming a physician or poet if that was how his inclination went. Ficino was not only one of the foremost scholars of the Italian Renaissance, but a priest, a doctor and a musician, entertaining Bishop Campano with Orphic hymns on the lyre just as Leonardo's performances later enthralled the Duke of Milan. Despite his short stature and slight stammer, Ficino made an impression of grace and refinement on all who met him, aided no doubt by his angelic wavy golden locks, his relaxed temperament and reputed ability to procure the finest wines.

Because Leonardo was in Florence as an apprentice in the workshop of Verrocchio while Ficino was active in the city, it is tempting to imagine that he may have acquired his Neoplatonism from, as it were, the horse's mouth. But although Ficino had close links with Florentine artists, being a friend of the Pollaiuolo brothers and apparently responsible for overseeing Botticelli's work on the *Primavera*, there is no evidence that he knew of Verrocchio's promising student. All the same, when Leonardo read the *Timaeus*, it was probably Ficino's version.

Hidden nature

The Neoplatonists' assertion of cosmic order – a mystical belief that had no real empirical basis – permeated early modern science, and has remained lodged in the scientific enterprise ever since. Most scientists continue to believe that the universe is fundamentally orderly and rule-bound, and moreover that these rules will be comprehensible, simple to state (if not to understand) and most probably beautiful to those with an aesthetic sensibility that can register it. There is today no logical reason to expect this to be the case at all, except for the fact that such an expectation of regularity underlying variety has worked rather well so far.

But the legacy of Neoplatonism at the start of modern science has another aspect, which conditioned the emerging culture of curiosity. For Plato, the world could not be taken at face value: superficial appearances are a mirage, a flawed and contingent screen behind which the eternal perfection of reality is concealed. The truth is then hidden from us – it is literally *occult*. To Neoplatonists, raw experience is replete with signs that only the adepts can interpret. Some of those who endorsed this philosophy followed the lead of the Corpus Hermeticum in thinking that the hidden structure was an alchemical one: the macrocosm and the microcosm were united through chemical analogies that entailed processes of transformation and rebirth. Others more closely observed the *Timaeus* in considering mathematics and geometry, not chemistry, to be the unifying scheme. But all Neoplatonists regarded nature as a repository of *secrets*, and considered that the aim of natural philosophy was to decode them. The answers would not be found in old books; nature had to be coaxed directly – by experiment – into revealing herself. In the late Renaissance, this viewpoint shaped the study of the natural world.

'Hidden nature' was explicitly gendered as a coy young woman who must be tricked into displaying what is normally concealed and private. The leering misogyny of that age, which turned nature into a maiden who must be unveiled either by cunning artifice or by brutal insistence, has been widely deplored by feminist historians, and is hardly exonerated by a defence which maintains that the language of coercion did not necessarily imply rape. It is as futile to deny the titillating role of this metaphor for the almost universally male audience who employed it, as it is to pretend that the fleshy females adorning the canvases of the High Renaissance were objects of purely intellectual appreciation. One can sometimes almost hear the panting excitement of the natural philosophers as they contemplate the removal of nature's garments and the penetration of her 'secrets'. This clandestine or sublimated sexuality is not pleasant to observe; neither is it incidental.

But the allure of secrets was not solely carnal. Access to secrets implies privilege, simultaneously a demonstration of power and a means to acquire more. The alchemist who possessed knowledge of the transmutation of metals would be rich and welcomed in any court; likewise the physician who knew the secrets of medicine. By uncovering the secrets of flowing water, Leonardo could claim an ability to control rivers and floods, to the advantage of princes and kings. This was more than theory: he and Niccolò Machiavelli collaborated on a project to divert the river Arno so that Florence's enemy Pisa would be deprived of water. Francis Bacon's dictum that knowledge is power has grown stale from exposure, but we must recognize that it rests on a tradition in which nature is a web of secrets.

This talk of secrets was more than rhetorical. The plain fact is that nature is, for the most part, hard to understand. Science exists at all because the way nature works is not obvious to everyone at a glance. It demands dedicated, patient and thoughtful study, and in one sense the discourse of secrecy merely reflects that. We still use this image today, in books or television programmes that promise to disclose the 'secrets' of plants or cells or the cosmos: all things that are hidden from everyday view. Yet there is an important distinction between the view that nature is hard to understand and that nature retains secrets, for the latter implies intentional withholding. Nature was either coy (which was exciting) or jealous (which was frustrating, and justified the use of force), and either way she (I will use but not condone the conventional gendering) would be revealed only with cunning. This promotes a very different view of the scientist from the one that prevails today, when he or she who discovers some particularly recondite aspect of nature is hailed simply as clever or ingenious. To be a scientist in the late sixteenth century demanded a degree of wiliness.

Secret society

In 1585 an Augustinian monk named Tommaso Garzoni published a book called *The Universal Piazza of All the Professions of the World*, among which he identified a recent category: the 'professors of secrets', who seek obscure and occult things. These secrets, he said, are things

> whose reasons are not so clear that they might be known by everyone, but by their very nature are manifested only to a very few; nevertheless they contain certain seeds of discovery, which facilitate finding out the way towards discovering whatever the intellect may desire to know.

While some of these secrets were 'good', and their pursuit an honourable trade, Garzoni clearly had a rather low opinion of these professors, who all too often sought 'ridiculous and vain secrets' that offered no benefits to humankind. And there are some, he said, who 'attend to this profession of secrets so zealously that they yearn for it more than for the necessities of life itself'. The stereotype on which Garzoni drew is the obsessive alchemist who lets his life go to ruin in foolish and futile pursuit of the philosopher's stone. Brueghel's caricature of this pathetic creature in his chaotic laboratory is the most famous of many sardonic depictions of the fool's quest. Like alchemists, the professors of secrets were a product of the Hermetic tradition, where grimy empiricism mixed with natural magic to offer what in the public gaze was apt to become a somewhat disreputable attempt to understand nature.

Among the people who Garzoni listed as 'professors of secrets' in Italy was Alessio Piedmontese, whose 1555 book *Secreti* can be described without too much anachronism as an international bestseller. It went through no fewer than seventeen editions in the first four years of circulation, and was still being published at the end of the seventeenth century, 104 editions later.

But no one knew who this 'Alexis from Piedmont' was. In the book's preface he claimed to be a devout man who had amassed his knowledge of 'secrets' outside of any academic tradition, by wandering the world gathering information from ordinary folk. His identity remains disputed. Of course, secrets are hardly that when they are published in a bestselling book, and Alessio's title is surely in part a marketing ploy. But since he pays lip service to the Hermetic convention of considering some knowledge too powerful to be appropriate for public disclosure, he is obliged to offer a moral justification for his decision to 'reveal' these precious secrets to the world. He explains that he had once withheld a remedy from a physician-surgeon, for fear that the man would use it 'for his own profit and honour'. As a result the physician's patient died, forcing Alessio to recognize that such valuable information should be disclosed to everyone.

What are these secrets? The reader hoping to find esoteric insights into the nature of the world was likely to be disappointed. On the other hand, Garzoni's disparaging remarks about the lack of practical value in secrets are here confounded, for *Secreti* was filled with useful recipes for everyday life: medicines, perfumes, soaps, lotions, cosmetics, cooking recipes and some more specialized craft procedures such as prescriptions for making metals or pigments. In the distance between the grandiose promise of the title and the mundane character of the contents, *Secreti* reveals the true span and significance of the tradition it represents. The first thing to notice is that, as a technical craft manual, the book is not connected to the academic disciplines studied at the universities. Even medicine, which was a scholarly subject, tended to be taught in terms of the vague and ineffectual theoretical principles expounded by ancient writers such as Hippocrates and Galen, while in practice the doctors relied on the practical skills of unlettered surgeons and the kinds of herbal and mineral concoctions that differed from folk medicines only in that they cost more. 'Secrets' were craft-based, largely devoid of theory, and scorned by the university professors as rudely empirical rather than the fruits of reason and logic. And yet, while books like Alessio's were nothing more than glorified versions of the popular recipe anthologies of the sixteenth century known as *Kunstbuchlein* ('booklets of the arts'), they gained prestige by asserting an association with the natural-magic tradition.

Alessio claims that one of his recipes, a cure for pleurisy, was given to him in Bologna by a gentleman called Girolamo Ruscelli. This Ruscelli was also on Garzoni's list, and what makes Alessio's story intriguing is that some historians now believe that Alessio was none other than Ruscelli himself. The basis for that claim seems at first rather secure, for Ruscelli made it himself in the introduction to his posthumous *Secreti nuovi* in 1567. Yet we should be wary of taking Ruscelli at his word – given his profession, he clearly stood to gain by being associated with such a popular title.

Ruscelli was one of the pre-eminent professors of secrets in the mid-sixteenth century. Born in Tuscany and educated at the University of Padua, he became a courtier to the Marquis of Vasto in the Abruzzo region of southern Italy, serving as a poet and ambassador. After the marquis's death, Ruscelli moved to Venice, where he became a proof-reader and writer for the publisher Vincenzo Valgrisi. His duties were those of a glorified hack: as a so-called *poligrafo*, Ruscelli would write anything the publisher demanded, from plays to travel books, all somewhat lowbrow and aimed at a general market. So it is quite possible that Ruscelli could have been involved in assembling a populist collection of 'secrets' for which the suitably mysterious author Alessio was concocted.

But Ruscelli may have played a more significant part in the establishment of a professoriat of secrets. He claims that he founded an 'experimental academy' called the Accademia Segreta (Academy of Secrets) in the kingdom of Naples. This was not only an institution for collecting secrets but also a secret society in itself: the twenty-four members took an oath to reveal the existence of the academy to no one unless approved by the others. Ruscelli declines to reveal the academy's patron, a prince of the city in which the members met every month, although one candidate is Ferrante Sanseverino, Prince of Salerno.

The aim of the society, according to Ruscelli, was 'to make the most diligent inquiries and, as it were, a true anatomy of the things and operations of Nature itself'. Although this knowledge was to be assembled in secret, the ultimate purpose was altruistic and egalitarian. 'In addition to our own pleasure and utility', says Ruscelli, 'we devoted ourselves equally to the benefit of the world in general and in particular, by reducing to certainty and true knowledge so many useful and important secrets of all kinds for all sorts of people, be they rich or poor, learned or ignorant, male or female, young or old' – a mission anticipating Francis Bacon's statement that science should strive towards 'the relief of Man's estate'. Ruscelli says that once the academy has collected all it seeks, the veil of secrecy would be withdrawn: 'it would be manifested and publicized to everyone as a thing most honoured, most virtuous, and most worthy to elicit the noblest rivalry from every true Lord and Prince in his state, and from every beautiful and sublime mind'. He takes care here to remind wealthy patrons what is in it for them.

The academicians of secrets in Naples did not just collect recipes, but tested them carefully before adding them to their list, so that only those that worked would eventually be published. Here, then, was apparently a systematic realization of Leonardo's method of trial and replication, distinct from the credulity of the much-maligned 'empiric' of the early Renaissance who would believe everything he was told. The Accademia Segreta allegedly based their programme on, if not exactly scepticism, then at least discernment. Crucially, their experiments were not tests of theories – on the whole the professors of secrets were wary of theorizing. Rather, they were tests of whether the claims espoused for particular procedures and concoctions were valid. To assist with this process of experimental sifting, the academy hired artisans including apothecaries, metalworkers, perfumers and gardeners.

It sounds all remarkably close to the practices of the scientific academies that flourished a hundred years later. Except for one thing: it is not clear that the Accademia Segreta ever existed beyond the imagination of Girolamo Ruscelli. There is no evidence of the society apart from his testimony, and it seems hard to credit that so substantial an enterprise could have really been kept otherwise entirely hidden. The ambiguities surrounding Alessio Piedmontese raise suspicion that Ruscelli is an unreliable witness, and his refusal to disclose any specific details of the academy's other members, patron or location only add to it.

All the same, his claim is not totally implausible. There may have been good political motives for maintaining strict secrecy. In Naples at this time, the Spanish rulers suspected private societies of fomenting social unrest and religious subversion. In 1543 the Spanish viceroy Don Pedro of Toledo closed down the humanist Accademia

Pontaniana on suspicion of heresy, and dismissed its leader Scipione Capece from the University of Naples. In retaliation, Ferrante Sanseverino brought Capece to Salerno and made him a professor of law, an act of defiance that supports Sanseverino's candidacy as patron of Ruscelli's academy. Four years later a public uprising in Naples provoked Don Pedro to close down all the philosophical and literary academies in the kingdom, and it is conceivable that this dangerous climate may have prompted Ruscelli to move to Venice around 1547, there to publish in the pseudonymous *Secreti* all that his academy of secrets had discovered so far.

Despite these obstacles, Naples remained a centre for the study of nature in the late Renaissance. The humanist scholar Bernardino Telesio spent much of his time there at the home of Alfonso Carafa, Duke of Nocera, and his son and heir Ferrante. In his home town of Cosenza in Calabria to the south, Telesio became the head of the Accademia Cosentina, a centre for the kind of natural philosophy that the schools spurned.

Telesio criticized the classical authorities Aristotle and Galen for their reliance on reason over the evidence of our senses, a position adopted in his great work *De rerum natura iuxta propria principia* (*On the Nature of Things According to their Own Principles*) (1586), which rejected the Aristotelian description of the material world in terms of matter and form. Telesio argued that our senses do not exist to support our sensitive human soul but are mere mechanical portals that mediate experience, conveyed by impulses in air and light. Understanding, in Telesio's view, stems not from analytical reasoning but from the senses.

In this cosmos there is no necessary role for the Aristotelian teleological agency that Thomas Aquinas equated with God's will. Things behave as they do of their own accord, motivated by an active force that ultimately derives from the opposing tendencies of hot and cold. Telesio never questioned that the world was made and shaped by God – he had spent time as a Benedictine monk, after all, and was even offered (but declined) the archbishopric of Cosenza by Pope Pius IV. But his God fashioned nature so that it might run without further intervention, governed by natural laws. These views would surely have got Telesio into trouble had he not cultivated good relations with the highest authorities of the Church.

In all of this, it is easy to appreciate Francis Bacon's assessment that Telesio was 'the first of the moderns'. But when we delve into the essence of what he believed, we find plenty that would make modern scientists uncomfortable with that association. For Telesio's vision borders on mystical pantheism: he believes that a corporeal spirit pervades all of nature and that to know nature we must unite with it. By learning what will help humanity survive and what will threaten it, the enquiry into nature becomes a moral quest.

Telesio's determination to ascribe rational, mechanistic causes to natural phenomena was the hallmark of the natural-magic tradition. The Italian physician and mathematician Girolamo Cardano, a native of Pavia, displayed the same impulse to account for things in terms of a balance in the opposed forces of sympathy and antipathy. This often left Cardano focused more on the mechanical explanation than on the veracity of the phenomenon in the first place, which was why he could end up seeking to rationalize such superstitions as the idea that eggs laid after the new moon in August would not rot or that if one spat on one's hand after hitting someone, it lessened the victim's pain. Such apparent capitulations to fantasy, along with Cardano's enthusiasm for astrology,* led historian Lynn Thorndike to be rather impatient with him: his 'voluminous works', Thorndike said, 'are very repetitious [and] ramble on and on without evidencing any inclination to stop'. But Thorndike's complaint that 'they contain much that would seem of no possible interest to anybody except apparently the author himself' is an indication that Cardano had the characteristically exhaustive curiosity of the natural magician. He sought specific explanations for parochial effects, rather than the generic reasons of Aristotelianism: maize grows without rain in the high Andes, he said, because the sun is weaker there and dries only the surface of the soil, while Ireland is snake-free because the 'bitumen' (peat) in the soil kills them with its bad smell.

Yet even Thorndike cannot deny Cardano's occasional acumen, especially in the field of mathematics. In his treatise on algebra *Ars magna* (*The Great Art*) (1545), he provides algebraic solutions for cubic and quartic equations (which contain variables raised to the powers

* Cardano's ill-advised publication of the horoscope of Jesus exposed him to charges of heresy in 1570.

3 and 4), and presents the first known reference to imaginary numbers (those containing the square root of −1). His *Liber de ludo aleae* (*Book on Games of Chance*) presents one of the first accounts of probability in relation to the rolling of dice – a topic that Cardano had ample cause to study, since his unconventional interests and theories excluded him from the universities and often forced him into gambling to pay his bills. He also came close to understanding the true nature of combustion two centuries before Antoine Lavoisier's oxygen theory, saying that a flame is merely burning air.

For Cardano, curiosity was spurred by wonder and an appreciation of the marvellous – which, he said, is more abundantly exhibited in nature than in art. In his *De rerum varietate* (*On the Variety of Things*) (1557) he provides a veritable taxonomy of wonders: 'wonders of the earth', 'wonders of water', and so forth. He argued that while some things are truly wonderful (and perhaps therefore beyond rational explanation), others are 'worthy of wonder, but not great wonder', and some simply not marvellous at all. In the first of these classes he places the 'blue clouds' said to be sighted in the Strait of Magellan off the tip of South America, and in the second, the foot jugglers of Mexico – indications that the discovery of the New World was expanding the categories of things that may exist and happen (see Chapter 6).

The magic man

In the 1560s Naples acquired its own Academy of Secrets: the Accademia dei Secreti, whose founder was Giambattista Della Porta, another of the 'professors of secrets' on the disapproving Garzoni's roster. It is possible that this group represented a revival of Ruscelli's programme after the political climate had become more settled, for Della Porta may have been a part of Ruscelli's Accademia Segreta too.

If it seems disturbing to place Telesio at the roots of modern science, locating Della Porta at that juncture will appear a whole lot worse. His 1558 book *Natural Magick* provided the definitive exposition on the view of nature as an occult storehouse of secrets. How could the author of a magical book be a predecessor of the scientist?

To understand that, we need to forget pretty much all we think we know about magic. The scorn that today's scientists reserve for the word is fully justified insofar as it relates to the modern, debased meaning, and is nicely illustrated in this recent exchange between Richard Dawkins and David Attenborough as Dawkins talks about his forthcoming children's book:

> *Dawkins*: It is called *The Magic Of Reality* and one of the problems I'm facing is the distinction between the use of the word magic, as in a magic trick, and the magic of the universe, life on Earth, which one uses in a poetic way.
>
> *Attenborough*: No, I think there's a distinction between magic and wonder. Magic, in my view, should be restricted to things that are actually not so. Rabbits don't really live in hats. It's magic.

We have perhaps already some intimation (we will acquire more later) that Attenborough's view of wonder is problematic when seen through the lens of history. The same is true in spades for his take on magic, although as an eminent natural historian he can be trusted for his knowledge of rabbits. That magic has come to connote 'things that are not actually so' is a story in itself, for in the mid-sixteenth century it was the closest thing that existed to experimental science. To its principal advocates, natural magic was the most promising avenue for eliminating the supernatural agency of demons or of God from the natural world. These supporters had constantly to refute charges of heresy by insisting that natural magic was in no way incompatible with the idea of God as omnipotent creator and prime mover. Like the modern scientist, the natural magician believed that natural phenomena are governed by invisible forces. And as historian William Eamon explains, natural magic was 'the science that attempted to give rational, naturalistic explanations' of those forces. Far from utilizing supernatural means, natural magicians knew how to harness the forces and tendencies inherent in nature. As the great early sixteenth-century proponent of the art, Cornelius Agrippa, explained:

> Magicians are like careful explorers of nature only directing what nature has formerly prepared, uniting actives to passives and often succeeding in anticipating results so that these things are popularly held to be miracles when they are really no more than anticipations of natural operations . . . therefore those who believe the operations of magic to

be above or against nature are mistaken because they are only derived
from nature and in harmony with it.

The natural magicians, says Eamon, believed that 'nature teemed with
hidden forces and powers that could be imitated, improved upon, and
exploited for human gain'. For this reason, there was no more potent
challenge to the old concept of forbidden knowledge – and therefore
no greater validation of curiosity as a virtue – than that offered by
natural magic.

Della Porta became regarded as 'the most diligent scrutinizer of
the secrets of nature' in his time, and before Galileo he was the dominant figure of Italian science. It was said at the end of the sixteenth
century that, alongside the famous Roman baths at Pozzuoli, Della
Porta was the greatest tourist attraction in the kingdom of Naples.
This is all the more remarkable when we find that his reputation today
rests as much on his skill as a playwright; he wrote at least thirty-three
plays, some of which no longer survive.

He was born into a wealthy and influential family of Naples (hailing
originally from Salerno), and he allegedly composed his *Natural Magick*
at the age of fifteen. Since we do not know exactly when he was born
(between 1535 and 1540), it is hard to know the truth of that claim,*
but Della Porta was surely a prodigy: the book was printed in 1558,
a year after Alessio's *Secreti*, when its author could not have been more
than twenty-three years old. *Natural Magick* had a much stronger link
with that intellectual tradition than Alessio's compendium of recipes,
but in other ways it was not so different. After an introductory chapter
explaining the causes of natural and artificial phenomena both 'manifest' and occult, according to such principles as the sympathy and
antipathy of forms and substances and how they are influenced by
the celestial bodies, it proceeds to list a collection of tricks, experiments and chemical procedures: how to write secret messages, how
to make cosmetics and love potions, what the special powers of gems
and stones are, and how to conduct a range of optical tricks with
mirrors, magic lamps and the camera obscura that paints images with
light. The book was expanded in later editions, incorporating all kinds

---

* This is partly Della Porta's doing: he was apt to lie about his age, as well as to
distort other facts, to the advantage of his reputation.

of useful agricultural lore such as how to crossbreed plants and animals and how to preserve fruits. The prosaic nature of much of the book's content no doubt helps to explain its popularity: this was 'magic' that anyone could use.

Natural Magick shows us precisely where curiosity was located in the mid-sixteenth century. The attraction of the natural-magic tradition for those of an inquisitive disposition is obvious: it is relatively undogmatic, open to experience, and can embrace just about any topic you like, from cryptography to apple-growing. Nothing is off-limits: as Della Porta put it, 'Magic is nothing else but the knowledge of the whole course of Nature.' You need not be a learned professor to participate in this exploration of nature, nor need you become mired in the rhetorical swamps of classical disputation. Natural magic is largely free from the proscriptions of religion, as well as from the imperatives of Aristotle, which try to tell nature what it can and cannot do. The testimony of antiquity counts for nothing if it is not borne out by your own direct experience. None of this amounts to true science, but it paves the way for it.

On the other hand, there are hazards. The most immediate is that you are liable to be accused of sorcery and necromancy. This frequently happened to Della Porta, who complains in the 1589 edition that a French jurist, Jean Bodin, had denounced him as a 'great Neapolitan sorcerer' and declared *Natural Magick* fit for burning because it included an unguent made by witches (Della Porta insists that he took the recipe instead from the writings of theologians). Often the mere inclusion of experiments, rather than the allusions to magic per se, was enough to arouse suspicion. Della Porta was hounded by the Inquisition, who investigated him in the 1570s after closing down his Accademia dei Secreti. It seems likely that he was forced to recant some of his views after being imprisoned, and he felt compelled to join the Jesuits and spend a day each week working on religious duties to allay suggestions of impiety. Even so, he struggled to gain Church approval to publish his books, on one occasion being threatened with excommunication if he did so without the permission of the Roman High Tribunal.

It is odd that conventional histories of science, so eager to leap from Copernicus to Galileo, ignore all of this: to be a martyr to science, it seems, you have to have been 'right' – or perhaps one must even say, to fit within a particular narrative. In some ways, Della Porta's natural magic was more threatening to the Church than Copernicus's heliocentric astronomical theory, since it sought to erode the superstitions on which ecclesiastical authority depended. The efficacy of holy relics and faith healing demanded an unquestioning acceptance of miracles that a naturalistic explanation undermined. And religious condemnation of demons, along with Church rites to quell their interfering ways, were rendered superfluous if the likes of Della Porta were going to leave these wicked spirits no role to play. As Eamon puts it:

> Just as the Church was intensifying its campaign against popular superstitions, Della Porta was developing a theory that would undermine its entire conception of superstition . . . the Church needed demons as much as it needed Aristotle . . . Witchcraft legitimized established authority and the instruments by which it exercised its power.

Perhaps a reluctance to afford natural magic its proper place in the beginnings of science stems from another of its dangers, glaringly evident in retrospect: it is credulous. If occult forces act in nature, how can you tell what they can and cannot do? If the compass needle can be oriented toward the north from anywhere in the world by some hidden power, is it so hard to believe that a 'marvellous water' can restore impaired vision (as Della Porta swears happened to him) or that certain gems can confer protection against disease, or that Ptolemy had a magical mirror (perhaps a lens) that could see ships 600 miles away? Della Porta was no more gullible than any of his contemporaries, including many of the acknowledged heroes of early science. Some of the ambitious claims and recipes in *Natural Magick* teeter on the brink of plausibility, such as drugs that will make a man think he is a bird or a fish, or the effect of diet on dreams, or a method for preventing dogs from barking. Other fantastical phenomena that the book describes, such as spontaneous generation, were universally accepted.

More importantly, Della Porta stands out not so much for what he believed but for why he believed it. He was convinced that even the most marvellous of events must have naturalistic explanations, being caused by the forces of nature rather than by God or demons. Jean Bodin's damaging diatribe was provoked by Della Porta's rational explanation for the salve that enabled witches to fly: Della Porta argued that they only *think* they do, because of the effects of a hallucinogenic drug rubbed on their bodies. It was this implied criticism of the witch-hunts that led the Inquisition to command Della Porta to account for himself in Rome.

Della Porta's book established experiment as the modus operandi of natural magic. He says that he personally tested claims found in older writings that seemed hard to credit, and ruled out ones that did not pass muster. Never mind that Albertus Magnus suggests that iron be hardened by tempering it in radish juice or 'water of earth-worms' – it doesn't work, Della Porta attested. (In fact, with characteristic exaggeration apt to undermine his empirical credentials, he says it makes the metal as soft as lead.) The 1589 edition of *Natural Magick* reports that the long-standing idea that the attractive force of a magnet is nullified by rubbing it with garlic does not pass this empirical test either (nor, for that matter, does the alleged enhancement in attractive power caused by goat's blood). This experimental disproof of garlic's 'anti-magnetic' property is sometimes attributed instead to the English natural philosopher William Gilbert in his 1600 work *De magnete* (*Of Magnets*), perhaps because Gilbert has been deemed a more respectable precursor to science. Della Porta himself grumbled that Gilbert 'took the whole seventh book of my *Natural Magick* and split it into many books, making some changes . . . the material which he adds on his own account is false, perverse and melancholy'. He does not help his case, however, by making Gilbert's claim that Earth is in motion one of these 'mad ideas'. But we should be wary of making into 'moderns' and 'medievalists' those at the turn of the century who were and were not Copernicans.

Della Porta is particularly insightful on the subject of optics. In the 1584 edition of *Natural Magick* he explains that:

Concave Lenticulars [lenses] will make one see most clearly things that are afar off. But Convexes, things near at hand. So you may use them as your sight requires. With a Concave Lenticular you shall see small things afar off very clearly. With a Convex Lenticular, things nearer to be bigger, but more obscurely. If you know how to fit them together, you will see both things afar off, and things near hand, both greater and clearly. I have much helped some of my friends, who saw things afar off weakly, and what was near, confusedly, that they might see all things clearly.

Although he is writing here not, as it might appear, of the telescope or the microscope, but of spectacles and other correctives to faulty vision, it is clear enough where the ideas are heading. If Telesio was right that our senses are mechanical, perhaps they can be supplemented and augmented by mechanical aids.

Della Porta's natural magic offered one of the main alternatives to Aristotelian natural philosophy during the late sixteenth and early seventeenth centuries. While it remained rooted in the Neoplatonism and Hermeticism championed by Ficino, exhibiting a (not altogether uncritical) enthusiasm for alchemy, astrology and the new 'chemical medicine', it took on different complexions in different hands. Perhaps the most well-known support er of the magical world view during this period is also one of the wildest and least influential. The Neapolitan friar Giordano Bruno had an arrogant and argumentative nature that was bound to get him into serious trouble eventually, although if he had not happened to promote Copernican cosmology it is doubtful whether he would command any greater fame today than the many other intellectual vagabonds who wandered Europe during the Counter-Reformation. It seems a vain hope that Bruno should ever cease to be the 'martyr to science' that modern times have made of him; maybe we must resign ourselves to the words spoken by Brecht's Galileo: 'Unhappy the land where heroes are needed.'

The fact is that Bruno's Copernicanism is not mentioned in the charges levelled against him by the Inquisition in 1576, nor the denunciation of 1592 that led to his imprisonment and lengthy trial. Of the heretical accusations that condemned him to be burnt at the stake in 1600, only two are still recorded, which relate to obscure theological matters. He held many opinions of which the Church disapproved deeply, on such delicate matters as the Incarnation and the Trinity, not to mention having a long history of associating with disreputable types. Bruno's death stains the Church's record of tolerance for free thought, but says little about its attitude to science. There is nothing in Bruno's espousal of a world soul, or his long discourses on demons and other spiritual beings, or his unconventional system of the elements, that makes him so very unusual for his times – but nothing either that qualifies him for canonization in the scientific pantheon.

An Academy of curious Men

Before incurring the wrath of Rome himself, Della Porta convened his Academy of Secrets in his own home in Naples. 'I never wanted [that is, lacked] also at my House an Academy of curious Men', he wrote,

> who for the trying of those Experiments, cheerfully disbursed their Moneys, and employed their utmost Endeavours, in assisting me to Compile and Enlarge this Volume [*Natural Magick*], which with so great Charge, Labour, and Study, I had long before provided.

The Italian writer Pompeo Sarnelli, who edited Della Porta's letters, says that these visitors 'vied with one another to add new discoveries to his researches'. They were wealthy men, and like Ruscelli's Accademia Segreta they paid artisans to help with the experiments.

Natural magic set the agenda for the scientific programme of the seventeenth century in the sense that its objective was not, as is often said of science, to banish occult qualities and phenomena, but to *accommodate* them by explaining them rationally. 'Hidden' qualities such as magnetism, gravity and light, far from being marginalized, supplied some of the central themes of the new philosophy: it was precisely these things that the scholastic tradition could not explain. Most importantly, natural magic regarded the natural world not as a logical puzzle that could be cracked with the bibliographic tools of antiquity but as a subtle, spontaneous, secretive structure whose rules needed to be teased out with cunning, guile and keen observation.

The magical cosmology of Della Porta, Telesio, Cardano and their followers supplied both a space to legitimize curiosity and a reminder of its disreputable image. This new philosophy was viewed with contempt and suspicion both by conservative academics and by the Church – largely because it challenged the cosy certainties of both traditions, with their hidebound interpretations and their rigid demarcations of what one might and might not acceptably ask. It refused to accept the complacent notion that what we (and in particular, what

Aristotle) did not already know about the world was not worth knowing. Della Porta put it like this:

> The ignorant philosophers, when they cannot give reasons to these things according to the principles of Aristotle (as if he knew all things), judge them as superstitious. But learned men know well that of the infinite number of things seen in this great machine of the world that they want to know, they scarcely know a particle of them.

However, sanctioning curiosity did not in itself guarantee a new natural philosophy worthy of the name. The books of secrets and of natural magic betray the consequences of over-permissive enquiry: they are apt to become vast, undisciplined lists of claims and recipes and marvels, of widely varying credibility, lacking any coherent explanation or structure. The aversion to detailed theory in the natural magic tradition is a strength, because it allows experience its full voice, and limits the hasty speculations which, in that age, were far more likely to be wrong than right. But it is also a weakness, because it constrains the scope and depth of understanding. Della Porta's explanations, rooted in a teleological Neoplatonism in which the universe is designed to display hints of God's handiwork, tend to be self-confirming dead ends. His occult influences are typically forces of attraction and repulsion, or kinship and enmity, with almost limitless versatility. They will 'explain', for example, why the plants rue and hemlock oppose one another, the inflammation caused by rue being soothed by hemlock juice and vice versa. These correspondences and interactions were considered to be evident in the nature of things via the so-called doctrine of signatures: little visual puns that God wove into the fabric of the world so that we might know what all these things are *for*. The game is to read the signs. And game it is, for nature does not just indulge in cryptic secrets but takes delight in elaborating them: in William Eamon's words, Della Porta portrays nature as 'ever changing, ever joking, subtle, ingenious, and prodigious'. Much as this is an invitation to the curious, it also implies that we cannot hope to explain too much: we are back with something resembling Aristotelianism, whereby a few broad principles suffice to convey the thrust of nature while the diversions and variations are mere caprice.

All the same, it is worth searching them out, for one never knows what they might reveal. This is evident in Della Porta's defence of apparently arbitrary gobbets of knowledge. It is true, he admits, that it might seem trivial to know that magnets are not nullified by rubbing with garlic – but a small truth is preferable to a great falsehood. In a motto that could read as an advertisement for the virtues of small science, he said that 'True Things be they never so small, will give occasions to Discover greater things by them.' If that was so, there was nothing that did not deserve our curiosity.

3 The Theatre of Curiosity

They say miracles are past; and we have our philosophical persons
to make modern and familiar, things supernatural and causeless.
Shakespeare, *All's Well That Ends Well* (*c.*1604)

The Heavens, the Seas, the Whole Globe of Earth (from the variously adorned surface, to the most hidden Treasuries in her
bowels) all Gods visible workes, are your Subject.
John Evelyn (1667)

Anyone who wanted to study the natural world in the sixteenth and
early seventeenth centuries needed, on the whole, to enjoy one of a
limited range of circumstances. To be both fabulously wealthy and
leisured (and of course male) was probably the best position. If you
could secure a university post, your salary was guaranteed but you
might have to abide by the conventional rules of enquiry – if not to be
an obedient Aristotelian, then at least not to rock the boat. Some priests
and other clerics were able to indulge their curiosity in the course of
their religious duties. Failing all this, you needed a patron: someone
who would not only pay for your keep, your books and (if you were
so inclined) your instruments, but would also champion your intellectual
authority and perhaps defend you if you fell foul of the Church.

The philosopher's need for patronage tends to get overlooked in
standard histories of science. We are told what the likes of Galileo
and Bacon said, but rarely the social and professional context in which
they said it. Yet the realities of patronage could condition not only
the manner in which the philosopher expressed himself but also the

nature of his aims and the questions he could ask. Patrons had to be impressed; they had to be flattered; they had to know that their money was being well spent. And the role of courtier demanded of the philosopher a certain attitude of mind and use of words. Early science was as much shaped as it was supported by patronage, just as it is moulded by funding agencies today.

Renaissance humanism encouraged the emergence of a courtly environment receptive to natural philosophy. In the Middle Ages the courtier was distinguished by prowess at hunting and warfare, but had little time for the arts or for learning, as caricatured by the Italian diplomat and writer Baldassare Castiglione in *The Book of the Courtier* (1528). He describes a typical courtier whom a lady asked for a dance, and who

> not only refused, but would not listen to music or take part in the many other entertainments offered, protesting all the while that such frivolities were not his business. And when at length the lady asked what his business was, he answered with a scowl: 'Fighting.' 'Well then', the lady retorted, 'I should think that since you aren't at war at the moment and you are not engaged in fighting, it would be a good thing if you were to have yourself well greased and stowed away in a cupboard with all your fighting equipment, so that you avoid getting rustier than you are already.'

Times, Castiglione implies, have changed. The new scholarly attitude of the nobles and courtiers was in some ways forced upon them, since they had time on their hands: as social mobility increased and power could be accrued from money rather than from hereditary titles, governance was no longer the sole preserve of the aristocrat. The Renaissance prince was now expected to be cultivated in many fields, able to turn from poetry and music to statecraft or philosophy. What is more, he should display these virtues whether among the nobles or the ordinary people, and without ostentation or apparent effort: a coolly calculated sophistication, devoid of the argumentative point-scoring that marred scholastic debate. Cosimo de' Medici's grandson Lorenzo, who was taught by both Marsilio Ficino and Leonardo da Vinci and who presided at the height of the Florentine Renaissance, was said to be the embodiment of this ideal.

The princes and dukes demonstrated their noble and honorable status not, as formerly, with lavish displays of wealth, but by attracting the best artists, craftsmen and scholars, and by being able to converse with them and even to rival their abilities in words, music and art. Those skilled courtiers, meanwhile, were expected to produce learned and beautiful works for the glory of their patrons: the equivalent, perhaps, of today's 'pressure to publish' for the prestige of one's academic institution. Without it, your funding was liable to evaporate.

Courtly etiquette determined the style of the product. It had to impress, and to do so with elegance, not vulgarity. Inevitably, that did not necessarily equate with intellectual profundity and could mean quite the opposite: court philosophers might engage in displays of empty virtuosity and showmanship, for example by supplying the prince with elaborate, intricate automata and other machines. If there was advantage in not having to justify one's work from its practical utility and economic benefit, that could ultimately risk making uselessness a virtue. If Della Porta was sometimes accused of excessive fascination with shallow (if ingenious) tricks and illusions, his showman's style was largely just the convention of the time.

It was a style that celebrated secrets and encouraged secrecy. 'If you would have your works appear more wonderful', Della Porta wrote, 'you must not let the cause be known.' Nothing might sound now more antithetical to the scientific spirit of transparency; but the aim here was not so much to jealously guard one's knowledge in the manner of the alchemists, but to maximize the impact of the demonstration and thus the observer's delight in it. In his book on mechanical marvels *Mathematicall Recreations* (1633), Henry van Etten* wrote that 'to give a greater grace to the practice of these things, they ought to be concealed as much as they may . . . for that which doth ravish the spirits is an admirable effect, whose cause is unknowne'. It is worth remembering that, for Della Porta and his contemporaries, this approach only mimicked how nature herself behaved: she too was a playful, cunning (in all senses of the word) guardian of secrets.

This concealment of 'how the trick is done' was itself a reflection

---

* Henry (or Heinrich or Hendrik) van Etten was the pen name of a Frenchman called Jean Leurechon. The first edition of his book seems to have appeared in or around 1627, but the 1633 edition was more widely read. It was translated into English in 1653 by the mathematician William Oughtred, who pioneered the use of logarithms.

of courtly norms. Like a painter making his brush strokes invisible, the courtier was supposed to display his learning and skill as if it cost him not the slightest effort. By unspoken agreement, courtiers agreed to pretend that the difficult performance they had to sustain was effected with ease and nonchalance (*sprezzatura*), and above all else with grace (*grazia*). Passion, enthusiasm and industry were unseemly.

To exhibit these characteristics was to show what the Italians (who excelled at it) called *virtù*. According to the science historian Alistair Crombie, 'A man of *virtù*, acting rationally in the image of his Creator, was a man with intellectual power to command any situation, to act as he intended, like an architect producing a building according to his design, not at the mercy of fortuitous circumstances.' Out of this environment came a fusion of the graceful courtier with the ingenious scholar, in the form of the virtuoso.

The virtuoso 'was a rational artist in all things', says Crombie – meaning the arts as well as the sciences, pursued methodically with a scientist's understanding of perspective, anatomy and so forth. (It is after all in the arts that the epithet 'virtuoso' survives today.) The virtuoso was permitted, indeed expected, to indulge pure curiosity: to pry into any aspect of nature or art, no matter how trivial, for the sake of knowing. There was no sense that this impulse need be harnessed and disciplined by anything resembling a systematic programme, or by an attempt to generalize from particulars to over-arching theories. In fact the virtuoso was more likely to be attracted to the rare and unusual than the common and mundane, and it was precisely from such examples that generalization was perilous. As Henry Peacham explained in his *Compleat Gentleman* (1634):

> The possession of such rarities, by reason of their dead costlinesse, doth properly belong to Princes, or rather to princely minds . . . Such as are skilled in them, are by the *Italians* termed *Virtuosi*.

As Peacham made clear, the fusion of courtier and scholar was emphatically a class-based affair: one could not be a virtuoso unless one was already a gentleman.

Here was a new model for curiosity: it became acceptable to wish to know about things, so long as that desire was pursued with dispassionate grace. As William Eamon says:

Among the qualities of the idealized 'learned prince' [and hence of the ideal courtier], two stand out as being particularly important for their influence upon courtly science: curiosity and virtuosity . . . the seventeenth-century rehabilitation of curiosity was in large measure a product of the virtuoso sensibility.

Opening the cabinet

The refined intellectual curiosity of the royal courts was aped by lesser nobles and wealthy households. Giambattista Della Porta was raised in such an environment; his youngest brother Giovan Ferrante gathered a large collection of crystals, gems and other geological specimens, while his elder brother Giovan Vincenzo amassed marvellous *artificialia*: marbles, busts and medals.

These were, in effect, what we now call cabinets of curiosities. Della Porta collected his own cabinet, which contained plants and botanical specimens, gems, stones and all manner of things odd and unusual. In the late sixteenth and early seventeenth centuries, sovereigns and scholars accumulated vast collections, housed in several chambers in their homes and palaces, often catalogued in obsessive detail. The collection of the Neapolitan apothecary Ferrante Imperato was stuffed with books, specimen jars, preserved animals, shells and plants. When the English writer John Evelyn visited in 1645, twenty years after Ferrante's death, he described among the 'incomparable rarities' a male and female chameleon, a great crocodile, a salamander and birds of paradise of both sexes. Evelyn was also impressed by the collection of Cassiano Dal Pozzo, a 'great virtuoso' whom he visited the previous year in Rome. The Danish physician and naturalist Ole Worm possessed another cabinet of international renown, with many trays of minerals, metals, woods, fruits, herbs, corals, strange gourds and skulls, antlers, giant turtle shells, ethnographic clothes, weapons and musical instruments.

These cabinets of curiosities or *Wunderkammern* ('chambers of wonders') were the material representation of what curiosity itself was becoming and of what function it was deemed to have in the study of the natural world. At first glance, they seem to speak of a voracious appetite for knowing. But how, if at all, did they really assist in the process of finding out about the world? How were the contents selected, and how organized? What did they mean?

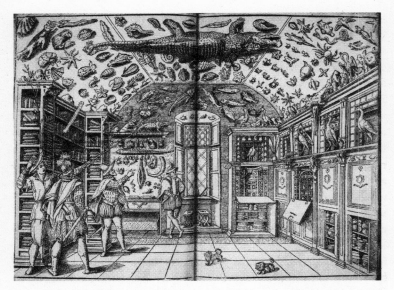

The cabinet of curiosities of Ferrante Imperato of Naples, as depicted in the frontispiece of his 1599 catalogue *Dell'historia naturale*.

The cabinets can be regarded as the descendants of the treasure troves collected by princes and churches in the Middle Ages: lavish hangings, gems, paintings and fine goldwork, like those described with somewhat improper delight by Abbot Suger at the royal abbey of Saint-Denis in France in the twelfth century. The Saint-Denis collection included not only religious relics, prized by all major medieval churches, and the paraphernalia of the French kings, but also rare, unusual and fantastical objects such as the Horn of Roland (made from an elephant's tusk), a griffin's claw and a unicorn's horn.

The ostentatious practice of displaying precious objects was a feature of the medieval French courts of Burgundy and Berry, and was later adopted by the Habsburgs. Yet in contrast to these rich hoards, characterized by expensive materials and the finest craftsmanship, the cabinets of curiosities were also stocked with purely natural objects. Since these items were also usually distinguished by their rarity, however, they too held financial value. These were not the kind of geological collections that will place basalt and granite next to diamond and amethyst: only the uncommon, the gorgeous, the wonderful, were generally considered worth collecting, and they were typically thrown together without any regard for 'scientific' classification. Thus some cabinets of curiosities remained little more than emblems of social status. In the late fifteenth century it became almost de rigueur for North Italian nobles to possess one; Leonello d'Este of Ferrara, Pietro and Francesco de' Medici in Florence, Isabella d'Este in Mantua and Federico da Montefeltro in Urbino (who commissioned a vast and famous library) all did so, and the tradition spread to Germany early in the next century. During the late Middle Ages, rich merchants and other urban elites joined the fraternity of collectors and connoisseurs.

But this was more than a case of 'look what I've got'. The power with which *Wunderkammern* were imbued was not wholly or even mainly symbolic. The ideal collection was *comprehensive* – not in that it contained an example of every object or substance in the world (although efforts were sometimes made towards such exhaustiveness), but in that it created its own complete microcosm: a representation of the world in miniature. While it would be too much to suggest that this representation is directly analogous to the simplified explanatory schemes that scientists now call models, by means of which they investigate experimentally and theoretically how the real world functions, the comparison is not entirely fatuous. The Flemish physician Samuel Quiccheberg, who administered the *Kunstkammer* of the Duke of Bavaria Albrecht V in Munich, called the cabinet a *theatrum sapientiae*, a theatre of knowledge,* amounting to a stylized representation of all that could be thought or seen in the world. A cabinet was not literally a single piece of furniture, but any enclosed space – the larger ones might occupy several rooms, usually crammed with objects from floor to ceiling, perhaps with annexes for conducting experiments. In his brief guide to how collections should be organized, *Inscriptiones vel tituli theatri amplissimi* (1565), Quiccheberg explained that they ought to contain 'authentic materials and precise representations of the whole of the universe'. By possessing this microcosm the collector-prince was not just symbolizing but also in a sense *exercising* his mastery of the

---

* The idea of the world as a theatre, often credited to Shakespeare – 'All the world's a stage'; 'The purpose of playing . . . is to hold, as 'twere, the mirror up to nature' – was (as with many of the Bard's most memorable images) in fact a familiar part of the intellectual scene of the late Renaissance.

world. The cabinet acted as a kind of mental laboratory within which the relationships between things could be contemplated via a process that shared elements of both experimentation and Gnostic revelation. Its organization often drew on the Neoplatonic idea of correspondence and analogy between disparate items, whose connections were revealed in resemblances of shape and form: a pre-scientific mode of finding or asserting order in the world. In short, the cabinet was a conduit to deep knowledge about the order of things.

It was acknowledged well before Francis Bacon that understanding promised worldly power: an epigram to Lorenzo de' Medici proclaims 'Because you know everything, you are all-powerful.' It was partly for this reason (and presumably partly just for purposes of sheer account-ancy) that most cabinets were carefully catalogued and their layout painstakingly planned – they were no crude piles of treasure. For you could not claim to be comprehensive in your collection unless you knew what was in it. Bacon himself advised princes and kings to establish a 'goodly huge cabinet, wherein whatsoever the hand of man by exquisite art or engine has made rare in stuff, form or motion, whatsoever singularity chance, and the shuffle of things hath produced, whatsoever nature has wrought in things that want life and may be kept; shall be sorted and included'. In this view, the collection confers power through knowledge by what seems to be a quasi-magical agency: 'Thus when your Excellency shall have added depth to knowledge to the fineness of spirits, and greatness of your Power, then indeed shall you be a new Trismegistus' – that is, the ruler becomes akin to Hermes Trismegistus, the founder of the Hermetic tradition.

But the contents of most cabinets were not literally comprehensive at all, nor even representative – they were *rare*, and they were wonderful, populating a magical pageant of power. That was the legacy of their courtly origins: the rarity flattered the narcissism and self-importance of the collector. Yet some cabinets were sober repos-itories for research and reference, and as such they did admit any item, no matter how humble. Such was the collection of Ulisse Aldrovandi, professor of philosophy and natural history at the University of Bologna, which was widely regarded as one of the finest in Europe even though Aldrovandi was no prince and held no courtly position. His collection included 11,000 animals, live plants and minerals, and several thousand pressed botanical specimens. If he could not get the originals, Aldrovandi employed artists to make paintings of them. 'His Museum', said a visitor in 1571, 'is like a compendium of natural things found below and above ground, in the air and in the water.'

Even those cabinets purporting to be collections in natural history, however, usually contained artifice too: virtuosic carvings, luxurious metalwork, fine paintings. Some of the most prized objects blended the natural and the artificial so that the distinctions were hazy. 'Figured stones' with complex shapes were common, many seeming to be poised ambiguously at the interface of the living and non-living worlds, and these might be further embellished by artists so that figures and faces materialized from the natural patterns. These adroit interweavings of art and nature leaned on metaphorical analogies of appearance, so that the billowing bands of agate were painted to become clouds, shells were made into eyes and lips, coral became the fingers and hair of statuettes. Here nature was an artist, a maker of luxury items that were of a piece with the work of the finest craftsmen. 'There are to be found in the sea such strange and diverse kinds of shells', wrote the French physician Ambroise Paré, 'that one can say that Nature, chambermaid of great God, plays in fabricating them.'

The collection could thus end up as a rather arbitrary and meaningless excess of undigested detail and virtuosity: a display of showmanship and acquisitiveness that was silent about the hidden workings of the world. To a man like Galileo, such an obsession with variety and multiplicity could obscure what he believed to be the unity and simplicity of underlying general principles. He contrasted 'a regal gallery adorned with a hundred classical statues with countless complete historical pictures by the most excellent masters and full of everything that is admirable and perfect' with

> a collection contained in the study of some little man with a taste for curios who has been pleased to fit it out with things that have something strange about them because of age or rarity or for some other reason, but are, as a matter of fact, nothing but bric-a-brac.

The caricature was doubtless sometimes warranted, but it speaks also to the perpetual anxiety in the natural sciences about the difficulty of distinguishing what is superficial and contingent from the real devils in the details.

It seems that collections *could* occasionally inspire the natural philosopher by suggesting new objects of study and new links between them. In his 1611 treatise on the snowflake Johannes Kepler, court astronomer to the Holy Roman Emperor Rudolf II in Prague, mentions that he saw in the palace of the Elector of Saxony in Dresden 'a panel inlaid with silver ore, from which a dodecahedron, like a small hazelnut in size, projected to half its depth, as if in flower' – a showy example of the metalsmith's craft which stimulated Kepler's seminal ideas about how crystals acquire their faceted forms.

The court of wonders

Kepler had another source of inspiration on his doorstep, because Rudolf II had one of the most renowned *Wunderkammern* in all of Europe. It was not merely a set of objects on shelves, but encompassed stables, aviaries and menageries, gardens and ponds, workshops and libraries, collectively embodying a miniature world in which nature, art and science were blended. Rudolf's cabinet proper was divided among four large rooms in his castle in Hradcany, organized into the categories of 'naturalia', 'artificialia' and 'scientifica' (instruments such as clocks and astronomical devices). To be offered a glimpse of these rooms was one of the highest honours that could be bestowed on visiting ambassadors and dignitaries.

By ordering nature and art in his galleries, Rudolf hoped he could grasp and control them beyond the castle walls. The fantastical portraits of Rudolf's court painter Giuseppe Arcimboldo, in which faces are constructed from arrangements of fruits, vegetables and flowering plants, were not as whimsical as they might seem, but flattered the emperor with the conceit that he could impose comparable order, harmony and form on the elements of the world, both natural and political.

But this sympathetic magic didn't work for poor Rudolf. An embodiment of the jibe that the Holy Roman Empire was neither holy, Roman nor an empire, Rudolf was rumoured to be an atheist, he made his court in Prague in Bohemia (it was previously and subsequently in Vienna), and his empire was a ramshackle affiliation of states who paid ever less notice to this reluctant and indecisive ruler, a man who some considered to be more comfortable with wizards and stargazers than with ambassadors and princes.

Towards the end of his reign, the rumours intensified that Rudolf was mad. Visiting Prague in 1609, the Tuscan traveller Daniel Eremita wrote that 'Disturbed in his mind by some ailment of melancholy, he has begun to love solitude and shut himself off in his Palace as if behind the bars of a prison.' Such gossip was useful to Rudolf's antagonists, but all the same it was not mere slander. An awkward man who suffered from depression, Rudolf would have struggled to manage his vast kingdom even in the most peaceful of times. But his accession occurred under impossible circumstances. Nominally the secular arm of an increasingly vicious and doctrinaire Church, the imperial throne was long affiliated by blood and by culture to the German rulers who, as Calvinists and Lutherans, were now locked into fatal conflict with Rome. Even when court artists depicted Rudolf in full armour, grasping a spear astride a rearing warhorse, his embarrassed gaze betrays a fervent wish to be squirrelled away in the library with his esoteric books. The hapless emperor wanted a peace that the rest of Europe was determined to refuse him.

Rudolf II on his warhorse, depicted in an engraving by Aegidius Sadeler (c.1603) after a sketch by Adrian de Vries.

The heir to Maximilian II, Rudolf was sent to be educated in Spain, the traditional seat of Habsburg power. He returned with his stiff disposition accentuated by the cold formalities of the Madrid court, and was crowned King of Hungary in 1572, aged twenty. After his father's death he became King of Bohemia in 1575, and was made Holy Roman Emperor the following year.

Collecting was in Rudolf's family. His uncle Ferdinand, Archduke of Tyrol, had a cabinet, and Arcimboldo was instructed to acquire artworks for Rudolf's father and grandfather. Rudolf was a generous patron of the arts – he commissioned Paolo Veronese and Tintoretto, and had over 3,000 paintings in his galleries, among them works by Bosch, Brueghel and Dürer. His court painter Arcimboldo was himself the archetypal Renaissance virtuoso who could turn his hand to anything from portraiture to engineering, making musical instruments or designing costumes for balls. He was considered something of a natural historian too; Aldrovandi contracted him to paint birds and other animals.

But it was the sciences that commanded Rudolf's principal allegiance. Whatever deficiencies he had as an emperor, he drew to Prague some of the finest philosophical minds in Christendom. His court astronomer was the Dane* Tycho Brahe, whose diligence in observing the movements of the heavens was unrivalled. And Brahe in turn employed the services of Kepler, who subsequently inherited Brahe's position and thereby his superb and voluminous astronomical data.

Rudolf set up laboratories for alchemists. Foremost among them was the German Michael Maier, one of the most formidable and dazzling scholars of nature in Europe, who was also the emperor's chief physician. Maier only began to publish after the emperor's death, when he left Prague; his first book, *Arcana arcanissima* (*Secret of Secrets*) (1614), shows that the tradition of 'secrets' was still thriving. More of an advocate than a theorist of alchemy, Maier is best known for *Atalanta fugiens* (1617), a curious collection of obscure but rather beautiful symbolic alchemical engravings along with poems and musical compositions. His *Themis aurea* is a Rosicrucian tract (see page 71), subtitled *The Laws of the Fraternity of the Rosie Cross* when it was published in

---

* Tycho was born on the Scandinavian peninsula in what is now Sweden, but was then part of Denmark.

English in 1656. Maier visited England after Rudolf's death, spending time at the court of James I and probably meeting with the like-minded alchemical mystic Robert Fludd, before becoming physician to Moritz, Landgrave of Hesse-Kassel.

Other renowned alchemists who came to Rudolfine Prague include Heinrich Khunrath and the adherents of Swiss physician Paracelsus's 'chemical philosophy' Oswald Croll and Michael Sendivogius. Giordano Bruno passed through, and John Dee, sometime astrologer and mathematician to Elizabeth I, travelled to Prague in search of better fortune and patronage when he fell out of royal favour in England. Rudolf tried, without success, to lure the great Della Porta to his court. And along with the jewels came plenty of dross: charlatans and quacks from all over made their way to Rudolf's high citadel, hoping to impress him with their trickery. As a Venetian observer wrote, 'he delights in hearing secrets about things both natural and artificial, and whoever is able to deal in such matters will always find the ear of the Emperor ready'.

While it would be too much to say that Rudolf was running a state-sponsored scientific programme, nonetheless his rule offered ample evidence of what could come from such a concentration of fine minds cushioned by the coffers of a wealthy patron. In particular, the combined efforts of Brahe and Kepler transformed the understanding of the heavens, not only by mapping out the paths of the planets in unprecedented detail but also by uncovering the mathematical laws that governed them. And Rudolf's passion for Neoplatonism and Hermeticism guaranteed that his proto-scientists had an experimental bias, driven by curiosity about nature. Yet Rudolf was not the ideal patron. He did not always pay what he promised; Kepler was constantly having to chase his wages, and both he and Brahe were diverted by their duties to provide regular astrological forecasts. Even if they had no ideological objections (astrology was almost universally accepted in principle, if not in practice), this was a wasteful drain on time. Moreover, Rudolf did nothing to foster a feeling of community among his scientific courtiers – he operated very much on the old model (shared by the Elizabethan court in England) of a ruler surrounded by advisers vying for attention and favour, so that any collaboration was ad hoc and hampered by competitive instincts.

The Winter King

Rudolf may have been a flawed emperor, but it seems possible that this in itself held Europe to a precarious peace. By prevaricating, he never precipitated the religious conflict that constantly loomed. By the end of the sixteenth century, Rome was beginning to doubt the commitment of this emperor who displayed such tolerance for other faiths. Yet Rudolf's clemency made Prague a burgeoning intellectual centre: the Jews were safe during his rule, and Protestants such as Kepler were welcomed at the nominally Catholic court. This was consistent with Bohemia's long tradition of nonconformism. The reformer Jan Hus was burnt at the stake for heresy in 1415, but his Hussite followers subsequently repelled armies of crusaders, and it was natural that the Moravian Church should find common cause with the Protestant Reformation.

By 1600 the papal ambassador in Bohemia, Cardinal Filippo Spinelli, felt compelled to advise Pope Clement VIII that 'It is generally agreed amongst Catholics in Prague that the Emperor has been bewitched and is in league with the devil.' In fact he seems to have had a nervous breakdown, eventually becoming so ineffective, reclusive and disturbed that his estranged brother Matthias (who had been bequeathed next to nothing by Maximilian II) marched against him and by 1611 had taken control of Bohemia, Austria and Hungary. Rudolf, stripped of any real power but permitted still to call himself emperor, was allowed to haunt Prague Castle, a pitiful and broken man until his death a year later. When Matthias died in 1618, the House of Habsburg came under the charge of Archduke Ferdinand of Styria, a fanatical Jesuit whose attacks on the Bohemian Church led to the famous Defenestration of Prague and eventually to outright revolt. Thus began the abominable Thirty Years War which devastated the German lands until the middle of the century.

These political and religious machinations are more than just the backdrop for the transformation of natural philosophy in the early seventeenth century, for they dictated the mood of those times. In 1619 the Bohemian rebels offered their crown to the Protestant Frederick V, Elector Palatine of the Rhine and head of the Protestant Union of German states, who six years earlier had become son-in-law to James I of England by marrying his daughter

Elizabeth, cementing the alliance between England and the German Protestant states that existed during Queen Elizabeth's reign.

This couple were no ordinary royals, but courtly scholar-rulers who embraced the Hermetically tinged experimental philosophy. Frederick's enthusiasm for the culture of wonders and ingenious mechanisms was reflected in the gardens that he commissioned for his new wife at his palace at Heidelberg. They were designed by the French Protestant engineer Salomon de Caus, former tutor to Elizabeth in the Stuart court, and they epitomized the allegorical character of late Renaissance landscape architecture: a world of wonders comparable to the *Wunderkammern* themselves. De Caus, thought to have been a member of the Rosicrucian movement, allegedly planned the gardens as a representation of the cosmos, ordered by geometrical and mathematical principles. They were populated with curious grottos and mechanical marvels, such as pneumatically driven mobile statues and bird automata that sang. De Caus made a water-organ like that described by the Roman engineer Vitruvius, and he recreated the statue of the legendary Ethiopian king Memnon said to have been built at Thebes by the Egyptians, which made musical sounds when sunlight fell on it. He dedicated his book on automata and fabulous machines, *Les Raisons des Forces Mouvantes* (1615) – a celebration of the noble, princely form of curiosity – to Frederick's bride.

Like the cabinets of curiosities, these wonders were regarded not as charming diversions but as expressions of occult wisdom. They showed how the forces of nature, such as those involved in hydraulics, could be harnessed 'out of sight' to create artefacts mimicking those in the natural world. Such apparently whimsical contrivances were one of the key uses to which science and technology were put in the late Renaissance. John Dee devised clever stage effects for the Elizabethan theatre, and Francis Bacon helped construct the entertainments for the wedding celebrations of Frederick and Elizabeth when they married in 1613. James I was beguiled by the Dutch inventor and virtuoso Cornelius Drebbel, who paraded before the royal court devices to 'make it rain, lighten, and thunder . . . so that you would have sworn it came in a natural way from heaven'. Virtuosic inventions and devices were described in popular books such as John Bate's *Mysteries of Nature and Art* (1634) and John White's *A Rich Cabinet, with Variety of Inventions* (1651), which seized on the latest advances in scientific instrumentation solely to concoct enthralling experiments. This wasn't exploitation – it was commonly thought that this was precisely what mechanical and instrumental ingenuity was *for*.

The liberal, humanistic climate of Rudolf's Prague spread to Germany during the early seventeenth century, and Heidelberg enjoyed a reputation almost equal to Prague's. Frederick, the most powerful of the German Electors, was therefore the obvious choice for the Bohemians when they sought to throw off the yoke of the new emperor. But it was a false dawn. Frederick was no match for the Habsburg army, which invaded the Palatinate in September of 1620 and crushed the Protestant forces at the Battle of the White Mountain, near Prague, in November. Frederick and Elizabeth ruled Bohemia for little more than a single winter of 1619–20, and were mockingly called the Winter King and Queen. They retired in exile to The Hague, where they created a modest court that welcomed intellectuals. René Descartes dedicated his *Principia philosophiae* (1644) to their eldest daughter, also called Elizabeth.

From secrets to lynxes

Given the reputation of Giambattista Della Porta (and how loudly Kepler sung his praises), it is no surprise that he was among those who Rudolf hoped to attract to Prague. But in 1610 the ageing Della Porta found another benefactor in Federico Cesi, the young Duke of Umbria. In 1603 the eighteen-year-old Cesi had convened a society of 'searchers of the arcane sciences', modelled on the academies of secrets and inspired by Della Porta's writings and example. At its inception the group had just four members: besides Cesi, it included the Dutch physician Johannes van Heeck (Heckius) and the Italians Francesco Stelluti, a mathematician, and Anastasio De Filiis, a mechanician – all of them young men. Taking as their credo Della Porta's declaration in *Natural Magick* that the philosopher should 'examine with lynx-like eyes those things which manifest themselves', they called themselves the Accademia dei Lincei: the Academy of Lynxes.

According to Stelluti, the Lincei aimed to 'penetrate into the inside of things in order to know their causes and the operations of nature that work internally, just as it is said of the lynx that it sees not just what is in front of it, but what is hidden inside'.* In other words, they maintained a belief that the true ways of nature were secret and occult, and needed to be teased out with skill, perspicacity and uncommon keen-sightedness: with an ability to 'see beyond'. Heckius called the group 'most sagacious investigators of the arcanities of nature and dedicated to the Paracelsan disciplines' – like the iconoclastic Swiss physician Paracelsus, they were unwilling to rely on the writings of ancient authorities but insisted that only direct empiricism could reveal what nature had hidden. They combed the hills and valleys of southern Umbria for specimens, they built instruments including a large and complex astrolabe for measuring the positions of the stars, and they studied Arabic in order to scour the commentaries of the medieval Muslims.

In contrast to the courtly philosophers, the Lincei were openly fervent about science, rejecting the insouciant mask of *virtù*; 'I hate the court and courtiers like the plague', Cesi declared. They agreed to exclude the divisive topics of religion and politics and to devote themselves to science alone. Like over-earnest teenagers, they adopted grand-sounding Latin names: Cesi was Coelivagus, the heavenly wanderer, Heckius was Illuminatus, De Filiis was Eclipsatus, and modest Stelluti, Tardigradus the 'slow-stepper'.

Academies that stirred up new ideas were still not welcomed by the leaders of the Counter-Reformation, and Cesi's father, the Marchese of Monticello and Duke of Acquasparta, would have none of this dangerous dabbling. Fearing that Heckius was a northern Protestant out to convert his son (which shows how little he knew of this almost rabidly Catholic Dutchman), the duke insisted that Cesi disband the group. He had Heckius framed for heresy and expelled from Rome.

But Cesi was not so easily deterred. He maintained correspondence with his colleagues, although this was far from easy in the case of Heckius, a fraught individual who gradually succumbed to paranoid mental decline as he wandered through Europe. Perhaps unsurprisingly, he ended up in the ferment of Rudolfine Prague, where he met Kepler. In 1605 Cesi sponsored the publication of Heckius's treatise

---

* The lynx's ability to see through solid matter was legendary: Plutarch said that it 'can penetrate through trees and rocks with its sharp sight'.

on the 'new star' that appeared the year before, on which Kepler published a much more widely read book a year later (see page 219). But Cesi felt it necessary to tone down Heckius's gratuitous diatribes against Tycho Brahe, whom he damned as a Calvinist and anti-Aristotelian – Cesi realized that the new astronomers were the natural allies of the Lincei, and that it would not do to alienate them. Heckius was furious at these changes made without his approval (and to which, as Cesi surely knew, he would never have acceded). But he kept in touch with Cesi all the same, united by an intense passion that probably had a sexual element.

Anastasio De Filiis died in 1608, and with Heckius now unstable and all but vanished from the scene, Cesi and Stelluti were all that remained of the Lincei when Cesi's father died in 1610. But Cesi had been encouraged by Della Porta's words of support when he visited the grand old man in Naples in 1604, and so the new young duke, finally at liberty to pursue his vision, now wasted no time in asking Della Porta to become the society's fifth member. The venerable philosopher was flattered and charmed by Cesi, and he agreed.

A year later the society acquired a sixth member. At a banquet held in Rome, Cesi met the astronomer who had just astonished the world with a pamphlet entitled *Sidereus nuncius* (*The Celestial Messenger*) that described telescopic observations of the rugged surface of the moon, and he recognized that this was just the person to add lustre to the Lincei. That was how Galileo found himself in a club with the natural magician Giambattista Della Porta.

It is sometimes said that this pairing of the old and fading Della Porta with the vigorous, ascendant Galileo represented the way science itself was changing: the natural magic tradition of secrets, codes, signatures and occult forces yielding to the Galilean model of observation, openness and scepticism. But Della Porta's world was not as moribund as the man himself. Many others shared his vision in some form – indeed, it was in the process of being reborn, in Prague and elsewhere. It was arguably Galileo, not Della Porta, who was the anomaly in this milieu. And yet Galileo and Della Porta did not find their approaches to natural philosophy so alien to each other; their differences made for a somewhat prickly relationship, yet it was one marked by mutual respect. When they briefly fell out (see page 223), it was over an issue of scientific priority rather than matters of ideology or methodology.

Nonetheless, Galileo's entry into the society did mark a shift from the old habits of secrecy and esotericism to the new ideal of transparency, and from the study of nature for personal edification to the application of knowledge for the good of humanity. In 1616 Cesi explained that the knowledge that the Lincei sought should be used not for profit, honour or reputation but to help humankind. For that reason, it should be made public. For his part, Galileo was evidently proud to be a Lynx, placing his affiliation on the title pages of the books, including *Letters on the Sunspots* (1613) and *The Assayer* (1623), that Cesi funded.

If it is mentioned at all in accounts of early seventeenth-century science, the Accademia dei Lincei usually appears as an incidental

ISTORIA
E DIMOSTRAZIONI
INTORNO ALLE MACCHIE SOLARI
E LORO ACCIDENTI
COMPRESE IN TRE LETTERE SCRITTE
ALL' ILLVSTRISSIMO SIGNOR
MARCO VELSERI LINCEO
DVVMVIRO D'AVGVSTA
CONSIGLIERO DI SVA MAESTA CESAREA
DAL SIGNOR
GALILEO GALILEI LINCEO
Nobil Fiorentino, Filofofo, e Matematico Primario del Serenif.
D. COSIMO II. GRAN DVCA DI TOSCANA.

IN ROMA, Appreffo Giacomo Mafcardi. MDCXIII.
CON LICENZA DE' SVPERIORI.

The frontispiece of Galileo's *Istoria e Dimostrazioni Intorno Alle Macchie Solari e Loro Accidenti Rome* (*History and Demonstrations Concerning Sunspots and their Properties*) (1613), bearing the crest of the Accademia dei Lincei.

footnote. It deserves better. The activities of the Lincei, and the efforts of Cesi in particular, have never been accorded their due. The aims and concerns of the society anticipated many of those that more celebrated figures would develop over the course of the century – in particular, the need not just to catalogue the world but to find ways of ordering, classifying and thus taming its extravagant variety. In this way Cesi hoped to advance beyond the random, chaotic profusion of medieval and Renaissance bestiaries and 'natural histories'. He didn't succeed – but as we shall see, neither did more illustrious like-minded souls such as Francis Bacon and Robert Hooke. It is an indication of the close affinity between the motivations of the Lincei and those of Bacon, discussed in the next chapter, that when one of their number, the resourceful Cassiano Dal Pozzo, went to France on a diplomatic mission with Cardinal Francesco Barberini, Pope Urban VIII's nephew, in 1625–6, he recommended that the society consider Bacon for membership. (Cassiano found no Frenchmen worth proposing; they are, he said, only interested in women.)

The Lincei were never a populous fraternity: only a total of thirty-one members were ever enrolled, and there were never more than twenty at any one time. When Cesi died in 1630 at the age of forty-five, they were left without funding or leadership. Stelluti hoped that Francesco Barberini, a friend of Galileo who had himself become a Lynx in 1623, might be persuaded to take over the helm, but he did not. Stelluti kept the flame alive until his death in 1652 brought an end to the Academy of Lynxes. But they had made their mark in many ways, and we shall hear more of them later.

Utopian brotherhoods

By the mid-seventeenth century, the notion of a society dedicated to the diligent investigation of nature in all her designs was well established. It grew from the soil tilled by the professors of secrets and the Neoplatonic advocates of natural magic. The fruits bore telltale traces of these origins.

These curious societies – some fictional, some actual – were generally envisaged as utopian projects. One of the most influential was described by the Italian philosopher and Dominican friar Tommaso Campanella. Here is another of the perplexing, inventive figures who, because they do not fit easily within the conventional narrative of the Scientific Revolution, are generally denied any role at all. Born in Calabria in 1568, the twenty-year-old Campanella came across Bernardino Telesio's *De rerum natura iuxta propria principia* (*On the Nature of Things According to their Own Principles*) in that author's home town of Cosenza. Campanella was gripped by this new philosophy of nature that gave precedence to experience over Aristotle and book-learning. Visiting Naples in 1589 he was drawn into Della Porta's circle. But like Giordano Bruno, he could not stay out of trouble. Campanella was imprisoned by the Inquisition in 1594 on suspicion of heresy, and had been out of prison for just two years before being thrown back inside for his part in a plot to expel the Spanish from Naples. Amazingly, he avoided the death sentence by feigning madness, but was instead incarcerated for a further twenty-seven years, during which time he studied and wrote. (The German papal doctor Johann Faber, a member of the Lincei, petitioned the viceroy of Naples unsuccessfully for his release.) He was eventually exiled to France, where he was welcomed in the court of Louis XIII and generally flourished until his death in 1639.

Campanella's philosophical sympathies were clear. Calling Aristotle the 'tyrant of minds', he studied Ficino and his commentaries on Plato, but also welcomed the boundless curiosity about the natural world in Pliny's *Natural History*. Campanella's *On the Sense of Things and On Natural Magic* helped to promote the tradition in northern Europe – he began writing the book after meeting Della Porta but it was published only in 1620 after the Frankfurt publisher Tobias Adami visited him in prison and took some of his manuscripts back to Saxony. While praising Della Porta's efforts to strip natural magic of superstition and association with demons, Campanella criticized him for not going far enough to offer rationalistic explanations of the host of bizarre and wondrous phenomena described in *Natural Magick*.

Campanella's explanations did not reach very much further towards any real understanding, however. His vision of the world was more or less animistic: every material, from minerals and metals to humans, has a kind of spirit or 'sense' that acts for its own self-preservation. Natural magic consists in the ability to alter and manipulate this tenuous spirit – stuff sufficiently vague and all-pervasive to justify stock wonders of sympathetic magic such as the 'weapon salve', in which wounds are cured with a lotion applied to the blade that caused them, and the shattering of a sheepskin drum from 'fear' when a wolfskin drum is sounded.

More influential in the long term than these somewhat derivative theories of natural magic was Campanella's notion of how science might guide society. Another of his books, published in Frankfurt but written much earlier, was *City of the Sun* (1623), which was framed as a dialogue between a Genoese sailor who had sailed with Columbus to America and a Knight Hospitaller. The sailor explains that on his travels he found himself in an unknown land in which there was a great walled city where the citizens were dedicated to a scientific understanding of the world. The city contains seven concentric walls, on which are painted pictures and diagrams that teach the citizens about the arts and sciences: an encyclopaedic mural that makes knowledge available to everyone. Here are images of the arrangements of stars and planets; geometrical figures; depictions of all the flora, fauna and minerals of the natural world; plans for ingenious inventions and descriptions of the mechanical arts. Among these inventions are some of those 'futuristic' machines that periodically excite breathless claims of foresight in populist accounts that cannot imagine why early technologists might have wished to be able to fly like birds, travel underwater like fish, or propel their boats without reliance on the whims of the wind. It's more striking that the mechanical arts are awarded such priority in the first place: whereas Aristotle felt that artisans and manual workers did not deserve full citizenship, getting your hands dirty and sweat on your brow was a positive virtue in Campanella's City of the Sun. By running around the city walls, the children of this utopian society would come to understand the sciences before they were ten years old, 'without effort, merely while playing'.

The city is a spiritual community but not an explicitly Christian one: Christ and the apostles are shown among the 'legislators' on the outside wall, but so are Moses, Osiris, Jupiter, Mercury and Mohammed. Although the citizens believe that the soul is immortal and that a deity governs the universe, Campanella's dangerously bold proposal is for a universal religion ordained in the stars. 'I shall make the heavens a temple and the stars an altar', he declared in one of his poems. The citizens offer prayers to the heavens, and their priests are astronomers and astrologers. The very city itself is organized along astrological lines, its seven walls echoing the seven planetary orbits. The cosmology of the City of the Sun is thus Copernican: its central altar has the form of the sun.

What most distinguishes Campanella's utopia from the fictional kingdom of that name described by Thomas More in 1516 is the emphasis on learning and natural history rather than political and social organization. While More's work contains elements of satire absent from the City of the Sun (but which returned to 'utopian' literature with a vengeance later, as we will see), it was concerned mostly to depict a peaceful and egalitarian society* in which there is no private ownership, the inhabitants pay their way with honest labour, and religious diversity – excepting atheism – is tolerated. Science and technology play little part in this largely late medieval vision, and while More's Utopians are contented and well cared for, they show no sign of being in the slightest bit curious.

The Rosy Cross

It was a mystico-scientific utopia like Campanella's that many in Bohemia had hoped to see materialize when they made Frederick V their short-lived ruler. At that time, intellectuals throughout Europe had been agitated by the appearance of the Rosicrucian manifestos, which issued a call for the members of this putative secret brotherhood to declare themselves. Known by their abbreviated titles of the *Fama* and the *Confessio*, these documents were published in 1614 and 1615 respectively, in Kassel in the Protestant state of Hesse. The authors were never disclosed, although the publication of a third Rosicrucian tract, *The Chemical Wedding of Christian Rosencreutz*, in 1616 by the Lutheran Johann Valentin Andreae of Württemberg, the state adjoining the Palatinate, raises the suspicion that he had a hand in them.

The *Fama* calls for a new philosophy to replace that of Aristotle and Galen, based instead on wisdom gained directly from God and nature: in other words, essentially the same Hermetic programme that was endorsed by Della Porta, Telesio and Campanella. Andreae's *Chemical Wedding* explains that such knowledge had been gained in

---

* That is, within the moral conventions of the age: More's citizens own slaves.

the East by a fifteenth-century Dutchman named Christian Rosencreutz, who had spread it through a secret network of adepts, the successors of whom – the Rosicrucian Order of 'learned men, magicians, cabalists, physicians and philosophers' – are now dispersed throughout Europe. The manifestos announced that it was time to renew this brotherhood and to assemble from the Book of Nature 'a perfect method of all arts'. The Rosicrucian texts drew heavily on the *Monas hieroglyphica* (1564) of John Dee: the *Chemical Wedding* prominently reproduces the occult symbol – a kind of cosmic cryptogram – that Dee introduced in the book, while an introduction to the *Confessio* quotes theorems from his book.

This was not just a philosophical manifesto but a political one, and it is no coincidence that it came just as Bohemia was looking to the Elector Palatine to lead them into a new golden age. The *Chemical Wedding* makes reference to the marriage of Frederick and Elizabeth Stuart, and the *Confessio* denounces the Pope as the Antichrist. The political message was clear to some of Frederick's opponents, who satirized aspects of the manifestos in images that were circulated to discredit the Winter King after his defeat.

The Rosicrucian movement provided a focus for supporters of a new philosophy. One such was the Neoplatonic English physician Robert Fludd, who considered that the teachings of the classical authorities should be replaced by natural magic, alchemy and the Paracelsian 'chemical medicine' that flourished in the court of the French king Henry IV. Responding to a criticism of the Rosicrucian

The Rosicrucian text *The Chemical Wedding of Christian Rosencreutz* reveals its debt to the Elizabethan mysticism of John Dee by reproducing his cryptic symbol called the 'hieroglyphic monad'.

manifestos published in 1615 by the Saxon physician Andreas Libavius, Fludd's *Apologetic Tractatus for the Society of the Rosy Cross* (1617) argued for a complete reformation of science and the scholastic curriculum, in which the traditional mathematical methods of axioms, definitions and theorems should be replaced by the Gnostic approach of the Pythagoreans, which could reveal the scheme of cosmic harmony that underpinned the world.

As Fludd's dispute with Libavius shows, his opponents were not all reactionaries eager to defend Aristotle. His ideas about universal harmony trod on the toes of Johannes Kepler, who was putting the finishing touches to his own work on that subject, *Harmonices mundi* (*Harmonies of the World*), published in 1619 (see page 199). It is not surprising, then, that Kepler sought to establish his precedence by adding to his book an appendix that criticized Fludd's Hermetic approach to the advantage of his own. Kepler asserted that Fludd's explanations were, as he put it, 'enigmatic, emblematic and Hermetic', while his own were based on clear mathematical principles. There was something in that, not least because, as Kepler pointed out, he but not Fludd was willing to change his hypothesis if quantitative observations of the real world – especially the movements of the stars and planets – did not accommodate it. Fludd responded by branding Kepler precisely the kind of mathematician of which he disapproved, who only 'concern themselves with quantitative shadows'. It is tempting now – especially in view of the posthumous reputations of these two men – to make Kepler the modern progressive and Fludd the muddled mystic. But in fact their dispute was, as many are, all the more bitter because the distance between the antagonists was so small. As historian of science Allen Debus says, 'No less than Fludd did [Kepler] believe in an *anima mundi* [a vital 'world spirit'], no less than Fludd did he argue for a near divine sun that belonged in the center of the world, and no less than Fludd did he believe in the stars as living entities.'

Despite these arguments, no one was quite sure who the Rosicrucians were. That's the problem with a secret society: you don't know who's in it and who isn't. The Brothers of the Rosy Cross were said to form an 'invisible college'; in Paris they were simply called the Invisibles. Rumours abounded of intellectuals eager to join but frustrated in their attempts to find the right contacts. Others were allegedly entreated by shadowy figures to enlist – the alchemist Michael Sendivogius apparently refused the advances of two Rosicrucians who sought him out in the remote castle in Poland where he worked after leaving Rudolf's court. Some Catholics considered the Rosicrucians to be evil magicians and subversives. One of these critics was the French Jesuit priest Marin Mersenne, who also became embroiled in a frustrating argument with Fludd, primarily about whether alchemists had any right to draw religious inferences from their work. Both Mersenne and his friend Pierre Gassendi in Paris were supporters of the mechanistic philosophy of René Descartes, which attempted to explain all physical phenomena in terms of the mechanical interactions and collisions of the tiny particles of which they were composed. This too was a 'new philosophy', but of a very different order from the Hermetic chemical philosophy founded by Paracelsus and expounded by Fludd and many other Rosicrucians, in which mathematics and indeed all logical reasoning tended to rest on analogy rather than deduction and proof.

But this secret society did not necessarily respect religious boundaries. After all, Descartes, who was also educated as a Jesuit, was rumoured to have became a Rosicrucian in Germany in 1619 (others say he tried but failed), even though he was there to join the forces of the Duke of Bavaria in suppressing Frederick V's claim to the Bohemian crown. Descartes in fact seems perversely capricious in his allegiances. He soon lost interest in the Bavarian enterprise, but nonetheless joined the Catholic army that fought at the Battle of the White Mountain, and he saw the subsequent entry of Emperor Ferdinand's victorious troops into Prague. Yet he became so dedicated to the tutorship of the Winter Queen's daughter Elizabeth that he later moved to Leiden to be close to her in The Hague, and in his dedication to her in the *Principia* he called her the daughter of the King of Bohemia, a title that Frederick's enemies would never award him. Elizabeth, the Winter Queen herself, retained her intellectual vitality, winning the affection of scholars of all persuasions.* Exiles from the strife in Europe, Protestants from the Palatinate and Bohemia and elsewhere, would come to her at The Hague, though she was in no position

---

* Her grandchild George become the first Hanoverian English monarch in 1714.

to offer financial succour. Among them were the Prussian scholar Samuel Hartlib, and his mentor the Moravian Johann Amos Komensky (known as Comenius), a bishop of the Hussite Bohemian Brethren. Hartlib fled from Prussia after the Catholic conquest of Elbing, where he had been at the centre of a mystical society akin to the Rosicrucians. He settled in England in 1628, where he became a prominent voice for social and scientific reform. Both Hartlib and Comenius expressed their visions of a better society in the form of utopian fables.

Even before the defeat of the Winter King, Johann Valentin Andreae seemed to conclude that the cloak-and-dagger conspiracy of the Rosicrucian Order was not the best way to reform learning.* Instead he began from around 1617 to promote 'Christian Societies' which would more openly promote charity and love as well as scientific knowledge. He laid out his plans in *Reipublicae Christianopolitanae descriptio* (1619), which describes a Christian society called Christianopolis. In a later letter to Hartlib in London he claims that he actually took steps to establish such a community somewhere in Germany, supported by an unidentified prince (perhaps August, Duke of Brunswick and Lüneburg) until it was broken up by the onset of the Thirty Years War. He attempted to resurrect this scheme at Nuremberg around 1628.

Christianopolis is another island utopia. Like the City of the Sun, its city-state is organized along geometric lines (here concentric squares), and the society venerates the mechanical and manual arts: 'Their artisans are almost entirely educated men'. Again,

---

* Andreae is one of the most perplexing figures in the 'history of curiosity'. His backing – some might even say invention – of the Rosicrucian movement seems to place him squarely within the tradition that encourages the investigation of 'secrets'. But in 1620 Andreae published *Treatise on the Pestilence of Curiosity*, which condemns this trait as 'that immodest thirst to know and do what lies beyond the customary cleverness of human beings' and accuses it of inhibiting the establishment of his Christian communities. Even more bafflingly, this tract attacks the Rosicrucians themselves as 'the inner core of curious people of our time.' Although Andreae never publicly admitted to authorship of *The Chemical Wedding* and the *Fama*, there seems every reason to attribute them to him. Did he have a change of heart, or is this another example of the early modern thinker's apparent ability to hold contradictory beliefs? Or is Andreae simply using the multivalent word 'curiosity' to make a moral point?

knowledge is conveyed partly by easily assimilated visual aids: in the natural history laboratory the walls are painted with depictions of animals and natural phenomena. Like Campanella, Andreae presents a civic union of religion with science, predicated on a Platonic concept of God as an architect of a divine mechanism governed by harmony.

Nothing better illustrates the corrosive effect of the Thirty Years War than the contrast between Andreae's optimistic vision at its onset and that written by Comenius four years later (but not published until 1631), titled *The Labyrinth of the World*. Again this describes a city divided up like an encyclopaedia, the districts and streets representing different branches of learning. But to the pilgrim who visits, it becomes clear that there is little order or harmony to be found. Religious sects and princes fight with one another in the streets, and Comenius implies that all knowledge is untrustworthy and futile.

Comenius's spirit seems to have rallied when he moved to England around 1641. In *Via Lucis* (*The Way of Light*, written 1641, published 1668), he foresees a time when all knowledge of the world has been assembled into a kind of universal wisdom: 'An Art of Arts, a Science of Sciences, a Wisdom of Wisdoms, a Light of Light'. He proposes the establishment of a sacred college or society devoted to seeking wisdom for the well-being of all mankind. But again worldly conflict got in the way: Comenius returned to Poland in 1642 as England slid towards civil war. All the same, Comenius took heart from the establishment of the Royal Society after the Restoration (see Chapter 5), dedicating his book to them and calling them (in a rather Neoplatonic image) the 'illuminati', 'Torch Bearers of this Enlightened Age'.

Thus the enthusiasm for a new philosophy to replace obsolete scholasticism, organized by a brotherhood of savants and reliant on experiment and experience to illuminate every wondrous corner of nature, was gathering momentum in the early seventeenth century. But had political events not turned out as they did, its eventual realization might have had a quite different character. The mystical philosophies presented by the likes of Campanella, Dee and Maier, cultivated in Prague, the Palatinate and to some extent the court of James I of England, and informed by Rosicrucianism and Hermeticism, could have been the dominant model, had the Thirty Years War not crushed this dream and scattered its supporters. Who knows what this model of science might have looked like as it matured? But even as it was, the vision exerted more influence than is usually appreciated. For it is within this context that we can now appreciate the origins of the new philosophy advocated by Francis Bacon, which is commonly considered to have sparked the Scientific Revolution. The traditional picture of Bacon as a modern empirical scientist who broke away from a mystical past is then no longer tenable. Indeed, according to the seventeenth-century writer John Heydon, Bacon was himself a Rosicrucian – the truth of which is far less important than its plausibility.

The Lord Chancellor's Utopia

Francis Bacon was raised for greatness, but what he achieved of it in his lifetime was repeatedly compromised and seemed to bring him little satisfaction. His father Sir Nicholas Bacon was Lord Keeper of the Great Seal of England for Elizabeth I, one of the key offices of state in the Tudor monarchy. Francis studied at Trinity College in Cambridge before enrolling as a law student at Gray's Inn in London. In 1589 he was appointed Clerk of the Council in the Elizabethan Star Chamber, although he was not at liberty to occupy that post until it became free under James I in 1604. In the meantime Bacon had managed to fall out with the Virgin Queen for opposing her request to the English Parliament for more funds. Some have suggested that Bacon's testimony against his former patron Robert Devereux, the second Earl of Essex, when Devereux was tried for treason in 1601 after open rebellion against Elizabeth, was an attempt to win back the queen's favour. In any event, Bacon's part in the trial, which led to Essex's execution, was widely considered dishonourable – and it seemingly did little to improve his standing at court.

The ambitious Bacon found better fortune with James I. He was knighted in 1603, and ten years later he began to rise up through the offices of state. He was made Attorney General, and then in 1616 became a member of the king's Privy Council. The following year he took up his father's former post, before being made Lord Chancellor (and accordingly, Lord Verulam of St Albans) in 1618. This made him one of the most powerful men in the country – which was why his fall from grace in 1621 was all the more spectacular. He was accused of abusing his office by taking bribes, and was imprisoned in the Tower of London. Although the price he paid was ultimately mild – through the king's intervention, he spent only three days in prison and the fine of £40,000 was remitted – he was forbidden to hold any office of state. Bacon had by this stage already set out the his agenda for a new form of natural philosophy, especially in the *Instauratio Magna* (*Great Instauration*) and the *Novum Organum* (*New Organon*), both published in 1620 after three decades of gestation. Now excluded from political life but still immensely wealthy, he had time enough to illustrate his ideals in the utopian tract *Nova Atlantis* (1624), published in English (*New Atlantis*) in 1627, a year after his death.

Popular legend has it that experimental science was Bacon's nemesis: he is said to have died after catching a cold while investigating whether meat might be preserved by freezing. Bacon's last amanuensis, the philosopher Thomas Hobbes, told John Aubrey – who revelled in such anecdotes in his *Brief Lives* – that Bacon was travelling by coach to Highgate in bitterly cold weather, when he alighted to buy a hen and stuff it with the snow that lay on the ground. He quickly became so ill that he was taken to bed at the Highgate house of the Earl of Arundel and died three days later. Although that was indeed where Bacon died, the veracity of the 'chicken' story has been questioned. He does seem to have been taken ill while on travel, and to have sought refuge in Arundel's house, but his fatal malady may have had other causes. Arundel was not there at the time – he had been taken to the Tower himself on the orders of Charles I, for reasons of subtle political intrigue – and Bacon sent him a letter explaining the imposition, saying that he had been taken by vomiting but 'knew not whether it were the stone, or some surfeit, or cold, or indeed a touch of them all three'. Historians Lisa Jardine and Alan Stewart suspect that the cause was a different kind of experiment: Bacon had recently been investigating the effects of opiates, and may have indulged too liberally.*

---

* Jardine and Stewart think that Bacon's letter to Arundel was another example of his political wiliness, even in that extremity: he was keen to establish an alibi for

Although *New Atlantis* is normally discussed as another vision of a technocratic utopia, a surprisingly large part of the book is devoted to scene-setting. (It is hard to know how much this is because the work was never finished, being published in its incomplete form.) Bacon's fictional land of Bensalem is said to have existed at the same time as the Atlantis mentioned in Plato's *Critias*, but it survived the flood that extinguished that legendary kingdom and has been protected from harm and decay thereafter by its mastery of science. Yet the subsequent decline in navigational skill in the rest of the world meant that the ships of the Phoenicians, the Carthaginians, the Chinese and others were no longer able to reach Bensalem, and its existence fell out of memory except for chance encounters like that which brings the book's narrator there. The island was converted to Christianity by a divine revelation twenty years after the Crucifixion.

The description of Bensalem's Christian foundations occupy Bacon at some length, for they are an essential part of his message: as with Andreae's Christianopolis, the perfection achieved by this society through mastery of nature is necessarily rooted in the Christian faith. This domination of nature is made possible by the knowledge gathered and utilized in Bensalem's 'temple of science', called Solomon's House. Here a caste of scientist-priests apply their understanding to make many wondrous things for the benefit of the citizens, while carefully regulating what information they make public and what they conceal. In describing the goals of Solomon's House, Bacon uses the old discourse of the 'secrets of nature': 'The End of our Foundation is the knowledge of Causes, and secret motions of things; and the enlarging of the bounds of Human Empire, to the effecting of all things possible.'

It is that 'all things possible' which draws most comment today, both from those who celebrate science's ingenuity and its capacity to relieve man's estate and from those who deplore Bacon's coercive attitude towards nature. While both points of view have their merits, Bacon's account of the wonders of Solomon's House loses much of its meaning when removed from its original context.

For one thing, the influence of the natural-magic tradition can be discerned in the way Bacon celebrates the manual, mechanical and

---

being at the discredited earl's house.

agricultural arts. Interest in questions of a technological nature had been growing since the early sixteenth century because of their economic, industrial and military implications. Rulers could appreciate that improvements in naval engineering, ballistics, mining, chronometry, printing and textile manufacture could have a significant bearing on their wealth and power. But serious treatments of these issues tended still to remain separate from the academic tradition. The German scholar Georg Agricola's authoritative text on mining, *De re metallica* (1556), was unusual in being the work of a university-trained physician, while just about the only other important manual on the subject, *De la pirotechnia* (1540), was written by a Sienese metalsmith and military engineer, Vanniccio Biringuccio, and was published in the Italian vernacular rather than scholarly Latin. To learn about arts such as dyeing, alchemy, metallurgy and printing, Rabelais' Gargantua in his sprawling mid-sixteenth-century tale had to go out with his tutor to visit the craftsmen at first hand (and only an iconoclast such as Rabelais would consider these things part of a proper education in the first place). By and large, written accounts of such topics were confined to the *Kunstbuchlein* of the tradition of secrets.

Yet mining and metallurgy is the first area of study shown to Bacon's narrator. He is regaled with a string of ingenious yet decidedly earthy marvels, not the least of which seems to be an account of composting: the production of new earths and soils 'for the making of the earth fruitful'. There are wells and fountains for making minerals and salts, and spacious houses where artificial snow, hail and rain are made. In orchards and gardens, plants are grown that may flower and fruit out of season, or faster than in nature, or which give fruits that are sweeter or differing in taste, smell, colour and shape from the natural varieties (all of which recalls Della Porta's interest in horticultural breeding techniques). Animals are bred or chemically altered to create new varieties and species, including 'commixtures' and strains that are more and less fertile than in nature. There are fabulous 'brew-houses, bake-houses, and kitchens': a reassuring amount of effort apparently goes into making superior wines and dishes. Nor are the decorative arts neglected: there are new dyes, papers, silks, tissues. Houses of sound produce music 'sweeter than any you have', as well as 'strange and artificial echoes', mimics of speech and the cries of birds and beasts. The optics laboratories ('perspective-houses') are a catalogue of wonders: 'glasses and means to see small and minute bodies perfectly and distinctly' (Galileo had only recently described his microscopes), artificial rainbows, 'all colourations of light: all delusions and deceits of the sight', and – it is difficult here not to track hindsight through to Newton's famous experiment on the colours in light (see page 344) – 'several colours: not in rain-bows, as it is in gems and prisms, but of themselves single.' Only after all these sometimes frivolous accomplishments are conventional scholarly expectations acknowledged with the perfunctory mention of 'a mathematical house, where are represented all instruments, as well of geometry as astronomy, exquisitely made'.

In none of this does Bacon disparage magic as such; rather, by welding magic's experimental attitude to the reason of natural philosophy, he hoped to delineate a true and useful approach to the study of nature. While he accused astrology, natural magic and alchemy of 'sway[ing] the imagination more than the reason', he felt that they needed to be reformed rather than abandoned. Magic in its 'ancient and honorable sense' means 'a sublimer wisdom, or a knowledge of the relations of universal nature' – it is, he said, 'that science, which leads to the knowledge of hidden forms, for producing great effects, and by joining agents to patients setting the capital works of nature to view'. The common natural magic found in books, he says, 'gives us only some childish and superstitious traditions and observations of the sympathies and antipathies of things, or occult and specific properties, which are usually intermixed with many trifling experiments, admired rather for their disguise than for themselves'. Even the alchemical transmutation of metals may not be impossible – but it needs to be practised not with obscure principles and cryptic terminology but through familiarity with 'the nature of gravity, colour, malleability, fixedness, volatility, the principles of metals and menstruums'.

In Campanella's City of the Sun, scientific knowledge is respected but present as a fait accompli: all knowledge is gathered together on the city walls, and all one has to do is read it. Where did this learning come from? Andreae's Christianopolis, in contrast, has an active research agenda: it contains laboratories and anatomy theatres, as well as cabinets that are starting to sound more like a natural history museum, filled with rare and precious specimens. But Bacon is even more systematic: a solid programme of research and innovation lies at the core of Bensalem, and his implication is that any ruler who wished to achieve similar technological mastery would be wise to follow this model.

No ruler did – Bacon was singularly unsuccessful in getting his vision of a state-sponsored institute for scientific research taken seriously. He had hopes that James I, crowned in 1603, would be more receptive to his schemes for experimental philosophy than Elizabeth I had been, for the Stuart king was known to take an interest in the new philosophy. He had visited Tycho Brahe in 1590 when he travelled to Denmark to claim his bride Anne, daughter of Frederick II, and Johannes Kepler was one of his correspondents (James even invited the German astronomer to settle in England). But the king was not convinced that he could use a Solomon's House. This was not just a monarchical lack of appreciation of the value of science, for Bacon failed to articulate how such an institution would actually operate. He took an almost adolescent delight in specifying the fine details of Solomon's House, down to the colour of the underwear of its savants (white), but he did not explain how it might be embedded in civic society: how it was financed, how it relates to the sovereign, how its members are recruited and trained. A king needs more than dreams.

The advancement of learning

New Atlantis is not a unique and revelatory vision, but just one in an evolving succession of utopian and often explicitly Rosicrucian* schemes for the procurement of universal knowledge of the world. It was the blueprint both for Comenius's *The Way of Light* and for another imaginary utopia described by Samuel Hartlib in *A Description of the Famous Kingdome of Macaria* (1641). Hartlib takes the name from one of the realms mentioned in More's *Utopia*, but in many respects it is modelled on Bensalem, albeit with a focus on social reform rather than technological capability. Hartlib seems to have wanted to advance

---

* John Heydon, for whom as we saw earlier Bacon was surely a Rosicrucian, more or less equates Solomon's House with the 'Temple of the Rosy Cross'.

a genuinely practical scheme, and to have hoped that the English Parliament would put it into action. His passion for reform of learning was shared by John Milton, who wrote his tract *Of Education* (1644) after discussion with the Prussian scholar. During Oliver Cromwell's Commonwealth after the civil war, Hartlib was appointed by Parliament as 'an Agent for the advancement of Universal Learning and the Publick Good'.

He was the right man for the job, for he was what we should now call a social networker, who knew everyone and where they might be found. As well as Milton, his associates included John Aubrey, John Evelyn, Robert Boyle, the eminent Cambridge Platonist Henry More, and the religious reformer John Dury, who was a part of the circle of Elizabeth Stuart and had been tutor to Charles I's daughter Mary, the Princess Royal and Princess of Orange, in The Hague. It was partly at Hartlib's invitation that Comenius came to England, and he was impressed by the Moravian's concept of Pansophia, which the historian Rosemary Syfret calls 'a collecting together of everything – eternal and temporal, spiritual and bodily, heavenly and earthly – that man ought to know, and a presentation of it to him in such a way that he could know it'.* The ultimate goal of this encyclopaedic collection of knowledge was a utopian one: as Comenius wrote, 'if all men understand each other they will become as it were one race, one people, one household, one School of God'. They would not, in other words, fight the kind of wars that had devastated his and Hartlib's homelands, or that ravaged England in the 1640s.

The way to this harmonious society would be shown, these men hoped, by a Universal College of scientists – one might call them priests of nature – modelled on Solomon's House and Christianopolis. Hartlib proposed the formation of a so-called Office of Address that would collect and disseminate scientific knowledge, perhaps by testing the various experimental claims made in earlier books much as the Accademia Segreta had done. Evelyn called for a brotherhood to be established near London that would conduct experiments and collect

---

* The pansophist's encyclopaedic attempt to systematize the variety of all the things in the world has been dismissed by historian Michael Oakeshott as a childish 'romantic obsession'. But it might equally be seen as an expression of the scientist's faith in a universe based on an orderly underlying scheme.

'rarities and things of nature'. Hartlib, immersed in the tradition of occult secrets, began to imagine this society as an 'Invisible College', an open reference to the name under which the Rosicrucians went in much of Europe. Inspired by the Lord Verulam, such a brotherhood seemed in the 1650s at last to emerge.

4 The Hunt of Pan

Men have entered into a desire of learning and knowledge, sometimes upon a natural curiosity and inquisitive appetite; sometimes
to entertain their minds with variety and delight; sometimes for
ornament and reputation.

Francis Bacon, *Advancement of Learning* (1605)

God has given enough for use, not for Curiosity, which is Endless.

John Evelyn, *Memoires for my Grand-son*

The theory that Francis Bacon wrote the plays of Shakespeare is one
of the sillier fancies of Victorian scholarship, but it inadvertently
stumbles upon a broader truth: the works of both men can be regarded
as contributions to a movement that sought to glorify the reign of
the Virgin Queen as the apogee of Neoplatonic Protestant mysticism.
Bacon's call for a new philosophy is of a part with the late Renaissance
tradition of liberal humanism that embraced curiosity and wonder
about the world and believed that it was a trove of secrets which
Hermetic philosophy could reveal. This tradition found its most vibrant
expressions in Rudolf's Prague and Elizabethan England, both of them
refuges from the conservative strictures of the Counter-Reformation.
When considered from the right direction, Bacon can be seen to keep
company not with Robert Hooke, Marin Mersenne and René Descartes
but with Shakespeare, Edmund Spenser, Walter Raleigh, Giambattista
Della Porta and John Dee.

As the historian Frances Yates says, 'the dominant philosophy of
the Elizabethan age was precisely the occult philosophy, with its magic,

its melancholy, its aim of penetrating into profound spheres of knowledge and experience, scientific and spiritual, its fear of the dangers of such a quest, and of the fierce opposition which it encountered'. *This* was the alternative to scholastic tradition, and most of the 'moderns' of the early seventeenth century felt some allegiance to it to some degree or another. The sceptics and the mechanists, whether they be the Royalists Thomas Hobbes and William Harvey, the Jesuit Mersenne, or Descartes joining battle against the Winter King, are more properly seen as the voices of reaction, and the science that ultimately emerged in the eighteenth century was a rude mixture of the old and new, into which the mathematical magic of Kepler and Newton is only now being cautiously readmitted.

It's for this reason that we can no more hope to make sense of Shakespeare's fairies than of Bacon's new experimental philosophy without reference to their religious and political context. Yates argues that the fairies of *A Midsummer Night's Dream* and *The Merry Wives of Windsor*, and the witches and ghosts of the Bard's great tragedies, are borrowed not from folk belief but from the occult tradition stretching back to the arch-magi of the early sixteenth century such as Cornelius Agrippa and Paracelsus, and that they function to celebrate the 'pure religion' of the Virgin Queen. It is widely thought that *The Merry Wives* was commissioned to celebrate the Garter Feast of 1597, perhaps by George Carey, patron of Shakespeare's company the Lord Chamberlain's Men, who received the Order of the Garter on that occasion. Carey was the archetypal Elizabethan 'occult courtier', a patron also of the well-known English alchemists James Forester and Simon Forman. *The Merry Wives* possibly contains an alchemical allegory in which Falstaff is purified of his grossness by being stewed under a pile of clothes in a laundry basket, like a lump of lead in the crucible.

Yates argues that Shakespeare's fairies are from the same stamp as those in Spenser's *The Faerie Queen*, a panegyric to Elizabeth I that brims with occult and astrological imagery. 'To read Shakespeare's fairy scenes without reference to the contemporary build-up of the Virgin Queen as the representative of pure religion', she writes, 'is to miss their purpose as an affirmation of adherence to the Spenserian point of view, a very serious purpose disguised in fantasy.'

The literary scholar Charles Nicholl proposes that alchemical themes infuse several of Shakespeare's plays, most notably *King Lear*, in which the tribulations of the king encode the transformations supposed to be borne by matter as it is brought to a state of perfection in the 'Red King', the philosopher's stone. Even if it is hard to believe that this is the play's primary allegorical meaning, the argument warrants consideration. But the most obvious of Shakespeare's debts to occult philosophy is found in the last and in some ways the most difficult of his plays, *The Tempest*, which was performed for Frederick V and his bride Elizabeth Stuart during their marriage celebrations. Here Prospero models the defence of 'white magic' – the devout and decorous form of magic – against accusations from scholastic reactionaries that it was necromancy and devilish conjuration. Several candidates have been advanced as the inspiration for Prospero, among them Shakespeare himself. Inevitably there is a degree of armchair speculation in all of this that neglects how literary figures are spun from a web of influences. But one of the more plausible suggestions for a formative model is John Dee.

The queen's sorcerer

The most famous Elizabethan magus and one-time mathematician to the queen's court, Dee provides a link between the courts of Prague and England and offers an alternative vision of what the early seventeenth-century virtuosi might be. He promulgated a philosophy seemingly so remote now from the careful and cautious experiments of the Royal Society only a few decades later that it is all too easy to brush him aside, like Della Porta, as a throwback to the mysticism of an earlier age. How else can we regard a man who claimed to commune with angels in a crystal ball?

But Dee's influence was immense – he was respected by Bacon – and he was in many ways more representative of the currents of proto-scientific thought in the early seventeenth century than was Galileo. Like Kepler, Galileo and later Newton, Dee held that the secrets of the world were at root mathematical and geometrical. In his preface to a 1570 translation of Euclid's *Elements* he quotes the arch-Neoplatonist of the Renaissance Pico Della Mirandola, saying 'By number, a way is had, to the searching out and understanding of every thyng, hable to be knowen.' What we struggle now to accept is precisely what was so central to the thinking of Dee and Kepler, and which sounds more subtly in the words of Bacon and Galileo: that mathematics is an art both mystical and practical, a hidden code of nature as well as a means of improving navigation and architecture. It was the key both to cosmic harmony and to quantitative science.

Dee seemed destined from birth for the courtly life. His father, a rich textile merchant, served as a 'sewer' to Henry VIII, and the young John developed his passion for mathematics while studying Greek at Cambridge. After serving as tutor for various nobles and weathering the storms of 'Bloody' Mary Tudor's reign, he found his way to her half-sister Elizabeth's confidence as an astrologer and alchemist. He established a famously grand library at his home in Mortlake, west of London, where he held meetings of like-minded souls that probably had something of the nature of the informal 'experimental academies' of Italy. Dee occasionally enjoyed the queen's favour, especially as she shared his passion for alchemy – for her, a promising way to swell the royal revenue. But he had little skill for navigating the dangerous currents of court intrigue, and failed repeatedly to secure a steady patron or source of income.

In 1582 Dee fell in with a strange man named Edward Kelley, a pathological fantasist who convinced Dee, and perhaps himself, that he possessed occult powers. Kelley claimed to be able to converse with angels through Dee's crystal ball, and together the two men set about trying to learn from them the 'Adamic' language spoken in Eden. Having fallen from royal favour, Dee and Kelley left England with their families in the retinue of Albrecht Laski, an impoverished and somewhat disreputable Polish prince with alchemical interests who was returning home. Dee and Kelley spent the next six years on the Continent, and in 1584 they travelled to Rudolf II's court in Prague, hoping to find a sympathetic patron. Dee was at first welcomed warmly into the house of Thaddeus Hajek, Rudolf's physician. But when he was granted an audience with the emperor, he delivered a disastrous message that his 'angels' had communicated, rebuking Rudolf for his sins. To make matters worse, he fell foul of the politico-religious machinations afoot in the imperial court and soon found himself accused of necromancy and other dark arts. He was peremptorily banished from Prague.

After that, everything fell apart in dramatic fashion. Kelley announced that his visions commanded him and Dee to share their wives with each other: a pact that they apparently fulfilled, with predictably dire results for domestic harmony. As Kelley's reputation as an alchemist and scryer waxed, he decided that he no longer needed Dee and left him stranded and all but penniless in southern Bohemia while he returned to Prague. There the cunning Kelley won over the emperor and briefly enjoyed the title of baron. But his fortunes continued to fluctuate. In 1591 he was arrested by Rudolf's agents on unknown charges, although heaven knows there was more than enough deception to charge him with. Yet by 1595, long after Dee had returned to England, Kelley wrote to his former colleague inviting him back to serve the emperor. What happened to him in the end is not clear – he was rumoured to be dead in late November, but there were reports that he was still practising his deceptive arts in 1598. Dee, meanwhile, had no option but to return to England at the end of 1589. He found his house ransacked, and although he clawed his way back to tenuous recognition at Elizabeth's court and was finally appointed warden of Manchester Collegiate Church (a mixed blessing, as the college had lost its grounds and property in the dissolution in 1547), he was plagued by debt and accusations of witchcraft until his death in London lodgings in 1609.

After such a chequered history, it might seem odd that Dee was held in esteem by anyone. Yet the vicissitudes of court patronage and ecclesiastical favour, not to mention suspicions of illicit magical enthusiasms, were suffered by all natural philosophers of the time – Tycho Brahe, Kepler, Galileo and Bacon all endured them. If it is hard for us to imagine that a man who communes with angels could also have made any useful contributions to the science of his time, that is because we have been encouraged to divorce mathematical and geometrical reasoning from its strong Renaissance associations with magic.

Dee is sometimes said to have inspired a character far less flattering than Prospero: the eponymous swindler Subtle in Ben Jonson's *The Alchemist*. At face value, Jonson's play is the rationalist antidote to the vapid fantasies of alchemy: Subtle's claims of gold-making are naked bluster and trickery. But in its time the play was not just a (much-needed) warning against alchemical charlatans; it was an attack on the entire tradition of Elizabethan occultism. When Subtle is disguised as an improbable fairy to fake a vision for one of his dupes, Jonson's audience was expected to recognize a swipe at Spenser. And it is not just alchemy, with its overblown allegorical language, that Jonson mocks, but the whole of Dee's mathematical science, in which 'algebra' becomes synonymous with 'black arts'. Subtle might be an outright scoundrel, but *The Alchemist* implies that serious practitioners of alchemy and experimental philosophy are no better, veiling their useless toil in elaborate words that signify nothing.

Because Jonson's satire, if competently staged, is still genuinely funny (frankly it has more comic value than *The Merry Wives of Windsor*), it is tempting to regard him as a voice of reason laying bare the absurd posturing and base deception of the Hermetic arts. But he scores his points by ridiculing a figure – the solitary and often deceitful gold-making alchemist, a Faustian trickster – who was already consigned to the fringes of natural philosophy. To the extent that the play is an attack on the new experimental programme as a whole, it is both reactionary and out of touch.

While Jonson fired the keen barb of mockery, Shakespeare's other great rival Christopher Marlowe assaulted the Hermetic Elizabethan Renaissance with moral probity. His *Doctor Faustus* was also quite likely to have been interpreted as an allusion to Dee – the play was probably first performed in 1594, when Dee was out of royal favour and stood accused of witchcraft. Marlowe shows where all this dabbling in magic must lead: to damnation. He drew on the Faust legend revitalized by the popular account of Johann Spiess, published in German in 1587, but played up the moralistic overtones in a way that set the template of narratives against science and curiosity for centuries to come.

The noble hunt

It's hard to see how the notion that Francis Bacon wrote Shakespeare's works can survive a reading of the Lord Verulam's books. It's not that they are poorly written, and to call them turgid would be going too far. But nowhere do they hint that Bacon was a man who could open up the human heart and set the soul soaring. His highly incomplete but nonetheless systematic agenda for the new philosophy at last gave a coherent voice to the approach that had been developing since the middle of the sixteenth century, and if it is occasionally heavy going, that is somewhat in the nature of the task that Bacon set himself. All the same, one finds here ample justification for William Harvey's famous quip that Bacon wrote natural philosophy 'like a Lord

Chancellor' – which is to say, as though it could be reduced to a quasi-legalistic set of rules.

Yet Bacon was enough of a Neoplatonist to know the power of analogy and metaphor. He devoted many pages to discussion of the allegorical meanings of ancient myths and fables, recognizing that they encode stories about the world and our place within it, whereby difficult concepts 'may find an easier passage to the understanding'. *Advancement of Learning* (1605) includes a discourse on the symbolism of the legend of the Greek god Pan that also reveals his affiliation to the tradition of secrets: 'This fable is, perhaps, the noblest of all antiquity, and pregnant with the mysteries and secrets of nature. Pan, as the name imports, represents the universe.' In particular, Bacon asserts that one interpretation of the Pan myth, which he says the Greeks borrowed from the Hebrews (perhaps via the Egyptians), is that 'it relates to the state of the world, not in its first creation, but as made subject to death and corruption after the fall'. Pan shows us the world as we now find it, banished as we are from the original knowledge of all things that Adam possessed in Eden. There is, then, a cautionary morality in the way this goat-footed god symbolizes nature.

Pan is one of the strangest and most ambiguous of the ancient Greek gods, doubtless an indication of the complex nexus of traditions that he represents. Excluded from the Olympian pantheon, he is an ugly, disruptive and impulsive creature in whose antics one can never quite distinguish the rascal from the outright villain. Even in a mythology that regularly conflates roguish seduction with rape, Pan's amorous exploits sail close to the wind. His legend clearly has its origins in fertility cults, but they share more of the riotous witches' orgy and the drunken frenzy of the Dionysian Maenads than the chaste maypole dance. It is clear enough whom this horny (in all respects), hairy man with cloven hoof came, in the Christian era, to represent. Indeed he does not seem so very far from the Devil of Cambridge theologian William Perkins's *Discourse of the Damned Art of Witch-craft* (1608), who has 'exquisite knowledge of all natural things, as of the influences of the stars, the constitutions of men and other creatures, the kinds, virtues, and operations of plants, roots, herbs, stones etc.' – which, as historian John Henry puts it, could be interpreted as making the Devil 'the first scientist'.

Although Pan kept a lazy eye on the flocks, herds and beehives of

Arcadia, he displayed a more animated aspect as the god of hunting. It was said that Pan, while out hunting, found the god Ceres, representing cultivated nature (grain or 'cereal'), after she had become lost.* In this tale Bacon found the perfect allegory for the way that an exploration of wild nature – the Hunt of Pan – can uncover things that serve humanity's welfare:

> That part of the fable which attributes the discovery of lost Ceres to Pan, whilst he was hunting, a happiness denied the other gods, though they diligently and expressly sought her, contains an exceeding just and prudent admonition: viz., that we are not to expect the discovery of things useful in common life, as that of corn, denoted by Ceres, from abstract philosophies, as if these were the gods of the first order – no, not though we used our utmost endeavours this way – but only from Pan, that is, the sagacious experience and general knowledge of nature, which is often found, even by accident, to stumble upon such discoveries, whilst the pursuit was directed another way.

The role of serendipity in scientific advance – responsible for, inter alia, the discovery of oxygen (some say), synthetic dyes, penicillin, Teflon, microwave ovens and the Big Bang – is now rehearsed to the point of cliché. It is used with justification as an argument for supporting so-called blue-skies research: curiosity-led investigations that have no fixed agenda (although in truth most serendipitous discoveries have come from research driven by other specific motives). In today's target-obsessed age it is no bad thing to keep banging this drum; but in Bacon's time this argument for an open-ended, curious exploration of nature was novel.

The most significant facet of Bacon's use of the Pan myth, however, is his evocation of the hunt. This too remains today a common metaphor for scientific research – we 'hunt' cancer genes, new particles, artificial chemical elements. But while these contemporary hunts usually seek out an entity or phenomenon whose existence is already supposed, Bacon envisaged something more like an unpremeditated

---

* Ceres, goddess of abundance, was a popular subject among early seventeenth-century artists, especially in Holland and Flanders. Rubens painted Pan's encounter with her around 1615.

Frontispiece of Francis Bacon's *The Great Instauration* (1620), showing the twin pillars of Hercules.

plunge into the thickets of nature to see what might be found there. This hunt was not the search for a missing sock in the bottom drawer, but an expedition into an uncharted and enchanted land. It was a quest for novelty: the ambition was not simply to explain what is already known but to find new things and effects that Aristotle and Plato never dreamed of. Constantly, the new philosophy of the seventeenth century insists on novelty: Bacon's *Novum Organum* and *New Atlantis*, the *Novum lumen chymicum* (1605) of the alchemist Sendivogius, Boyle's *New Experiments Physico-Mechanical* (1660). The frontispiece of Bacon's *The Great Instauration* shows a ship sailing beyond the Pillars of Hercules, symbolizing the limits of ancient learning.*

---

* There is another, not incompatible, way to interpret this imagery: the literary scholar Mary Baine Campbell suggests that the book can be read 'as an allegorical account of colonial exploration and conquest'. Indeed, we will see that the 'new worlds' opened up by the experimental philosophy were often spoken of by the virtuosi in these terms, which were likely to chime with the acquisitive temperaments of their (actual and wished-for) patrons.

Furthermore, the natural philosopher must hunt for new knowledge not just because it is hard to find but because it is actively concealed – nature guards her secrets jealously. This was not simply an anthropomorphic metaphor, for nature was genuinely regarded by Bacon and his contemporaries as an autonomous, teleological and in some sense intelligent agency. Whether or not it was taken literally, this imagery was still current among the members of the Royal Society in the later seventeenth century: its leading propagandist Joseph Glanvill wrote that '*Nature* works by an *Invisible Hand* in all things', while its foremost experimentalist Robert Hooke spoke of nature's 'secret and subtle Actors' and her 'abstruse and hidden Instruments and Engines'. There was an implication here that, in her opacity, nature was more than a little perverse – that she might in fact be wilfully deceitful. She seems, said Hooke, 'to make many *doublings* and *turnings*, and to use some kind of art in indeavouring to avoid our discovery'. Faced with such a slippery opponent, the philosopher could not rely on deductive reason alone to penetrate to the truth.

As a result, the scientist is required to display not the cool, methodical pedantry of the scholastic but the lively vigilance of the hunter. He needs to be a lynx, sharp-eyed and alert to every sign of his prey. The hunter must be patient, gathering and committing to memory every trace of his quarry, finding his path from the slightest hints, the faintest of scents. He must be prepared to spend hours wandering on the margins, guided as much by intuition as by logic. He sees and understands things that others do not: he can read the hidden codes and clues. The historian Carlo Ginzburg has argued that the evidence-based reasoning that distinguishes science from other modes of thought has its roots not in the classical, geometric and mathematical proofs of Copernicus and Galileo but in the imagery of the hunt.

Diligence alone will not suffice for the hunter, who needs also a certain guile, or what the Greeks called *mêtis*: cunning intelligence. This entails a readiness and boldness to leap ahead of what the facts will reliably support, yet with judgement and perspicacity rather than mere hopeful guesswork. These characteristics are indeed usually possessed by the greatest scientists. Yet because they cannot be reduced to a technique, far less be taught, they are rarely acknowledged as a part of the so-called scientific method, which focuses on the verification of deductions rather than the process of deducing. It is of course true that a bold and cunning mind does not count for much if it does not produce ideas and theories that accord with the evidence; yet it seems odd to talk as if science were only what comes in the aftermath of these leaps of imagination and intuition.

Bacon seems to ask for even more than *mêtis*: to endorse a martial or violent attitude to the gathering of data. He has been reproached for his apparent recommendation of force, even torture, in extracting nature's secrets. One might infer that he is thinking of the thumbscrew and the rack when he says:

> For like as a man's disposition is never well known till he be crossed, nor Proteus ever changed shapes till he was straitened and held fast; so the passages and variations of nature cannot appear so fully in the liberty of nature as in the trials and vexations of art.

But that is not the only interpretation. 'Trials and vexations' might mean nothing more than rigorous questioning by means of experiment and instrumentation. There is a difference (even if it is not in the quality of the information obtained) between the lie detector and the foot roaster. But just as it was obvious that the answers a man gives under duress are not necessarily the true ones, so it was open to debate whether one could trust what nature might reveal in such straits. Perhaps it was for this reason that Robert Hooke praised microscopy as a way of getting nature to 'exhibit herself' without recourse to undue force. With the same sexual metaphors of violence and voyeurism that Bacon deploys, he says that instead of 'pry[ing] into her secrets by breaking open the doors upon her', the microscope permits one to 'quietly peep in at the windows, without frighting her out of her usual byas'.

Bacon's description of science as a hunt (*venatio*) was one of the central conceits of the new philosophy. But it did not originate with him; it was introduced in Della Porta's *Natural Magick*. The *venatio* was thus already a familiar emblem in the late sixteenth century for those sympathetic to the experimental philosophy, as illustrated by the (short-lived) scientific society founded in Venice in 1596 called the Accademia Cacciatore: the Academy of Hunters. Of course, it was no coincidence that this image was immediately meaningful and appealing to the courts of the late Renaissance in which philosophers sought their patrons.

The analogy was still used in the eighteenth century. In a discussion of curiosity in his *Treatise of Human Nature* (1739–40), David Hume wrote that 'there cannot be two passions more nearly resembling each other, than those of hunting and philosophy, whatever disproportion may at first sight appear betwixt them'. Both, he said, require attention and dexterity if they are to overcome the inherent difficulties and uncertainties. And he perceptively notes that for these pursuits to excite feelings of passion and satisfaction they must have apparent utility, even if it is only a convenient fiction. The rich man does not need personally to go hunting for his evening meal, yet he finds pleasure in shooting partridges and pheasants that he will not feel by bagging crows and magpies. 'The utility or importance of itself causes no real passion, but is only requisite to support the imagination', says Hume – just as the remotest link to practical applications will justify recondite research in the mind of today's scientist. Even such pretence at utility can be superfluous once the hunt is underway: 'tho' in both cases the end of our action may in itself be despis'd, yet in the heat of the action we acquire such an attention to this end, that we are very uneasy under any disappointments, and are sorry when we either miss our game, or fall into any error in our reasoning'. We need a story only in order to begin the quest, not to sustain it.

Machine learning

For Galileo, nature was an 'open book': anyone who knows its language (which is, he said, mathematics) can read it plainly enough. But it is not to deny the genius of Galileo's insights into the arrangement of the heavens or the principles of motion to point out that he tended to focus his efforts where mathematics would be effective. He could afford this comforting vision because he had never tried to understand how plants grow or how minerals are formed. To Bacon, for whom no question was too narrow or exotic, nature looked very different. He believed that it was fundamentally ordered, but this design was not lucid and geometric – it was concealed within the dense forest of the particular and the idiosyncratic. 'The universe to the eye of the human understanding', he wrote, is framed like a labyrinth, presenting as it does on every side so many ambiguities of way, such deceitful resemblances of objects and signs, natures too irregular in their lines and so knotted and entangled. And then the way is still to be made by the uncertain light of the sense, sometimes shining out, sometimes clouded over, through the woods of experience and particulars; while those who offer themselves for guides are (as was said) themselves also puzzled, and increase the number of errors and wanderers.

As we shall see, it is Bacon's picture (derived from the natural magic tradition), and not Galileo's (which drew as much on scholastic deduction from theorem and axiom as it did on observation), that conditioned the emergence of experimental, empirical science – what has been called the Scientific Revolution. For Galileo, experiment was still as much a demonstrative tool as it was a search for new things to explain. But the professors of secrets were never quite sure what they would find in their lenses and crucibles.

Yet if the world is so obscure, how do we make progress?

First, we must accept that the ancients didn't know everything, and that there is still much of which we are ignorant. 'That wisdom which we have derived principally from the Greeks is but like the boyhood of knowledge', Bacon wrote. The schoolmen's obsession with the past meant that 'philosophy and the intellectual sciences . . . stand like statues, worshipped and celebrated, but not moved or advanced'.

Besides, he asserted, what has been found so far are the low-hanging fruits: those aspects of nature that present themselves most obviously and accessibly. 'The discoveries which have been hitherto made in the arts and sciences', he said,

are such as might be made by practice, meditation, observation, augmentation, – for they lay near to the senses, and immediately beneath common notions; but before we can reach the remoter and more hidden parts of nature, it is necessary that a more perfect use and application of the human mind and intellect be introduced.'

There is no one, Bacon concludes, 'who has dwelt upon experience and the facts of nature as long as is necessary'. We need to hunt harder. And we must be prepared to look everywhere, for we simply do not know where the most productive paths lie, nor where the spores are genuine and where they are false trails. As Della Porta had said, even the most trivial and humble phenomena might lead to great ideas – or as Bacon puts it, 'the magnetic virtue of iron was not first discovered in bars, but in needles'. He tells a familiar tale of the philosopher

> who, while he gazed upwards to the stars, fell into the water. For had he looked down, he might have discovered the stars in the water; but looking up to heaven, he could not see the water in the stars; for mean and small things often discover great ones, better than great can discover the small.

So a universal curiosity becomes not only permissible but essential: we can rule out nothing a priori.

This raises the question of what *kinds* of data are the most valuable. To the Aristotelians, the wonders celebrated in the cult of curiosities and the *Wunderkammer* offered little insight precisely because they were aberrant: examples of nature behaving abnormally and even whimsically. It was not that Aristotle exactly scorned experience as a guide to philosophy; but Bacon accused him of filtering and moulding it 'to suit his opinions' and to make it 'captive and bound', ignoring things that did not fit. That the sun rises in the east is a useful observation to an Aristotelian because it speaks of something that always happens. But an oddly shaped rock is silent for this sort of philosopher, because it is not like any other. To put it another way, ancient and medieval science was really about using observation to demonstrate *what was already known*: to fit observations within a pre-existing scheme of axioms. When experience deviated, that made it by definition unworthy of careful scrutiny, for it then obeyed no apparent order.

The enthusiasts of the *Wunderkammer* went to the other extreme in fixating on exceptions and anomalies – the stranger the better – and did not seem unduly concerned to fit these oddities into any system of nature. Bacon takes the middle ground: deviations were of interest precisely because they offered clues and trails to new secrets, new types of order. And Bacon's impulse to turn scientific understanding towards technological application led him to regard natural marvels as *innovations*, revealing new ways in which the fabric of the world might be arranged, and which could therefore offer inspiration to the craftsman. 'From the wonders of nature', he wrote, 'is the nearest intelligence and passage towards the wonders of art.' Nature was a warehouse of possible designs. Thus Bacon offered a formal justification for an all-embracing curiosity that could previously have been dismissed as a sign of mere dilettantism or a childish infatuation with wonders. As William Eamon puts it, 'All the bizarre objects and rarities that had fascinated and delighted visitors to the Renaissance curiosity-cabinets became urgently relevant to the Baconian scientific enterprise' – they became the vital clues pointing to the underlying mechanisms.*

The price of this concession was that marvels were to be regarded with sober reflection and analysis: in fact, *sans* wonder at all. It was not necessary that wonders should cease, but just that we should discipline ourselves not to wonder at them. 'For we are not to give up the investigation until the properties and qualities found in such things as may be taken for miracles of nature', Bacon wrote, 'be reduced and comprehended under some form or fixed law.'

In collecting data for the compilation of exhaustive descriptions of phenomena, Bacon insisted that it was no use conducting experiments at random. 'Our steps must be guided by a clue', he said, 'and the whole way from the very first perception of the senses must be laid upon a sure plan.' He believed that experiments should investigate particular properties and characteristics, which he listed as follows:

1. *Variation*: vary the experimental conditions. For example, will a piece of amber still pick up straw if it is heated rather than rubbed?
2. *Production*: repeat the same experiment in different situations: in cold and warm rooms, for example.
3. *Translation*: use experiments developed to investigate one phenomenon in the context of another.
4. *Inversion*: look for the effects of opposites. If a magnifying glass can make things hot, can it also make them cold?
5. *Compulsion*: test things to destruction. 'In other hunts the prey is only caught', he explained, 'but in this it is killed.'

* Attention to anomalies remains a central principle of science – that is, after all, how quantum theory emerged in the early twentieth century, forced upon us by a few niggling problems with what otherwise seemed to be a satisfying and complete scheme of classical physics.

6. *Application*: apply the results of some experiments to other situations. For example, by knowing the weights of equal volumes of wine and water, one can deduce whether a sample of wine has been watered down.

7. *Conjunction*: see if a chain of experiments produces different results than each one alone. For example, roses are known to bloom late if their early buds are plucked off or if their roots are exposed in the spring. What happens if one does both?

8. *Chance*: for after all, there is still a place for haphazard 'try and see'. 'This form of experimenting', he admitted, 'is merely irrational and as it were mad, when you have a mind to try something, not because reason or some other experiment leads you to it, but simply because such a thing has never been attempted before . . . the very absurdity of the thing may sometimes prove of service.'

This last principle is rather delightful, not least because it exposes the madness in Bacon's method. There is of course no end to the crazy experiments one might dream up, although in general they will stand to tell us rather little if we do not at least have some prior expectations about them. But before we dismiss this as an uncharacteristically quirky flourish from Bacon, let's just observe that a 'what the heck' approach is still indulged from time to time in modern science, and is occasionally fruitful.

Bacon was always very clear that these accumulations of raw data and experience ('natural histories') are merely the means to an end – which is to find 'an unbroken route through the woods of experience to the open ground of axioms'. His criticism of the natural magicians is not that they were obsessed with wonderful and fantastical phenomena but that they were content, even determined, to leave them as such and not to *explain* them:

> The mark of genuine science is that its explanations take the mystery out of things. Imposture dresses things up to seem more wonderful than they would be without the dress.

But how are raw data to be boiled down to a deeper understanding of their causes? This is the fundamental question for Bacon's inductive science: how to filter important clues from incidental effects. Aristotle never had to confront this problem, for while he paid close attention to nature, he did not believe that observations guide us towards fundamentals; rather, the philosopher's task was to work out how a priori axioms could be marshalled to account for what he saw. For Bacon (at least in principle), it should work the other way round: one should scrupulously avoid theorizing or generalizing until all the evidence was in. That, of course, was an open-ended affair, since neither Bacon nor anyone else could provide criteria for deciding when enough was enough – as Leonardo had discovered, one can ask questions for ever.

Nonetheless, Bacon tried gamely to devise a scheme for turning observation into understanding. His method is revealing precisely because it is so ineffectual (not to mention incomplete). Bacon's noble effort in the end shows us that this approach of proceeding from pure, hypothesis-free observation isn't the way to do science.

What he had in mind was a kind of machine for digesting experimental facts into theory. Aristotle's six books on logic, which formed the basis of his axiomatic, deductive philosophy, were known as the Organon, meaning tool or mechanism. Bacon proposed to replace Aristotle's logical machinery with his own new organon (*Novum Organum*), a description of which he published in 1620. 'What the sciences stand in need of', he said, 'is a form of induction which shall analyse experience and take it to pieces, and by a due process of exclusion and rejection lead to an inevitable conclusion.'

The first task, Bacon said, was to tabulate all related data and observations, and to look for similarities and differences in a systematic way. He drew up a list of twenty-seven such bases for comparison, which he called 'Prerogative Instances'. For example, if we are comparing quadrupedal animals, we might consider how an elephant differs from a cat. Yet even this much barely establishes the grounds for beginning the process, since we have first to decide on some classification scheme that determines what it is we must compare. Should we, say, group together all substances that are hard, or combustible, or which fracture in a certain manner? Do we group bats with birds or rodents? The similarity of form between, say, a tree and the branches of the lung or circulatory system was noticed by Leonardo, and was considered (rightly in this case) to reveal a deep connection between their functions. But does that mean we should group the tree with the flower, or with 'branching things'? In short, there is no obvious and unique way to carve up nature's diversity into unique and non-overlapping categories.

No matter; let us continue with the Lord Chancellor's knowledge-distilling machine. The basic principle is that, having collected all the data pertaining to a phenomenon, including experimental manipulations to alter its circumstance and expression, one must make comparisons between different observations and situations so as to sift what is contingent from what is fundamental. It is the latter that will ultimately furnish a hypothesis about the causes of the effect in question. This method is known as eliminative induction.

Bacon illustrated the procedure with reference to investigations of heat. In a table called 'Essence and Presence', he lists every situation in which heat seems to be present or produced: the sun's rays, flames, warm liquids, rubbed bodies, quicklime sprinkled with water, animals and so forth. Then he prepared a table of 'Deviation, or Absence in Proximity', which enumerated all phenomena that seem to be related to these but which do not produce heat – moonlight, for instance.

The causes of heat may then, Bacon believed, be deduced by figuring out what seems to make heat become manifest, or to wax and wane, in one circumstance and not in a similar one. This is in principle a good approach. Scientists today often seek to vary their experimental conditions so as to identify which of the conditions or variables are essential for producing the effect under consideration. For example, the sun's heat seems to depend on its position in the sky, it being hotter when it is overhead. And small bodies warm up more quickly than larger ones, or as Bacon put it, matter seems to 'oppose' heat.

Having made these comparisons, we are ready to start drawing conclusions – but at this point only to advance to the 'Commencement of Interpretation', or as Bacon puts it in an elaborate oenological metaphor, to the 'First Vintage'. For heat, this conclusion may be expressed as follows:

> Heat is an expansive motion restrained, and striving to exert itself in the smaller particles. The expansion is modified by its tendency to rise, though expanding towards the exterior; and the effort is modified by its not being sluggish, but active and somewhat violent.

Read in the light of the modern view that heat is a manifestation of the agitation of a body's component atoms and is transferred from one place to another by their collisions, this seems to be a pretty good definition for its time. But it can't really be considered an inevitable corollary of the observations on which it is based. For one thing, it already assumes that bodies are made of 'smaller particles' – an atomistic or corpuscular model of matter, which was commonly held by the experimental philosophers but was at that time no more than an article of faith. And what can it mean to say that heat is (uniformly) expansive and yet has a 'tendency to rise'?

However, this is just the First Vintage. There is much more to come. The conclusions are to be refined through a sequence of nine steps, of which the 'Prerogative Instances' are just the first. Typically, Bacon explains only this first step in any detail, although doing so requires the one hundred or so remaining pages of the *Novum Organum*. Still to come, presumably at comparable length were he ever to have finished the job, are accounts of the 'Supports of Induction', 'Rectification of Induction', 'Prerogative Natures' and so on. We have to take it somewhat on trust that Bacon himself ever had a full vision of the workings of this knowledge machine; even if he did, one has to wonder if this arcane mechanism would ever have succeeded in processing any raw data into true understanding. The whole business begins to sound unhappily legalistic – or, far worse, scholastic. As John Henry says, 'the more details of his method he provides, the less appealing it becomes', and you can rather see what William Harvey was getting at. James I was said to have been magisterially dismissive of Bacon's system, wittily comparing it with the will of God in that it 'passes all understanding'.

There is no evidence that Bacon's method was ever used by him or anyone to discover anything; indeed, Bacon made not a single significant scientific discovery himself. The important point, however, was not the details but that it is a quasi-mechanical step-by-step (algorithmic) process, so that anyone could conduct it, without recourse to inspiration or intellectual genius. 'Our method of discovering the sciences', wrote Bacon, 'merely levels men's wits, and leaves but little to their superiority, since it achieves everything by the most certain rules and demonstrations.' You don't need to be clever, but merely careful and thorough. This – and not Bacon's own quasi-vitalistic, occult conception of nature – is what separates his philosophy from that of the natural magicians such as Paracelsus, Cardano or Della Porta, for all of whom scientific understanding was a privilege conferred by the possession of exceptional intellect or insight which made one a conduit for the almost Gnostic revelation of truths.*

In other words, the *Novum Organum* democratizes science. Indeed, it positively demands that discovery be assigned to legions of plodders who patiently collect all the data while lacking the imagination to leap to conclusions. 'I am emphatically of the opinion', Bacon wrote, 'that man's wits require not the addition of feathers and wings, but of leaden weights. Men are very far from realising how strict and disciplined a thing is research into truth and nature, and how little it leaves to the judgement of men.' The implication is that science must be institutionalized and professionalized: there should be what amounts to a 'knowledge industry' dedicated to churning out scientific understanding just as bakers make bread. That sentiment was echoed in Thomas Sprat's *History of the Royal Society* in 1667, in which he adduces a role for 'plain, diligent and laborious observers [who] bring not much knowledg, yet bring their hands, and their eyes uncorrupted'. With its hierarchical delegation of jobs to various types and grades of researcher – to the colourfully titled Merchants of Light, Depredators, Mystery-Men, Pioneer-Miners, Compilers, Dowry-men, Inoculators and Interpreters of Nature – Bacon's scheme could be seen as the blueprint for the modern military-industrial complex.

The corollary is that curiosity too should be institutionalized. Then it is not so much curiosity at all, but rather, thoroughness: one asks comprehensive questions simply to ensure that Bacon's 'philosophical tables' have no important gaps. If one must become encyclopaedic in canvassing nature, that should not reflect a voracious appetite so much as a diligent attitude. When Bacon's successors used 'curious' as a term of praise for natural philosophers, it was often with this

---

* It's ironic, then, that this image of scientific understanding coming into the minds of great men in a flash of inspiration is precisely the one promoted in old-fashioned Whiggish histories of science, such as Bernard Jaffe's popular history of chemistry *Crucibles*.

connotation: the *cura* here was 'care' rather than inquisitiveness or zeal.

The central flaw in Bacon's system is evident even from a cursory reading: there is no obvious way to prescribe the limits of investigation. How do you know when the philosophical table is full? Although Bacon clearly wants to avoid *indiscriminate* accumulation of data, he offers no rules for how one might be discriminating about it, and so the Baconian approach was apt to be interpreted as mindlessly encyclopaedic, as exhausting as it was exhaustive.

One of the central motivating forces of Bacon's new philosophy was that it should yield *useful* (as he put it, 'profitable') knowledge. The motivation in searching for axiomatic principles is not simply to know them but because, as Bacon is convinced, 'axioms once rightly discovered will carry whole troops of works along with them, and produce them, not here and there one, but in clusters'. According to historian Stephen Gaukroger, Baconian natural philosophy 'manifests its worth in enabling us to secure control over our environment'. This implies that knowledge about nature is not neutral but is accompanied by questions of how it ought to be used: that natural philosophy acquires something of the character of moral philosophy. Bacon disparages those who study nature out of sheer pleasure: that was the frivolous indulgence of the dilettante. The only pleasure he will permit is a utilitarian one: that of discovering new ways of mimicking natural processes in bountiful 'art'.

It also means that, from Bacon's point of view, some of the most pressing questions are those in which nature is altered and manipulated. In Aristotelian science it was not clear that these situations can be considered a part of natural knowledge at all. Since it is in the nature of heavy objects such as stones that they should fall to the earth, the act of raising them up by artificial means is *contra naturam*, against nature, and as such is an anomaly that cannot tell us about how the world works. But for Bacon, it is precisely these motions that are most useful, for example in the operation of machines. Aristotelian philosophy, by implication, was looking at (or for) the wrong things. In modern science it is now taken for granted that highly artificial and constrained laboratory manipulations are not just useful but essential for uncovering natural laws; but Bacon's proposal to make them central to science was new and controversial, and as we shall see, it was not accepted lightly.

Is it proper?

Confronting the religious proscription of curiosity, supporters of Bacon's experimental philosophy usually echoed the curious-minded iconoclasts of the Middle Ages by declaring that an interest in the natural world was an act of reverence and homage rather than hubris. For Robert Boyle, the foremost experimental philosopher in England before Newton, 'whatever God himself has been pleased to think worthy of his making, its fellow-creature man should not think unworthy of his knowing'. This defence of science as a form of natural theology – the claim that knowing more of God's creation can only increase our awe, admiration and wonder of his power and wisdom – remained the attitude of many Christians until at least the nineteenth century, and is still invoked today.

But if God had meant us to understand all the secrets of nature, why did he strip Adam of this knowledge after the Fall? Bacon argued that this was not exactly what had happened:

> For it was not that pure and uncorrupted natural knowledge whereby Adam gave names to the creatures according to their propriety, which gave occasion to the fall. It was the ambitious and proud desire of moral knowledge to judge of good and evil, to the end that man may revolt from God and give laws to himself, which was the form and manner of the temptation. Whereas of the sciences which regard nature, the divine philosopher [Solomon] declares that 'it is the glory of God to conceal a thing, but it is the glory of the king* to find a thing out'.

Thus, says Bacon, it is fine to be curious so long as 'we do not presume by the contemplation of nature to attain to the mysteries of God', but instead 'use our knowledge so as to give ourselves ease and content'. There are some modern scientists who would do well to remember his injunction that 'if any man shall think, by his inquiries after material things, to discover the nature or will of God, he is indeed spoiled by vain philosophy'.

Here, then, Bacon is making two striking claims for science. First, it has the potential to return us to an Edenic state in which we enjoy again the dominion over nature that God awarded to Adam. That

---

* Flattery of the monarchy was never far from Bacon's thoughts.

idea was repeated by Robert Hooke in the introduction to his *Micrographia* (1665), where he suggests that instruments such as the microscope might repair the defects of vision ('mischiefs and imperfections') that mankind has drawn on himself partly because of 'innate corruption' or original sin:

> As at first, mankind *fell* by *tasting* of the forbidden Tree of Knowledge, so we, their *Posterity*, may be in part *restor'd* by the same way, not only by *beholding* and *contemplating*, but by *tasting* too those fruits of Natural knowledge, that were never yet forbidden.

Second, Bacon rejects the medieval belief that God has hidden nature's workings because he does not want them to be discovered, in favour of a view in which God has constructed an intricate puzzle to the challenge of which he hoped mankind will rise. That venture should be undertaken not timidly like a monk but boldly and passionately like a noble huntsman. It was Solomon, says Bacon, who declared that:

> God has framed the mind like a glass, capable of the image of the universe, and desirous to receive it as the eye to receive light; and thus it is not only pleased with the variety and vicissitudes of things, but also endeavours to find out the laws they observe in their changes and alterations.

Other theologians and moralists objected not so much about the propriety of interrogating God's creation as about that of expending so much effort that ought better to be devoted to pious contemplation. In his *Reflections upon the Conduct of Human Life* (1690), the Cambridge Platonist John Norris railed at how 'absurd and impertinent' it was to find 'a Man, who has so great a Concern upon his Hands as preparing for Eternity, all busy and taken up with *Quadrants*, and *Telescopes*, *Furnaces*, *Syphons*, and *Air Pumps*'. This complaint is still current, voiced today in secular terms as the sheer, profligate waste of time, energy and money spent investigating questions of no conceivable value – a pursuit that appears to be a dereliction of social responsibility. Bacon admits that some people consider it damaging to the dignity of the human mind to spend long and frequent intercourse with experiments and particulars, which are the objects of sense, and confined to matter; especially since such matters generally require labour in investigation, are mean subjects for meditation, harsh in discourse, unproductive in practice, infinite in number, and delicate in their subtilty.

He recognized also that some politicians felt an excess of learning would make men 'too curious and irresolute' and 'too unsociable and incapacitated'. But on the contrary, he says, learning confers protection against such inactivity, since it 'defends the mind against idleness and pleasure'.

All the same, Bacon does retain the definition of curiosity that the ancients insisted on: an obsession with 'things of little use'. Curiosity, he explained, 'consists either in matters or words, that is, either in taking pains about vain things, or too much labour about the delicacy of language'. The latter, he said, is in fact precisely where the scholastics are led astray, for he accuses them, shut up in their cloisters with only a few dogmatic old authors as their guides, of spinning 'cobwebs of learning, admirable indeed for the fineness of the thread, but of no substance or profit'. He is dismissive of knowledge gathered simply 'out of a natural curiosity and inquisitive temper' or 'to entertain the mind with variety and delight', or as 'a walk for a wandering mind'.

One might say then that while Bacon issued a defence of curiosity, he was not yet ready to call it that, but reserved the word as a pejorative for enquiries that were idle, aimless and without fruit.

A change of mind

Bacon was critical of the tradition of secrets and natural magic, but only of its practices and not its principles. He endorsed its view of nature as an interplay of occult forces, but argued for a new context in attempting to understand them. No longer was the natural philosopher to be a lone figure or member of a shadowy sect, masking his enquiries in esoteric language and guarding his discoveries jealously. He was to be a public servant, acting within an organized and institutional programme for the public good, which is to say, in the service of the state.

A state-sponsored experimental philosophy processed through Bacon's new organon would, he said, give the monarch a powerful source of intellectual authority, and by implication, of moral authority too. But it was still more than that, for Bacon's constant refrain about the importance of *useful* knowledge shows his keen awareness of the military, industrial and economic benefits of science. This was the bait with which many virtuosi, most obviously those within the Royal Society, tried (with mixed fortunes) to hook their sponsors. Today we witness the case being made afresh with every generation.

Bacon felt that not only should the scientist be working in a new institutional setting but he needed also to reform his character. This was more than a matter of good intentions and attitude; he proposed that philosophers must become actively aware of their assumptions and prejudices in order to overcome them. That demands a retraining of the mind to expunge its habitual blindness. Although differing in their methods and approach, both Galileo and Descartes shared with Bacon this desire to portray the new philosopher not just as a person who follows different principles from the scholastics but as one who has a wholly different mentality.

This marks a profound shift in the status of knowledge itself. A man like Paracelsus, or indeed Galen, would have rested his claim to speak authoritatively about nature on a superior intellect, or greater diligence, or perhaps a better method of pursuing knowledge and a sounder judgement of its sources. To Bacon, in contrast, one did not have to possess any special abilities to contribute to science, so long as one underwent a programme of self-cleansing to prepare the mind. His aim was not merely to make genius or privileged insight dispensable but to dehumanize the process of science entirely, so as to eliminate the pitfalls of subjectivity. This, ultimately, is the significance of the Baconian knowledge-machine: being 'artificial, chaste, severe', it offers a route to understanding that is disinterested and independent of the sensitivity and perceptiveness of the investigator. It is in this respect that it can be seen to articulate the aspirations of modern science. (I consider in the final chapter whether science really can and does work in this manner.)

The Baconian philosopher needed to be an iconoclast – not in exhibiting bold or radical thought, but literally in dismantling 'idols' of the mind, habitual ways of thinking that hinder perception. There are four ways in which our judgement is led astray, says Bacon, for which he selected perversely opaque labels. The 'idols of the tribe' are the characteristics that hamper all men. Among them is the tendency to anthropomorphize, to imagine that 'nature acts as man does'. Each of us has also learnt misleading habits of thought, or 'idols of the cave'. The ambiguities and traps of language and communication create 'idols of the market', while 'false theories, or philosophies, and the perverted laws of demonstration' are 'idols of the theatre'.

At the same time, Bacon maintained the long-standing view that the natural philosopher should comport himself in a virtuous manner. In ancient Greece and the Middle Ages, the value of a scholar's work was related to his personal decorum, and the enemies of early sixteenth-century mavericks such as Cornelius Agrippa and Paracelsus could still effectively undermine their ideas by portraying the men themselves as wicked, decadent and wild. The scientist-priests of Solomon's House, in contrast, were reliable and trustworthy because they had mastered their passions, exhibiting self-control and self-respect at all times. Zealots, says Bacon, tend to be 'men of young years and superficial understanding, carried away with partial respect of persons', who 'leap from ignorance to a prejudicate opinion'.

This treacherous zeal was deemed by Baconians to manifest itself as an urge to theorize, to advance with unseemly haste from a collection of facts to an explanation of them. Bacon praised those

> who aspire not to guess and divine, but to discover and know, who propose not to devise mimic and fabulous worlds of their own, but to examine and dissect the nature of this very world itself.

Such aspirations, he said, demand that one 'must go to the facts themselves for everything'. For facts were facts and incontestable by any competent observer. Theories and explanations, on the other hand, were contentious and hard to prove at best. It is for this reason that even the cool and calculating Descartes could be denounced as an unreliable 'enthusiast' because he was not content to make observations but insisted on developing theories to explain them. As the classicist Meric Casaubon* wrote of Descartes in 1668:

* Casaubon's orthodox Anglicanism, it must be said, led him to regard all the new philosophy as a hotbed of both atheism and necromancy. He accused John Dee of dealing with the Devil, and considered the Royal Society to be awash with the fanatical subjectivity of 'enthusiasm'.

But for his *Method*: I tooke him for one, whom excessive pride and self-conceit (which doth happen unto many) had absolutely bereved of his witts. I could not believe that such stuffe, soe ridiculous, soe blasphemous (as I apprehended it, and doe still) could proceed from a sober man.

What might have been

Francis Bacon and John Dee were united not only in their view of nature's character but in their objectives for prosecuting a programme of natural philosophy: it was for the glory of the state, specifically for the state represented in the cult of the Virgin Queen. Knowledge found its virtue in the pursuit of power. The reason these two men look so different now is that Dee's model was that of the early Renaissance magus: a lone figure possessed of privileged knowledge which he communicated to the cognoscenti in cryptic ways. Bacon, on the other hand, advocated something like a national project of research, itself a recognition of the magnitude of the task of preparing 'histories' of all phenomena, both natural and artificial: the job was too great for one man. For Bacon there were no short cuts, no divine revelations of forgotten wisdom or codes that, once unlocked, would open the gates to a flood of understanding. He rightly perceived that nature was subtle to an almost disheartening degree, and that in consequence natural philosophy needed manpower as much as genius.

In this much, Bacon was absolutely right. But that wasn't obvious at the start of the seventeenth century, and there were some – such as Robert Fludd – who preferred to follow their boundless curiosity in Dee's example as an exploration of the Hermetic mysteries of the macrocosm and the microcosm. As we saw, Fludd's approach was criticized by some of the 'new philosophers', including Kepler and Mersenne. Indeed, the astronomer Seth Ward argued that 'There are not two waies in the whole World more opposite, than those of the L. *Verulam* and D. *Fludd*, the one founded upon experiment, the other upon mysticall Ideal reasons.' But the distinctions were less clear-cut than the likes of Ward might prefer to imagine. Both men after all, conducted experiments and preferred 'experience' over the ancient authorities. Neither is it impossible to distinguish moderns from mystics on the basis of which of them has been vindicated by posterity. Fludd wrote of the circulation of the blood before William Harvey, and although his argument was Hermetic rather than experimental, drawing on an analogy with the circulation of the planets, it was one that Harvey himself later cited. What is more, there was certainly no consensus at the time about which of these 'new philosophies' was the more reliable. Frances Yates asserts that Dee's 'Mathematicall Preface' (see page 87),

> has been said to be of greater importance than Francis Bacon's *Advancement of Learning* published thirty-five years later, for Dee fully understood and emphasized the basic importance of mathematical studies for the advancement of science, whereas Bacon underestimated mathematics. Dee's mathematical preface had a great influence and was widely read until well on in the seventeenth century.

It is only when we recognize that the likes of Fludd and Dee flourished and were powerful intellectual forces at the same time as Boyle, Hooke, Newton and Christiaan Huygens were crystallizing the methods of modern science that we can appreciate why the rise of curiosity was not a simple displacement of religious constraint with a single form of open-minded experimental science. The Hermetic philosophers came adrift from the current of early modern science not because of *what* they believed but because of what they *did* with those beliefs. They did not recognize, as Bacon did, that the curious pursuit of knowledge needed to become less a personal Gnostic voyage and more a state enterprise.

5 Professors of Everything

Is it not evident, in these last hundred Years (when the Study of
Philosophy has been the business of all the *Virtuosi* in
Christendom) that almost a New Nature has been revealed to us?

John Dryden (1668)

These two Subjects, *God*, and the *Soul*, being only forborn, in all
the rest they wander at their Pleasure.

Thomas Sprat, *The History of the Royal Society* (1667)

In 1648 Reverend John Wilkins became warden of Wadham College
in Oxford and published a book called *Mathematical Magick*, which
became a favourite of the young Isaac Newton. We might expect
from such a title that this was a compilation of numerical parlour
tricks, or perhaps instead a serious work of numerological mysticism. But it is neither. Wilkins' book is an exposition on mechanical
devices, from simple levers to the sort of complex automata with
which Salomon de Caus had populated the royal gardens at
Heidelberg. The title shows that mathematics was linked both to
the magical tradition and to the practical marvels of mechanics –
with just a hint that there was an element of trickery involved in
that art. Wilkins alluded to the way ignorant folk attribute mechanical and mathematical ingenuity to magic, understood in its modern
sense of something superstitious rather than rational. John Aubrey
attests that in the previous century people 'burned Mathematical
bookes for Conjuring bookes', and even in 1651 John Rawley, a friend
of Francis Bacon, was accused of conjuring by a country parson

when he measured the height of his church steeple using geometrical calculation.

John Dee's reputation as a conjuror itself rested in part on his mechanical wizardry. In 1547 he staged Aristophanes' *Pax* at Trinity College in Cambridge, in which there was, as Dee boasted, 'a performance of the Scarabeus his flying up to Jupiter's palace, with a man and his basket of Victualls on her back; whereat was great wondring, and many vaine reports spread abroad of the meanes how that was effected'. It was all done with weights and pulleys, but magic seemed to some folk a more likely explanation.

Dee helped to cement the union of mathematics with mechanical ingenuity in his 'Mathematicall Preface' to Euclid's *Elements*, where he asserted that the art of geometry included such engineering feats as the construction of water pumps and automata, the measurement of time, architecture and navigation. In fact, historian Peter Zetterberg suspects that 'after reading Dee's exciting preface many were disappointed by the *Elements* itself and its pedestrian tone and topics'. Henry Peacham stressed this wondrous aspect of mathematics in his *Compleat Gentleman*, advising that 'geometry' was 'more than a dull study required for building engines of war', but might also be used to make marvels such as the flying wooden dove allegedly constructed by the Pythagorean philosopher Archytas in the fourth century BC, or the 'iron fly' of the fifteenth-century German mathematician Regiomontanus.

Dee's contemporaries considered that the mechanical mathematicians of antiquity had almost divine powers. The Welsh mathematician Robert Recorde wrote apropos Archimedes' devices for defending Syracuse from Marcus Claudius Marcellus that he knew 'how to steale fire from Heaven . . . [he] transformed himselfe into *Jupiter*, thundering downe from the highest Towers of the Town, his thunder-bolts of lightning'. Archimedes was 'a petty God, more than a man', according to Robert Burton in *Anatomy of Melancholy* (1621). After all, as Wilkins pointed out, did not Archimedes claim to be able to move Earth itself with a suitably placed lever? Wilkins attached Archimedes' name to the section of his book that dealt with machines.

Wilkins was no armchair mathematician. According to John Evelyn, he filled his rooms at Oxford with the kinds of ingenious devices that had appealed to Archimedes and Hero of Alexandria:

an hollow Statue which gave a Voice & utterd words, by a long & concealed pipe which went to its mouth, while one spoke thro it, at a good distance, & which at first was very Surprizing: He had above in his Gallery & Lodgings variety of *Shadows*, Dyals, Perspectives . . . & many other artificial, mathematical, Magical curiosities: a Way-Wiser [for measuring distances as you walked], a *Thermometer*, a monstrous *Magnes* [magnet], *Conic* & other *Sections*, a Balance on a demie Circle.

Wilkins had seen Salomon de Caus's creations for himself. He acted as chaplain to Prince Charles Louis, the second son of Frederick V and Elizabeth Stuart, who, exiled in England, became heir to the title of Elector Palatinate following the death of his elder brother. When the conflict in the German states drew to a precarious close via the Peace of Westphalia in 1648, Charles Louis was able to reclaim part of his inheritance in the Lower Palatinate, and when he returned to the marvellous palace and gardens of Heidelberg, Wilkins accompanied him.

The warden of Wadham was thoroughly familiar with the works of Dee and Fludd and with the tradition they represented. His 1641 book *Mercury, or the Secret and Swift Messenger*, was explicitly Hermetic in its title (Mercury being the Roman equivalent of Hermes), and its subject matter of cryptography connected it both to Dee's work and to that of the German natural magician Trimethius, the mentor of Agrippa in the early sixteenth century, whose *Stenographia* is one of the first works on the subject. Agrippa himself referred to mechanical inventions as a kind of 'mathematical magic', and Wilkins' book on the subject seems to draw on Fludd's encyclopaedic *Utriusque Cosmi Historia* (1619). Wilkins was associated with the circle of Comenius and Hartlib, and was linked to the Rosicrucian movement; in *Mathematical Magick* he mentions an underground lamp that is similar to one 'related to be seen in the sepulchre of Francis Rosicrosse, as is more largely expressed in the Confession of that Fraternity'.

All this explains why Wilkins was such an important figure in the development of experimental science. He is a link between the virtuoso culture of curiosity and the new natural philosophers, such as Robert Boyle and Robert Hooke, who grew increasingly important in the Royal Society. In its earliest days the Royal Society was dominated by

men like Wilkins: virtuosi who revelled in wonder more than explanation and who paid lip service to Francis Bacon while ignoring his
injunctions against 'fabulous experiments, idle secrets, and frivolous
impostures'. John Evelyn, another archetypal virtuoso who, while no
scientist himself, lubricated the wheels of the new philosophy with
elegant words of advocacy, exposed the gap between courtly curiosity
and scientific understanding in his description of a marvel that he saw
in a cabinet during a continental sojourn: 'a large pendant candlestick,
branching into several sockets, furnished all with ordinary candles to
appearance, out of the wicks spouting out streams of water, instead
of flames'. There is evident disappointment in his comment that this
device 'seemed then and was a rarity, before the philosophy of
compressed air made it intelligible'. He clearly preferred Della Porta's
approach of hiding the trick to enhance the wonder.

Ingenious gentlemen

According to John Aubrey, Wilkins 'was a very ingeniose man, and
had a very Mechanicall head'. He was 'much for the Trying of
Experiments', to the extent that Aubrey credits him as 'the principall
Reviver of Experimental Philosophy at Oxford'. From 1649 Wilkins
convened at his college lodgings each week an 'experimental club',
which Aubrey identifies as the 'Incunabula [cradle] of the Royall
Society'.

It has been claimed that the origins of the Royal Society predate
Wilkins' Oxford club, springing instead from meetings he had with
like-minded souls in London in the early 1640s. One of these, the
mathematician John Wallis, called its members 'Persons inquisitive
into Natural Philosophy, and other parts of Humane Learning: And
particularly of what hath been called the New Philosophy or
Experimental Philosophy.' Wallis attributed this 'new philosophy' to
both Francis Bacon and Galileo. The London meetings, he said, were
suggested by the German theologian Theodore Haak, an exile from
the Palatinate and friend of Comenius, who may have begun them
several years before Wallis joined in 1645. Another member was
Oliver Cromwell's doctor Jonathan Goddard, who had also attended
Charles I while he was held captive by Parliament. The group would
gather in Goddard's rooms in Wood Street, or 'at a convenient place

in Cheapside' (the Bull-head Tavern) – or else at Gresham College in Bishopsgate.

Established in 1596, Gresham College was housed in the mansion of its patron, the wealthy merchant Thomas Gresham, who endowed professorial chairs of law, medicine, rhetoric, music, divinity, geometry and astronomy.* It is not clear now what motivated Gresham to set up this centre of learning, but he stipulated that the professors were to give weekly public lectures for the educated gentlemen of London. Those gentlemen often did not seem terribly interested in this provision, but nonetheless for the professors Gresham College offered a comfortable simulacrum of a university in a city that otherwise lacked such an institution. It seems possible, indeed natural, that Gresham was the key locus of the London group of new philosophers, and by the time Wallis joined them they were in the habit of meeting after the weekly lecture by the Gresham professor of astronomy Samuel Foster.

The idea that the London group, and not the later Oxford club, gave birth to the Royal Society was promoted by Wallis when, as a Fellow of the Royal Society (FRS), he wrote a rather self-serving account of the society's origins in order to challenge the one put forward by his scientific rival William Holder, who located its genesis squarely in Oxford.† But the London meetings clearly had much of the same flavour, as well as some of the same members. According to Wallis, their interests included:

the circulation of the blood . . . the Copernican hypothesis, the nature of comets and new stars, the satellites of Jupiter . . . the inequalities and selenography of the Moon . . . the improvement of telescopes . . . the weight of air, the possibility or impossibility of vacuities [vacuums] . . . the descent of heavy bodies, and the degree of acceleration therein.

In the end it is a matter of taste whether one decides that the Royal Society started in London or Oxford. It has been said with some cause

---

* It is perhaps a little surprising that Francis Bacon did not think of Gresham College as a possible location for the English Solomon's House that he desired, given that Gresham himself was Bacon's uncle.
† John Aubrey took Holder's side – he clearly did not like Wallis at all, accusing him of being 'so extremely greedy, that he steales flowers from others to adorne [his] owne cap'. Aubrey's account of the matter is thus clearly partisan.

that the Oxford group undertook experiments whereas the London group did not. But many of the Fellows of the Royal Society weren't experimentalists either – they went along merely to hear the latest about the new philosophy. Awarding too much importance to the 'experimental' activities at Oxford runs the risk of implying that this method of investigation was invented there. And we should not give too much weight to the 'official' history of the society by Thomas Sprat, an Oxford protégé of Wilkins, who was commissioned to write it in 1667. Such accounts were inevitably as much propaganda as they were history, and Frances Yates suggests that Sprat – a non-scientist who may have been admitted to the Royal Society solely as an eloquent propagandist – was encouraged to downplay the role of the London meetings in the mid-1640s because these were deemed in retrospect to have rather too much of an odour of Rosicrucian utopianism about them.*

In any event, the formation of the Royal Society was also indebted to a *second* group of London-based experimental philosophers besides the Wilkins/Wallace group. Devoted Baconians all, they were led by Samuel Hartlib, and included the German philosopher and theologian Henry Oldenburg and the reformer John Dury, whose daughter became Oldenburg's wife. Robert Boyle was also one of Hartlib's regular correspondents. For Hartlib science was not a self-motivated exploration of nature but should instead dedicate itself to the Baconian goal of improving the lot of humankind.

The Cromwellian purge of Royalists at Oxford left many vacant posts, of which several of these men took advantage. Among those who moved to Oxford were Wilkins, Wallis, Goddard, and later Boyle and Oldenburg, who became tutor to the sons of Boyle's sister Lady Ranelagh and was employed by Boyle as an editor and translator. All of them attended Wilkins' club in Wadham College, where they were joined by the Savilian professor of astronomy Seth Ward, the astronomer Laurence Rooke and the mathematician and polymath William Petty – and by an undergraduate named Christopher Wren.

---

* The reliability of Sprat's account has been much contested. The historian Michael Hunter calls it 'as much a confession of faith as a factual record'. Even if it is not simply 'the story according to Wilkins', the book was not closely supervised, and several members of the Royal Society were unhappy with the picture it gave both of the society's inception and of its aims. There was much talk of a revised version, or a follow-up by another author, but nothing ever materialized.

It is sometimes implied that these gentlemen convened for the purpose of inventing modern science. But as is clear from the descriptions of Wilkins' crafty 'mathematical magic', the club entertained a culture of virtuosity acquired from the courtly scene of the late sixteenth century, occupied as much with demonstrations of skill and wonder as with understanding the world. The Royalist physician Walter Charleton describes it thus:

> It is their usual recreation, to practise all Delusions of the sight, in the Figures, Magnitudes, Motions, Colours, Distances, and Multiplications of Objects: And, were you there, you might be entertained with such admirable Curiosities, both Dioptrical and Catoptrical, as former ages would have been startled at, and believed to have been magical . . . Were Friar [Roger] Bacon alive again, he would with amazement confesse, that he was canonized a Conjurer, for effecting far lesse, than these men frequently exhibit to their friends in sport.

As befits a man wedded to the Baconian enterprise of experimental philosophy, Wilkins had a wide-ranging curiosity: he wrote on botany, agriculture, hydraulics, navigation, heat, magnetism and mechanics. He did not share Bacon's semi-mystical reverence for the monarchy: he was (like Prince Charles Louis, surprisingly) a Parliamentarian during the English Civil War, and in 1656 he married Cromwell's widowed sister Robina. This meant that, after the Restoration, Wilkins was substituted by a Royalist in the academic post he then held, Master of Trinity College in Cambridge, which forced him back to London in 1660. Here he resumed his 'philosophical club', first back at the Bull-head Tavern in Cheapside. When it grew too big to meet at a pub, it looked to the other regular venue of the earlier Wallis group: Gresham College. Some of Wilkins' fellow exiles from the universities had become Gresham professors: Laurence Rooke took the chair of astronomy in 1648 and subsequently moved to geometry, William Petty held the chair of music, Jonathan Goddard was made professor of physic in 1655, and Rooke's successor as professor of astronomy was Christopher Wren.

As will be apparent to those who know anything of Petty and Wren, their Gresham titles were rather arbitrary, prompting the suspicion that anyone of sufficient wit and education could flit at will between disciplines that today are separated not just by chasms but at times by fortifications. What's more, these positions may have demanded little more than tokenistic effort, for their incumbents typically pursued other occupations at the same time. While holding forth on music at Gresham, Petty was at the same time professor of anatomy at Brasenose College in Oxford, yet he gave rather little time to either when, in 1651, he was appointed chief physician to the English forces occupying Ireland. Even there Petty found himself tasked with a quite different job from looking after the health of the troops, for he was placed in charge of surveying the country. Quite aside from all this, Petty is now best known as a pioneer of social statistics, arguing in his book *Political Arithmetick* (posthumously published in 1690) that the governance of a nation can be improved by being conducted as a science based on quantitative measures.

If Petty was a polymath, how should we describe Wren, the astronomer who built St Paul's Cathedral, and whose efforts to reconstruct much of London after the Great Fire of 1066 look from this distance almost like something he fitted into his spare time? Wren had been a protégé of Wilkins ever since his father (also Christopher), the Dean of Windsor, befriended Wilkins during the latter's spell at the royal castle as chaplain to Charles I's nephew. When he was granted the Gresham chair of astronomy in 1657, Wren was only twenty-five. The previous incumbent Rooke apparently moved to geometry simply because the rooms were better – another intimation that any of these positions was up for grabs once you were in the right circles, and that the successors to the 'professors of secrets' can be reasonably called 'professors of everything'. Polymathy – or as it was often called at the time, pansophism or universal wisdom – was the order of the age. 'He can hardly be a good scholar, who is not a general one', wrote Isaac Barrow, Isaac Newton's predecessor as Lucasian Professor of Mathematics at Trinity.

When the London philosophical club met in Rooke's rooms at Gresham on 28 November 1660 after one of Wren's astronomy lectures, they decided it was time to learn from the example that the Italians had already set and formalize their endeavours to promote the new experimental philosophy. 'It was proposed', says the Royal Society's official record, that some regular course might be thought of, to improve this meeting to a more regular way of debating things, and according to the manner in other countries, where there were voluntary associations of men in academies, for the advancement of various parts of learning, so they might doe something answerable here for the promoting of experimentall philosophy.

Thus the plan for the 'Royal Society of London for Promoting Natural Knowledge' was hatched. Among those present was Lord William Brouncker, an able mathematician who was ultimately more useful to the fledgling Royal Society for his connections than for his intellect: he was a Royalist who became Chancellor to Charles II's queen, Catherine of Braganza, in 1662. Another link to the king was supplied by the Scot Robert Moray, who deserves rather more credit than posterity has afforded him for his role in establishing the society. Moray led a colourful life during the reign of Charles I, acting as a French spy for Cardinal Richelieu and marching with the Scottish Covenanters when they defeated the English king's troops at Newcastle in 1640. Yet Moray played so crucial a role in the negotiations between the rebellious Scots and the English that Charles knighted him in 1643. Fighting with the Scots Guard in France during the Thirty Years War, Moray was captured and imprisoned in Bavaria; eventually he returned to Scotland during the Protectorate where he took part in the uprising in support of Charles II in 1650, and subsequently went into exile in France. After the Restoration Moray returned to England in the late summer of 1660, and by November he had already established himself as a regular attender at Gresham. A close confidant of the new king, Moray helped to bring the plans of the Gresham group to Charles II's attention. Samuel Pepys records that Moray and Brouncker sometimes carried out chemical experiments in the king's laboratory in Whitehall. Moray was also the patron of the Welsh Paracelsian alchemist Thomas Vaughan, who produced the first English translation of the Rosicrucian manifestos in 1652.

Of those others present at the November meeting of the Greshamites, Alexander Bruce, Paul Neile, Jonathan Goddard, William Balle, Laurence Rooke and Abraham Hill were surely able fellows, yet all can without unkindness be described as bit-part actors in this historic episode. Not so the last member of the group, Robert Boyle.

While natural philosophy in the Middle Ages depended largely on the existence of curious clerics who had both access to books and the leisure to think about what they read, in the transitional period from the Renaissance to the Enlightenment it was sustained by the attention of secular gentlemen of independent means who needed no patron in order to be able to pursue their interests. Boyle's English father was a canny investor who took advantage of the confiscation and redistribution of Irish land by the Crown in the early seventeenth century to acquire a sizable estate, which grew larger still as he rose in the administration of Ireland from privy councillor to Lord High Treasurer and ultimately to become the first Earl of Cork in 1620. By the time Robert Boyle was born in 1627 in Lismore Castle, twenty-five miles from Cork, his father was fabulously wealthy. In Boyle's adult years the Irish estates that he inherited brought him £3,000 a year, which enabled him to construct private laboratories and employ staff to assist him in his experimental studies. When Boyle encountered the experimental philosophy in the late 1640s, it was a revelation – he called it 'a kind of Elizium'. His first laboratory was built at his manor house in Stalbridge, Dorset, and there he began his 'Study of the Booke of Nature'.

Unlike many gentlemen who dabbled in philosophy, Boyle's interest was not only sustained and profound but also guided by one of the most astute minds in England. His investigations were conditioned and motivated by strong religious beliefs: for Boyle, understanding nature's mechanism was a devotional act, a demonstration of the wisdom with which God had made the world. Boyle felt that science could supply a counter-argument to the atheism that extreme rationalism seemed to threaten. His name is now most commonly associated with his investigation of the behaviour of gases* and with his book *The Sceptical Chymist* (1665), which has often been misinterpreted as an attack on alchemy. But Boyle was interested in virtually every facet of seventeenth-century science, and by the 1660s his opinion was sought and esteemed on all matters connected to natural philosophy. An endorsement from Boyle came almost to have the authority that a corroboration from Aristotle had borne in the Middle Ages, and it would have

---

* The somewhat inaptly named Boyle's law (see page 271) states that the volume of a gas is inversely proportional to its pressure.

been unthinkable to launch an institution like the Royal Society
without Boyle being included. Like most of the experimental philosophers, he had little time for Aristotle, discerning that his tautological
explanations of phenomena were all but useless for predicting the
outcomes of specific experiments. The Aristotelians, he wrote,

> make it very *easy* to solve All the Phaenomena of Nature in Generall,
> but make men think it *impossible* to explicate almost Any of them in
> Particular.

A description of Boyle after he left Oxford to live at his sister's palatial
house on Pall Mall in London gives a lucid indication both of his
comfortable circumstances and the extent to which his experimental
science was indebted to Bacon and the old tradition of secrets:

> His greatest delight is chymistrey. He haz at his sister's a noble labora-
> tory, and severall servants (prentices to him) to looke to it. He is char-
> itable to ingeniose men that are in want, and foreigne chymists have
> had large proofs of his bountie, for he will not spare for cost to gett
> any rare secret. Experiment, he declared, is the interrogation of Nature.

It was said that Boyle's residence was 'constantly open to the Curious'.
According to John Evelyn, Boyle learned 'more from men, real experiments, & in his laboratory (which was ample and well furnished)
than from books'.* He himself professed to have gained more understanding of stones from 'two or three masons, and stone-cutters, than
ever I did from *Pliny*, or *Aristotle* and his commentators'.

Curating curiosity

To the Royal Society's energetic publicist Joseph Glanvill, a philosopher
who styled himself 'apologist of the virtuosi', the group represented the
fulfilment of Francis Bacon's 'mighty Design': an organization devoted
to collecting scientific knowledge in the service of the state. Moray
appealed to Charles II for a royal charter, which was granted in 1662,

---

* Such remarks have encouraged the false idea that Boyle had no library. In fact he
had a large one, although its contents have now been lost.

but the king showed little serious interest in the society's activities and was more bemused and amused than impressed by them. Its royal status meant nothing more than that the society was exempt from certain taxes, was permitted to employ staff and buy premises, could publish its own literature, correspond with foreigners and bear its own coat of arms. The king evinced no faith in Bacon's argument that a scientific society could serve the nation's political interests, for he never gave the Royal Society a penny, and it often struggled through lack of funds. It was unable to take on many staff, and often could not pay their salaries in any case – Oldenburg was secretary from 1662 but received no recompense until seven years later. And it was in no position to take advantage of its property purchasing rights: the members continued to meet gratis at Gresham College once a week, where they were afforded a public room for their discussions, another for their instruments, books, papers and rarities, and 'several of the other Lodgings, as their occasions do require'. A sustained effort to raise funds and draw up plans for the society's own building, a truly Baconian Solomon's House, in 1667–8 came to nothing.

In order to finance itself, the Royal Society levied a hefty subscription fee, which meant that its membership was confined mostly to the gentry. At the beginning it was in truth more important that members could pay than that they could understand or contribute to the discussions. Although the society accrued around 260 members by the late 1660s, many were aristocrats who attended the meetings mainly in the expectation of a diverting spectacle. This constrained what the Royal Society could achieve, for it was forced to cater to the whims of its dilettante members.* The experimental demonstrations could not be too obscure, and it was best if they yielded dramatic or surprising results. There was an emphasis on clever and entertaining gadgets and curiosities of the sort that appealed to John Wilkins.

These weekly demonstrations were devised and performed by a curator, who was obliged to take great care (*cura*) in preparing them. In its early days the society had only one curator, an overloaded fellow

---

* In the seventeenth century, 'dilettante' had not in fact acquired quite the pejorative implication that it has today – it retained its literal meaning simply as 'one who delights'.

who was expected to produce 'three or four considerable Experiments' each week. It would probably have been an impossible task for anyone less inventive and ingenious than Robert Hooke, who was informally appointed to the post in 1662 (it was formalized three years later), and soon thereafter took up residence for four days a week at Gresham.

A poor scholar from a humble family on the Isle of Wight, Hooke was studying at Christ Church in Oxford when his prodigious mechanical talents came to the attention of the Wadham club, which took him on as a paid technician. He was introduced to Boyle, who employed him as an assistant in his laboratory in Oxford High Street. The relationship, while mutually respectful, was very much that of master and servant – Hooke described himself as 'belonging' to Boyle even in 1663, after Boyle had 'released' him for the benefit of the Royal Society. Yet for all his deference to Boyle and the society's other Fellows, Hooke had begun by then to gain a sense of his own abilities as both an experimenter and a philosopher. His duties compelled him to be interested in everything, but such curiosity seemed to come naturally to Robert Hooke. If the social hierarchy would never permit him to become a professor of everything, he would have to content himself with being a technician of everything instead.

Here is a list made by Hooke in the 1660s of some of the topics he intended to explore:

Theory of Motion
 of Light
 of Gravity
 of Magneticks
 of Gunpowder
 of the Heavens

Improve shipping
 — watches
 — Opticks
 — Engines for trade
 — Engines for carriage

Inquiry into the figures of Bodys
 — qualitys of Bodys

Today this looks like an absurdly ambitious plan. Scientists spend their lifetime studying magnetism alone, or gravity, and if they make a single really significant contribution to their chosen topic then they can consider their career fruitful. Moreover, the expertise needed to understand gravity bears very little relationship to that needed to effect improvements in vehicle technology or the chemistry of explosives. That's not to say that a person capable of one could not do the other; but these things are recognized as distinct endeavours that, for reasons of sheer time and effort, are likely to be mutually exclusive.

That is partly because we know how hard most of these problems are – or perhaps one should say, how hard they have become now that the most accessible spoils have been claimed. It is one thing devising, say, better suspension for a horse and cart, quite another making a hydrogen-powered car. Discovering the inverse-square law of gravity was no simple matter, as we'll see, but the maths can now be learnt at school; reconciling gravity with quantum theory probably requires maths that hasn't yet been invented. All the same, it would be quite wrong to suggest that doing science was easier in Hooke's day, when there was still so much to be discovered about even (perhaps especially) the most mundane aspects of the world. The ambition apparent in Hooke's list speaks of a confidence that nature will yield her secrets to a universal approach – the experimental philosophy – once we have mastered it. There is also here something of the tone of a kid in a candy store: once curiosity was sanctioned among a group of like-minded individuals, the best thinkers could not refrain from asking everything at once and attempting to grab all the most appetizing questions for themselves. Hooke, probably more than any other of his contemporaries, shows what happens when Bacon's encyclopaedic research programme collides with human nature, which is rarely content just to beaver away at the finer points of this or that problem in natural philosophy. Individuals tried to lay claim to the grandest or most lucrative problems, and (especially in Hooke's case) were prepared to defend their priority ferociously.

Remarkably, however, Hooke's list does not simply reflect bravado or naïve enthusiasm, for he made genuine contributions in many of these areas, and in others too. He pioneered the design of watches and of microscopes; he designed and built new carriages; he quantified the manner in which rods bend when loaded; he helped Boyle to make new discoveries about the behaviour of gases. Most famously, he helped Wren rebuild fire-ravaged London, designing several of the city's monumental buildings, including the Monument itself, the great column (and disguised astronomical observatory) that marks the site where the conflagration began. For John Aubrey, writing in 1680, Hooke had shown himself to be 'the greatest Mechanick this day in the world'.

Hooke's imagination verges on the incontinent, which would not be so bad if he were not determined to assert priority for every fleeting fancy. In a book of 1675 he dashed off ('to fill the vacancy of the ensuing page', or perhaps, one suspects, to plant his flag) a list of 'a *decimate* of the *centesme* [a thousandth] of the Inventions I intend to publish . . . as I can get opportunity and leasure'. Among them were:

A Way of Regulating all sorts of *Watches* or *Timekeepers*
The true Mathematical and Mechanical form of all manner of Arches for Building
The true Theory of *Elasticity* or *Springiness*
A new sort of Object-Glasses for *Telescopes* and *Microscopes*
A new *Selenoscope* [for surveying the moon]
A new sort of *Horizontall Sayls* for a *Mill*
A new way of *Post-Charriott* for travelling far
A new sort of *Philosophical-Scales*

Again this list is not mere bluster; he managed much of this, and more. Yet occasionally one has to exercise a little scepticism about Hooke's claims. His 'sprung shoes' that enable the wearer to jump twelve feet into the air conjure up an amusing sight in the quadrangle of Gresham College, but probably owe something to Hooke's penchant for exaggeration.

It's no surprise, then, that Hooke was constantly struggling to get anything completed before the next matter diverted his attention. The excessive demands came not only from his own ungovernable impulses but from others who clamoured for his services, especially the haughty and unreasonable Fellows of the Royal Society, who tended to treat their curator as a lackey at their beck and call. Yet his efforts to investigate everything under the sun were less a perceived obligation to Bacon's comprehensive knowledge-generating scheme and more an insatiable wish to *know*. Hooke nevertheless managed to avoid the kind of aimless collection of isolated, quirky observations that sometimes bedevilled the Royal Society, and although he did not exactly develop a coherent research programme – nobody did at that time – he did sometimes manage to prune his questions in favour of those most likely to bear fruit.

Hooke's achievements surely rival those of the better known Wren and Boyle. If Hooke had had his way, his renown would also exceed that of his nemesis Isaac Newton, whose superior mathematical skills and ruthless political acumen (not to mention his having outlived Hooke) gave him the upper hand. Hooke was a strange man, at once crabby and clubbable, liable to be conceited but tortured with feelings of inadequacy. But these idiosyncrasies do not alter the fact that he perhaps best personifies the changing face of curiosity during his time.

Omnivorous curiosity of the kind displayed by Boyle and Hooke was a defining feature of the entire enterprise. A manuscript apparently in the hand of Robert Moray makes the intended scope plain enough:

> The business & designe of the Royall Society is, To improve the knowledge of Naturall things and all usefull Arts, Manufactures, Mechanick practises, Engynes & inventions by Experiments. Not meddling with Divinity Metaphysi[cs,] Moralls, Politicks, Grammar, Rhetorick, or logick . . .
>
> To Examine all Systemes, Theories, principles Hypotheses, Elements, Histories & Experiments of things Naturall, Mathematicall, & Mechanicall, invented, recorded, or practised, by any considerable Author Auncient or Moderne: In order to the Compiling of a Complete System of Solide Philosophy, for explicating all phenomena produced by Nature, or Art . . .

This eclecticism is evident from the outset in the Royal Society's journal of record, the *Philosophical Transactions*, which was the brainchild of the society's secretary, Henry Oldenburg. Another of the singularly energetic individuals without whom the Royal Society could hardly have survived, Oldenburg possessed a unique combination of diplomacy, zeal, multilingualism and journalistic flair that made him the hub of a network of new philosophers both in England and abroad, whose observations and researches swelled the journal's pages. The first page of the inaugural volume of the *Transactions*, published on 6 March 1665, announces that one will herein find 'some accompt of the present Undertakings, Studies, and Labours of the Ingenious in many considerable parts of the World', and lists among its contents the following:

> An account of the Improvement of Optick Glasses at Rome. Of the Observation made in England, of a Spot in one of the Belts of the Planet Jupiter. Of the motion of the late Comet predicted . . . An Experimental History of Cold . . . A Relation of a very odd Monstrous Calf. Of a peculiar lead-ore in Germany . . . Of an Hungarian Bolus . . . Of the New American Whale-fishing about the Bermudas. A Narrative concerning the success of the Pendulum-watches at Sea for the Longitudes . . . A Catalogue of the Philosophical Books publisht by Monsieur de Fermat, Counsellour at Tholouse, lately dead.

Something for everyone, then, and illustrative of the society's purpose, as avowed by Thomas Sprat: 'they confin'd themselves to no order of subjects; and whatever they have recorded, they have done it, not to compleat Schemes of opinions, but as bare, unfinish'd Histories'. To judge from the *Natural History of Oxford-shire* (1677) by the Oxford chemist and FRS Robert Plot, the first keeper of the Ashmolean Museum (see page 161), it is not surprising that these histories were often unfinished. Plot's contents list makes it immediately clear that there is little prospect of the reader remaining for long in Oxfordshire. It treats of topics ranging from 'the Heavens and Air' to 'Formed Stones' (those with odd, quasi-crafted shapes), 'Men and Women', the 'Arts' and 'Antiquities'. It is soon apparent that the role of Oxfordshire here is to act as a microcosm of the whole world. So it is, for example, that the maximum age of child-bearing is discussed with reference both to Pliny and to one Catharine Tayler of Shetford, near Banbury, who told Plot that she birthed a son ('living and lusty') at the age of sixty. As a result, this 'natural history' is not only a delightful compendium of local tales and gossip but a discourse on all of creation as viewed from, as it were, Plot's backyard. The implication was striking, at the same time both liberating and daunting: to understand everything, you could start from anywhere.

The character of the virtuoso

For Francis Bacon, Descartes and Galileo, reshaping natural philosophy was largely about reimagining the philosopher. While the 'professors of secrets' promoted an empirical approach by demanding that phenomena and recipes be tested and reproduced before being pronounced real, that did little to banish either credulity or wilful speculation. If there was one thing that the Baconian, Galilean and Cartesian systems had in common, it was an attempt to instil discipline into the experimental method: to refrain from premature hypothesis and to filter experience carefully. If none of these schemes always lived up to their claims of sober objectivity, nonetheless the intention was clear enough that natural philosophy should be conducted with what amounted to a new moral attitude. For Bacon this entailed a purging of impulses and habits, almost a kind of spiritual and intellectual purification that would conquer the misleading idols of the mind. There was thus a whiff of asceticism about the new philosophy, which blended well with the Protestant and Puritan currents of the early seventeenth century. But its apparent denial of classical authority, evident for example in the Royal Society's motto *Nullius in verba* – take no one's word for it – aroused distrust in the Catholic Church and among conservatives and Royalists. And its ostensible disengagement from the passions meanwhile bemused and, in truth, offended the libertarian sensibilities of Restoration England. The talk of brotherhoods and societies of privileged cognoscenti even disturbed some Anglicans, especially the Puritans themselves, who saw there the spectre of religious enthusiasm.

Moreover, to be an experimental virtuoso entailed not just a denial of certain states of mind but the conscious creation of new, artificial ones. It required a kind of unknowing, such that events and objects that seemed familiar would not be ignored because of any prejudice that there was nothing more to be known about them. Everything had to be, in a sense, made extraordinary, everything rendered uncertain. So that the scientist should not fail to make his search rigorous and diligent, one had to force oneself to find a subject fascinating, perhaps by means of consciously *deciding* to consider it wonderful. Here is how Hooke put it:

> In the making of all kinds of Observations or Experiments there ought to be a huge deal of Circumspection, to take notice of every least perceivable Circumstance . . . And an Observer should endeavour to look upon such Experiments and Observations that are more common, and to which he has been more accustom'd, as if they were the greatest Rarity, and to imagine himself a Person of some other Country or Calling, that he had never heard of, or seen the like before; And to this end, to consider over those Phenomena and Effects, which being accustom'd to, he would be very apt to run over and slight, to see whether a more serious considering of them will not discover a significancy in those things which because usual were neglected.

In other words, the banal had to become the mysterious. The world was to be re-enchanted, but that enchantment then swiftly denied. As historian Catherine Wilson says, 'science destroys the image of the familiar world and substitutes for it the image of a strange one, wonderful to the imagination and at the same time resistant to the projection of human values'.

This strangeness was awakened by the assumption that nature hides her mechanism – it is encoded in patterns of behaviour, but only indirectly, in ways that measurement and observation must tease out and separate from what is apparent and 'obvious'. In this sense there is not so much space between the web of forces and corpuscles investigated by Hooke and Newton and the occult influences and virtues that underpinned the Paracelsian philosophy. To discern the threads of this web, one had to be forever watchful. 'Fearful of excluding anything', say Daston and Park, the natural philosophers 'strained every nerve to catch everything. The feats of concentrated attention this required were herculean.' They needed to be sustained by a curiosity that was itself to some extent artificial: a manufactured, chosen state of mind. This position was revealingly described by the Cartesian philosopher Nicolas Malebranche, who wrote that:

> It is necessary that we *deceive our imagination* in order to awaken our

spirits, and that we represent the subject upon which we wish to meditate in a new way, so as to excite in us some moment of wonder [my italics].

In practice, the attitude of unprejudiced observation advocated by Hooke is impossible. Just as it is not feasible to make ourselves forget how to speak, so we cannot hope to eliminate all assumptions and suppositions. Neither, even if we could, may we put that attitude to good use, for everything would then become as incomprehensible as an Arabic text to an Englishman. And the questions would never cease: will the same behaviour we see on the lab bench be found in the dark, or in the cold, or during a full moon, or in Africa or at the poles?

Nonetheless, Hooke's point that the experimenter must become aware of what he takes for granted is an important aspect of the scientific enterprise. Perhaps a more immediate problem with adopting this highly constructed perspective, however, is that to the outsider it can seem absurd. What nonsense, to insist to oneself that rotten meat is a source of deep wonder! To reinvent oneself as a mere visitor to the everyday world, a naïf who gasps at fleas and specks of dust! We will see later what mockery this invited.

Facts first

Bacon's stricture that theories should be deferred until all the facts were at hand sometimes became, with the Royal Society, an apparent determination to avoid causal explanations altogether. Instead the Fellows were apt to collect facts much as some collectors and antiquaries stocked their shelves with curiosities. And like the antiquary, the collector of 'facts' often had a dubious preference for those items most titillating to the senses: the most colourful, the most gaudy or entertaining, irrespective of their informative nature. As Daston and Park say, these facts, 'too bizarre or singular to be classified, much less theorized . . . often seem to have been chosen with an eye to thwarting any explanation or generalization'. Thomas Sprat may have criticized Pliny for considering only 'the greatest Curiosities' and not 'the *least*, and the *plainest* things'; but one could easily turn the same complaint on the eminent Fellows. Their baroque reports – take, for instance, 'An Account of *Four Suns*, which very lately Appear'd in

France' – allowed readers to do little more than shake their heads and mutter 'Well, there's a thing.' Even when they concern more prosaic matters, it is as if the observations have to be rendered peculiar to justify their presence: the colours of wine or the shape of a hanging chain become 'new', 'singular', 'uncommon', and inevitably, 'curious'.

Why was theory so distrusted? Bacon was right to warn that explanations were prone not only to rush ahead of the facts but also to ossify into dogma that could be taught in schools. However, it may be that the main objection to causal explanation is that, being much more conjectural and open to dispute than observation, it was likely to arouse discord. And the Royal Society was still sufficiently wedded to the courtly tradition of curiosity that its gentlemanly members shunned the thought of something so vulgar as argument.

No amount of self-discipline could entirely suppress the instinct to ask 'why?' Instead, the Fellows tried to justify their reticence with humility, or to phrase hypotheses so tentatively that it was not entirely clear they had been advanced at all. To the issue of interpretation, Thomas Sprat averred, 'the *Society* approaches with as much circumspection, and modesty, as human counsels are capable of: They have been cautious, to shun the overweening *dogmatizing* on causes on the one hand; and not to fall into a *Speculative Scepticism* on the other.' Robert Boyle prevaricates thus over his observations of bodies that emit their own light (see Chapter 10): 'It is not easy to know, what phaenomena may, and what cannot, be useful, to frame or verify an hypothesis of a subject new and singular, about which we have not as yet (that I know of) any good hypothesis settled.' Hooke, meanwhile, advanced any theorizing with such caution that it could be read as nothing more than a subjective opinion: 'Whenever [the reader] finds that I have ventur'd at any small Conjectures, at the causes of the things that I have observed, I beseech him to look upon them only as *doubtful Problems*, and *uncertain ghesses*, and not as unquestionable Conclusions, or matters of unconfutable Science.' Isaac Newton even stated boldly that 'Hypotheses are not to be regarded in experimental Philosophy.' All the same, human nature prevailed, and a reluctance to leap from observation to explanation did little to prevent Hooke, Newton and their contemporaries – but those two especially – from being plunged into bitter disputes about interpretation.

The Royal Society was not alone in proclaiming an aversion to theory. Bernard de Fontenelle, secretary of the French Académie Royale des Sciences (of which, more below), assured readers of the first volume (1702) of his annual *Histoire de l'Académie des Sciences* that they will find in the academy's deliberations no theories or hypotheses but only 'detached pieces', bare facts kept at arm's length by 'a kind of violence'. 'Hitherto', wrote Fontenelle,

> the Académie des Sciences has grasped Nature only in small chunks. There is no general system . . . Today one fact is established, tomorrow an entirely unconnected one. Conjectures about causes continue to be haphazarded, but they are just conjectures. So the annual collections which the Académie presents to the public are composed only of detached pieces, independent of one another.

Daston and Park describe the empirical science of the late seventeenth century as 'grainy with facts, full of experiential particulars conspicuously detached from explanatory or theoretical moorings.' Most importantly, these facts were often not motivated by a question of the sort 'I wonder why/if/how . . .' Their proliferation encouraged a suspicion that *anything* might happen – and if so, how could one hope to construct an explanation? There was a danger that the whimsy of God, which had in earlier times undermined confidence in reliable explanatory scientific principles, might be replaced by the whimsy of nature. Certainly, facts have a stubborn insistence that could unseat the cosy certainties of Aristotle; but if they are too strange, rare and irreproducible, how can they help us construct anything better?

One curious aspect of this readiness to grant nature the capacity to invent and elaborate was a hesitation to attribute too much importance to precise measurement. Even so careful a scientist as Robert Hooke, whose law of springs and elasticity provides one of the early examples of a precise mathematical relationship between cause and effect, admitted that 'Nature it self does not so exactly determine its operations, but allows a latitude to almost all its workings.' All scientists today recognize that their experiments are generally subject to random influences that may create small differences between the measured values from one experimental run to the next – but Hooke's comment is not quite speaking to that circumstance. Instead of regarding natural phenomena as being quantitatively determined by the combination of precise forces plus random disturbances, he seems to consider nature as an autonomous agent who is simply not concerned to ensure that she always acts in exactly the same way. The result may be the same in either case – experimental measurements that exhibit a range of values for ostensibly identical conditions – but Hooke's remark tells us something important about how he and his contemporaries regarded natural laws. These were mathematical, to be sure, but that did not oblige nature to observe them precisely. The idea of nature as a rule-bound mechanism coexisted with the picture of a quasi-intelligent, self-determining agency.

Empiricism and reluctance to theorize mark the Royal Society apart from the deductive and starkly rationalistic new philosophy developed in continental Europe by René Descartes and his followers Pierre Gassendi and Marin Mersenne. The Cartesians sought for explanations of phenomena that were not just mechanistic but materialistic: they believed forces could only be transmitted by the mechanical impinging of particles of matter. On this basis Descartes purported to explain tides, rainbows, magnetism, freezing and boiling, digestion and nerve action – all, it must be said, according to hypotheses that were more or less inaccessible to experimental tests.

Descartes' mechanistic system has sometimes been presented as an advance over the 'occult' views of earlier times. But as magnetism and gravity show, sometimes an occult explanation was more productive, and indeed more progressive. When Alistair Crombie describes Descartes as aspiring towards 'a completed natural science, following from the first principles of nature as infallibly as in a mathematical demonstration', he unwittingly exposes it as being not so far removed from the old axiomatic model of Aristotle, albeit with different axioms. Descartes' *cogito ergo sum* demonstrates the hazards of a self-contained system of reasoning that cannot be reached and modified by experiment. Even if informed by observation, this rush to find first principles runs the risk of becoming tautological: the principles, allegedly deduced from experiment, are then 'supported' by appeal to the very experimental demonstrations that furnished them. Evidently, neither an endless accumulation of 'facts' without theory, nor a premature leap from 'facts' to explanation, is likely to sate a curiosity for reliable and deep understanding of the world. Neither could by itself offer a firm base for modern science.

The world wide web

Gathering information for compiling 'histories' was a collective endeavour in which the Royal Society sought to enlist the whole of Europe. Oldenburg established lines of communication with trusted sources abroad, such as Christiaan Huygens in Holland and the physician Marcello Malpighi, who became the first Italian FRS. A letter of 1667 from Oldenburg to Malpighi on Sicily typifies the spirit of this fact-finding enterprise:

> We earnestly beg you to be so good as to let us know of all that is noteworthy – of which there is so much in your island – concerning plants, or minerals, or animals and insects, especially the silkworm and its productions, and finally concerning meteorology and earthquakes, known to you or to other ingenious men.

Italy was, as we have seen, the cradle of the proto-scientific society, and these continued to spring up. In the mid-1660s the astronomer and mathematician Geminiano Montanari, a supporter of Galileo's experimental science, founded the Accademia della Traccia (Academy of Traces, a hunting allusion in the tradition of the *venatio*) in Bologna, which mutated in the 1690s into the delightfully named Accademia degli Inquieti (Academy of the Restless Ones). The Tuscan Accademia del Cimento (Academy of Experiment) predated the Royal Society and served as something of a model for it. It was founded in 1657 by Leopold de' Medici, who had been tutored by Galileo, and its members included Galileo's last pupil Vincenzo Viviani and Viviani's later mentor Evangelista Torricelli, one of the finest experimentalists of the age. The academy lasted just ten years, but its philosophy, recounted in the *Essayes of Natural Experiments made in the Academie del Cimento* (a 1684 English translation of the academy's records by Richard Waller FRS), displayed an affinity with that of the Royal Society. The *Essayes* refute the religious objections to curiosity, saying that 'the sovereign beneficence of God when he creates our souls, [does not deny] them to pry, as we may say for a moment into the immense treasure of his eternal wisdom'. Yet despite

Galileo's legacy, the academy admitted that mathematical reasoning alone has its limitations and that 'we can rely on nothing with greater assurance than the faith of experience, which . . . by adapting the effects to the causes, and again the causes to the effects . . . at last succeeds so happily'.

Like its predecessor the Accademia Segreta, this Tuscan academy proposed to test both the claims of the ancient authorities and those made in present times: 'it is our principal intent to incite others also to repeat with the greatest severity and niceness, the same experiments, as we have now adventured to do with those of any other person'. And as with its counterpart in London, the academy proposed to regard all speculations as personal and fallible, and called for a 'free correspondence with those several societies that are dispersed throughout the more illustrious and noted parts of Europe'. Perhaps, however, it was as well that the opinion of the illustrious scholars of London was not conveyed to Italy, for the experiments on pneumatics, mechanics, freezing, physiology and chemistry described in Waller's translation were considered by them to be rather old-fashioned and unexceptional.

Unlike the Royal Society, the French Académie Royale des Sciences benefited from the active support of the state. It was founded in Paris in 1666 at the instigation of the Controller-General of Finances Jean-Baptiste Colbert, who sought the approval of Louis XIV. The academy's beginnings were modest – thirty or so scholars met twice a week in the king's library – but as a government-affiliated institution it could employ professional scientists, thereby resembling even more closely the Baconian Solomon's House. Bacon was explicitly the inspiration, as Christiaan Huygens explained when he first petitioned Colbert to set up the academy:

> The principal and most useful occupation of this group, in my opinion, is to work at a natural history, pretty much according to Bacon's plan. This history is composed of experiments and observations and is the sole method of arriving at the knowledge of the causes of all that is perceived in nature. As to learn the nature of weight, heat, cold, magnetic attraction, light, colours, the particles which compose the atmosphere, water, fire, and all other bodies, the purpose of animal respiration, how metals, stones, and plants grow – all things of which

little or nothing is yet known, although nothing in the world is more desirable or useful to know.

In Germany, the Rosicrucian movement lent impetus to the early foundation of societies with an interest in natural philosophy. One of them, the Fruchtbringende Gesellschaft (Fruitbearing Society), was formed in 1617 by a group of nobles at Castle Hornstein, near Weimar. It was funded by Prince Ludwig of Anhalt-Köthen, who modelled it on the Florentine Accademia della Crusca, a society of linguists and philologers. The Fruitbearing Society was itself linked to the mysterious Orden der Unzertrennlichen (Order of the Inseparables), founded in 1577, which concerned itself with alchemy and mining technology and has been proposed as a forerunner of the Order of the Rosy Cross itself. Despite this, and notwithstanding the society's use of botanical imagery and the fact that Johann Valentine Andreae was a member, the Fruitbearers were primarily interested in linguistics and literature.

The mathematician and physician Joachim Jungius, a friend of Andreae, Hartlib and Comenius, established the short-lived Societas Ereunetica sive Zetetica (Society for Research or Investigation, also known as the Collegium Philosophicum) in 1622 in Rostock, an explicitly Rosicrucian organization modelled on Andreae's Christianopolis. In 1672 a professor of mathematics and physics at Altdorf, Johann Christoph Sturm, launched the Collegium Curiosum sive Experimentale, although it did not survive his death in 1695. And a group of physicians formed the Academia Naturae Curiosorum (the Academy of Those Curious about Nature) in Schweinfurt in 1652, which concerned itself with medicine and related disciplines: botany, pharmacy, anatomy, chemistry and physics. From 1670 this institution published its own journal, the *Miscellanea curiosa*, modelled on the *Philosophical Transactions*, and it evolved into the German Academy of Sciences, called the Leopoldina, when it was formally recognized in 1677 by the Holy Roman Emperor Leopold I. Gottfried Leibniz proposed in 1675 that a scientific academy be created to exhibit the greatest scientific instruments and devices of the age: calculating machines (he invented one himself), the air pump, pendulums, and also a catalogue of marvels such as 'all sorts of optical wonders . . . unusual and rare animals [and] artificial meteors'. This would have the dual purpose of stimulating interest in science among the public ('All respectable people would want to see the curiosities in order to be able to talk about them', he said) and attracting potential patrons and sponsors of the putative academy. Clearly the propaganda value of wonders was still appreciated by the new philosophers. In the event Leibniz's scheme came to nothing, but twenty-five years later he helped to found the Prussian Academy of Science in Berlin and became its first president. Although supported by the Prince-Elector Frederick III of Brandenburg, it was funded through the strange ruse (Leibniz's idea) of being granted a monopoly on the selling of calendars in the principality.

The magical Royal Society

While these and other institutions began to make science an organized, formalized and collective enterprise, they did not exactly fulfil the conventional narrative of creating a revolution in how science was practised and how it conceptualized the world. Like any historical change or tendency, the so-called Scientific Revolution was an interleaving of many ideas and approaches, ancient and new, in which seemingly disparate ideas and practices coexisted, often within individual minds, in ways that look baffling in retrospect, provided that we do not care to edit and prune them into a tidier form. Catherine Wilson points out how this tendency to mould facts into a unified narrative of progress has in the past distorted our view of how science developed:

> The history of science was formerly presented as an incremental process, in which facts were unearthed and stored in the treasure-house of human achievement, from which the corresponding errors were carted away. The emphasis now is on the dependent and relative status of facts and observations themselves, which are said to have meaning only within a wider theory, and to achieve their status only in relation to other assumptions and background conditions.

Some scientists today, who feel that the history of their subject should be about handing out medals for discoveries rather than about how people in former ages thought, will complain that this latter viewpoint looks perilously relative, tantamount to the postmodern claim that all scientific knowledge is a social construct that has nothing to do with the objective world and is no more privileged than any other system of knowledge. But that is not what is being claimed. Rather, it is nothing more than the self-evident fact that knowledge is not formed in a vacuum, nor does it arrive as unalloyed nuggets of truth. It is distilled from a turbulent brew of observation and conjecture, filtered through the preoccupations and prejudices of its times. Certainly, the pressure and volume of gases really *are* related in the manner Robert Boyle described; there really *are* moons around Jupiter, and microscopic animals in pond water. But not everyone accepted these facts when they were discovered, and no one who accepted them really understood what they meant. And many of those who knew and believed these things did so for social and philosophical reasons, and not because they were more clever and less prejudiced than those who did not.

Nothing makes this more evident than the enthusiasm that Boyle and Newton shared for alchemy. This has widely been deemed perplexing: how can someone be so right and so clear in one field of science, and so misguided in another? In contrast to the old view of Newton as the great modernizing force of science – Alexander Pope's famous bringer of light – it has become popular more recently to deal with this dilemma by portraying him as a Janus figure, part progressive mathematical physicist, part medieval alchemist and chiliastic astrologer. And we have been encouraged to think that he hid the latter face from his colleagues for fear that they might ridicule it. This is no more than a clumsy attempt to reconcile what still strikes us, in thrall to the positivistic view of the history of science, as a baffling and contradictory combination of attributes, although it was in fact a common one at the time. Robert Boyle was equally convinced that transmutation of metals might be achieved and no less vulnerable to crude confidence tricks as a result. He was wholly taken in for some years by a mysterious French visitor named Georges Pierre des Clozets who claimed to have links with a European circle of adepts that knew how to make the philosopher's stone. Boyle wasted much time and money in lavishing gifts on Pierre before he finally accepted the truth that the man was a charlatan. And when Boyle died, Newton sent desperate letters to their mutual friend John Locke in the hope of getting his hands on a red elixir with fabulous properties, which he thought Boyle had bequeathed to Locke.

There was more magic and demonology afoot in the entire Royal Society than we are wont to believe – not because the members were superstitious or credulous (or at least, not all of them) but simply because that is how people thought in the seventeenth century. When the clergyman John Webster expressed scepticism about the reality of witchcraft in his book *Displaying of Supposed Witchcraft* (1677) – to which the Royal Society gave its imprimatur – Glanvill attacked its attempt to rationalize away such beliefs.* Boyle was of the same mind – he actually hoped that the existence of demons might be proved, because this would strike a blow at atheism. To that end he encouraged the publication of, and wrote an introduction to, a short treatise called *The Devil of Mascon* (1658), which reported the activities of that demon in the house of a French minister of Burgundy named Francis Perraud.

Joseph Glanvill, who believed that at least some of the legends of demons and spirits in England were well founded, proposed that among the Baconian 'natural histories' that the Royal Society ought to compile there might be one on the 'World of Spirits'. Robert Plot investigated whether fairy rings were produced by the dancing of witches and their familiars (he decided probably not). Again, there was nothing unusual in this belief in spirits; the question was rather to what extent they intervened in the world. Natural magic provided a relatively systematic and even mechanistic way of thinking about such effects. As historian Charles Webster argues:

> The idea of harmony in nature, parallelism between the macrocosm and microcosm, the pervasiveness of forces akin to sympathy and antipathy, the application of animistic explanations to directional processes, and reference to emanations and hierarchies that bridged the gulf between the material and non-material world, remained viable explanatory options which were actively drawn upon by forward-looking thinkers throughout the seventeenth century.

This is not simply to say how hard it was for these scientists to escape the superstitions of their age. For the fact is that their system of thought was not infected by these things but encompassed and embraced them, and did so because of its origins.

All the same, magic was never a significant aspect of the society's

---

* Michael Hunter believes that this book, publication of which had been refused by the church censors, slipped through the Society's system largely unnoticed, on the recommendation of the physician and Fellow Martin Lister. He notes that there is now no copy in the Royal Society's library.

collective agenda, remaining merely an interest of some of its members. Glanvill's call for a natural history of spirits was ignored, and Michael Hunter has argued that, without taking an explicit position on alchemy, astrology and witchcraft, 'the Society's corporate policy simply sidelined such pursuits'. As a whole, they simply weren't that interested, perhaps because they deemed it wisest to avoid these troublesome topics on which the members had a diverse range of views. What is perhaps more telling, as Hunter has pointed out, is that this cautious and pragmatic avoidance of magical matters was reconstructed, from the eighteenth century onwards, into a myth that the Royal Society was actively responsible for banishing such 'superstition' from science. There was a perceived need to present science as antithetical to the magical tradition. This process began early: the literary historian Michael Winship asserts that in his 1718 book *Historical Essay concerning Witchcraft* the bishop Francis Hutchinson 'rewrote Restoration intellectual history by crediting the decline of witchcraft to the influence of the Royal Society'. Despite the best efforts of today's science historians, this sort of rewriting still plagues the public face of the history of science.

Bacon's House

In its early days, the Royal Society made much of its debt to Francis Bacon's vision, both in its commitment to empirical data-gathering rather than the premature construction of hypotheses and theories and in its status as a research institution along the lines of Solomon's House. Some older histories of the society's role in the genesis of modern science, such as that of historian Margery Purver occasioned by its 300th anniversary in 1960, imply that, by adopting a Baconian programme of inductive enquiry, the English scientists in the mid-seventeenth century totally transformed, in a matter of a few decades, the way science was thenceforth to be done. It should become clear that this is far too simplistic a view.*

* Among other things, Purver betrays her obsolete positivistic bias by hurriedly dismissing the alchemical studies of Boyle and Newton, saying that they 'had no place in Royal Society policy'. In fact there was plenty of 'alchemy' (better called by the transitional term chymistry) going on in the activities of the Fellows, and Boyle published some of this in the *Philosophical Transactions*.

For one thing, the Royal Society was not a monolithic enterprise with aims and methods that were universally agreed. The members had diverse social origins – some were members of the landed gentry, others from the families of the yeoman or artisan classes, others from the clergy – and their attitudes, priorities and expectations reflected this. We've seen already that it was the society's burden to be obliged to admit wealthy dilettantes whose membership fees were sorely needed (though not always paid, or not promptly), yet who made no contribution to research or scholarship but came looking for diversion. Others, such as John Evelyn, John Locke and Samuel Pepys, were energetically committed to the institution but were not trained in natural philosophy and participated more as observers and propagandists than as researchers. Indeed, several of the key figures in the society's foundation, such as William Petty, Robert Moray and Henry Oldenburg, left little if anything in the way of scientific achievements that obviously justify their being regarded as 'scientists' (the word is anachronistic in any event) at all. And even among the serious scholars there was no consensus about, say, the value of research in areas such as alchemy, astrology and natural magic, nor about the proper methodology for scientific study or the relative weights that one should assign to experiment and theoretical interpretation. Few if any of them were rigidly Baconian in their approach. Indeed, as Michael Hunter has pointed out, Bacon's natural philosophy could act as a convenient smokescreen for those members 'lacking in imagination to make a virtue of their abstention from philosophical speculation' – they could justify the dull amassing of 'facts' as an adherence to the principles of the Lord Verulam rather than a necessity occasioned by their intellectual limitations.

The Royal Society's claim to focus on Bacon's 'Experiments of Fruit' – on practical applications of natural philosophy, which is to say, on technology – was also problematic. Robert Boyle avowed that this was always his goal: 'I shall not dare to think my self a true naturalist', he said, 'till my skill can make my garden yield better herbs and flowers, or my orchard better fruit, or my field better corn, or my dairy better cheese, than theirs that are strangers to physiology.' But did Boyle ever achieve any of this? He did not, and neither, on the whole, did his colleagues. For all that the Fellows sought to emphasize Bacon's advocacy of science as a productive servant of the state, improvements in medicine, in timekeeping and shipbuilding, agriculture and brewing and dyeing were notably absent in the harvest of the Royal Society's researches. This is not because the scientists were incompetent, but because they tended to underestimate the challenge. It remains a well-kept secret even today that turning discoveries and good ideas into viable applications is one of the hardest parts of science, and also one of the aspects in which scientists themselves are (for generally good reasons) poorly skilled. The gentlemen of the Royal Society fondly imagined that their superior intellect would enable them to master any craft quickly and improve it beyond what ignorant artisans had been able to achieve. They soon found out that it is not so easy, and indeed that the artisans might not be so ignorant. Sometimes they explained away their failure by saying that it was the ignorance and conservatism of tradesmen and craftsmen which prevented their proposals from being adopted. This was not entirely untrue – guilds were notably protective of and slow to change their methods – but often the technological proposals of the scientists were rejected because they were plainly impractical or inefficacious. It was probably in part the paucity of useful applications that led to a decline in the attention the Royal Society gave to technology towards the end of the seventeenth century.

Ultimately, then, the Royal Society proved to be a disappointment to some of its founders. It never became a high-powered research institution capable of wonders to match those of Solomon's House, nor was it even agreed that this *ought* to be the aim. The lack of a clear focus and programme was felt even in the 1660s, and by the 1670s, when Wilkins, Oldenburg and Moray died, it was in serious crisis. Various plans for its restoration and reform were drawn up. They were usually shelved; the society survived nonetheless, but hard times returned in the ensuing decades. And yet what it achieved – somewhat haphazardly, patchily and gradually – was something else, in the long run far more valuable and significant than the establishment of a scientific institution. Perhaps more than any other body or movement, the Royal Society helped Bacon's incomplete and confusing scheme for an empirical method of knowledge generation to permit the emancipation of curiosity.

6 More Things in Heaven and Earth

The variety of nature, is a stumbling block to most men, at
which they break their heads of understanding, like blind men,
that run against several posts or walls: and how should it be
otherwise, since nature's actions are infinite, and man's understanding finite?

Margaret Cavendish, *Observations Upon Experimental Philosophy*
(1666)

Multi pertransibunt et augebitur scientia [Many will go to and fro,
and knowledge will be increased]

Motto in the frontispiece of Francis Bacon's
The Great Instauration (1620)

In 1520 Albrecht Dürer saw golden ornamental artefacts of the Aztecs,
which had been shipped to Spain the previous year by the conquistador
Hernán Cortes. 'In all my life', he wrote,

I have seen nothing that made my heart rejoice so much as these
things. Here I have found wonderful, costly things, and I have marvelled
at the subtle ingenuity of people in strange lands.

It was one of the first intimations to Europeans that there were lands
they never knew existed. Uncatalogued in the encyclopaedias of the
ancient world, these exotic kingdoms yielded marvellous arts of strange
races, as well as new fruits and plants and animals, shells and minerals.
Some of the novelties were less welcome. Six years after Dürer gaped

at Aztec gold, the Spanish writer Gonzalo Fernández de Oviedo y Valdés suggested that among the things Columbus's crew brought back from the New World was the 'Spanish disease' that first appeared in the port of Naples: syphilis, named by the Italian doctor Girolamo Fracastoro in 1530. Whether or not that is true is still debated, but there was certainly no remedy for this dreadful affliction to be found in the books of Hippocrates or Galen. Indeed, Fracastoro fostered the common belief that local afflictions have local cures when, in his mythologized account of the origin of syphilis, he explained how a nymph called Ammerice found a remedy among the plants of the West Indies. For European sailors had since 1517 been importing from Haiti (Hispaniola) a dense wood called guaiac, which was boiled up to make an unpleasant decoction that was eagerly consumed by syphilitics who could afford it. (Those who couldn't swallowed mercury.) The New World offered up other new medicines too, the most prized being an extract of the cinchona or *quina* tree of Peru, which was used to treat malaria. This plant was the sole source of the antimalarial drug quinine until synthetic substitutes were found in the twentieth century.

Encountering such things for the first time, explorers who voyaged to Africa, Asia and the Americas could not help but be curious about them. When he reached Cuba in October 1492, Columbus wrote in his journal that:

> There are trees of a thousand kinds and all with their own kinds of fruits and all smell so that it is a marvel. I am the most sorrowful man in the world, not being acquainted with them. I am quite certain that all are things of value, and I am bringing samples of them and likewise of the plants.

Columbus's voyage was the curtain-raiser to a century and more of wonder about the world far beyond Europe's borders. By the late seventeenth century, curiosity was invoked as an almost default response to new horizons in the burgeoning genre of travel writing. Claude Biron described his *Curiosities of Nature and Art Brought Back from Two Voyages to the Indies* in 1703, while in the previous decade the Capuchin missionary Dionigi Da Piacenza Carli offered a 'curious and true narration and description of all the curiosities and noteworthy things' he found in Africa, Asia, America and Europe. It was abundantly clear that, as Walter Raleigh put it, 'there are stranger things to be seen in the world than are contained between London and Staines'.

These discoveries not only tempted the curiosity of the philosopher and natural historian but also expanded the limits of what they might legitimately consider plausible. There was a long tradition, going back at least to medieval accounts by travellers such as John Mandeville (who journeyed to China), of suspending disbelief for marvels that would be considered too unlikely on one's own doorstep. In his *History of a Voyage to the Land of Brazil* (1578), the explorer Jean de Léry admitted that his experiences in the New World had forced him to revise his former scepticism about the descriptions of foreign lands by Pliny and other classical writers, 'because I have seen things as fantastic and prodigious as any of those – once thought incredible – that they mention'. Elizabethan writers exploited this elasticity of the imagination: locating tales of fictitious marvels in foreign lands elevated them beyond the status of mere invention, allowing the reader to toy with the notion that they *might* be true. When seventeenth-century scientists seem gullible about the wonders of faraway places, we must remember that the golden age of exploration had undermined any basis for placing restrictions on the kinds of creatures, plants and phenomena that might exist in the world. After all, as historian Madeleine Doran puts it, 'For myself, I find the unicorn much less improbable than the giraffe.'

Travel, trade and exploration were more than introductions to new experiences and new classes of object, however. They brought some urgent practical focus to the experimental philosophy. In observing the heavens and theorizing about Earth and its place in the cosmos, men like Galileo, Kepler, William Gilbert and Thomas Harriot were addressing real problems in navigation, and their instruments were the tools of the adventurer. Great voyages created a demand for improved star charts, for telescopes and timekeepers, for meteorological devices such as thermometers and barometers. To map the stars and planets precisely, one needed to understand the laws of optics and the nature of the atmosphere. It was necessary to ask what controlled the weather, the tides, the currents of the oceans. Travel invited curiosity, but many of the questions it raised were far from idle, and virtuosi such as Robert Hooke and the astronomer Edmond Halley were aware that practical problems of exploration, empire and trade could provide a powerful justification for the experimental philosophy.

Travellers' tales

It is significant that these new sights and sensations were being experienced by men like Columbus – merchants, artisans and adventurers who stood outside any academic tradition and who therefore encountered the novelties unencumbered by preconceptions about how nature should be. Before long they were joined by scientists and other curious gentlemen eager to catalogue and examine the abundance that lay beyond the horizon. At first the bounty of the New World was considered *sui generis*, a collection of marvels that lay outside of conventional categories. But that would not for long satisfy natural philosophers.

The English virtuoso Thomas Harriot accompanied his patron Walter Raleigh on an expedition in 1584 to Roanoke Island in what is now North Carolina: a place that Raleigh christened Virginia in honour of his queen. Harriot returned the next year on a voyage funded by Raleigh and led by the explorer Ralph Lane, with the aim of establishing a colony on the island. Raleigh was the central figure in a network of poets and philosophers interested in esoteric knowledge, which included Harriot and probably also his friend Christopher Marlowe, as well as another of his patrons, the alchemically inclined Earl of Northumberland (the so-called Wizard Earl) who was later imprisoned in the Tower of London by James I on (false) suspicion of being involved in the Gunpowder Plot.*

Harriot was the archetypal Elizabethan polymath: a mathematician and astronomer, he was also an able linguist who learnt the language of the Roanoke natives after two of them were brought back to England on Raleigh's earlier voyage. This made him an indispensable interpreter on Lane's Virginia venture. Keen to attract investors and settlers to the new settlement (which was in the end small and short-lived), Raleigh encouraged Harriot to write a description of the potentially valuable resources in Virginia, and helped finance its publication in 1588. This work, *A Briefe and True Report of the New Found Land of Virginia*, was the first thorough description of the flora, fauna and

---

* This web of acquaintances has sometimes been said to constitute a secret society, the School of Night, but that is no more than the fanciful invention of Victorian scholars, based on a passing phrase in Shakespeare's *Love's Labour's Lost*. Giordano Bruno is said to have been a member of this allegedly atheistic sect too, a notion that merely fuels its romantic allure.

mineralogy of America by an authoritative first-hand observer.* It
listed plants with European analogues, such as what the natives called
okindgíer (beans that 'in taste . . . are altogether as good as our English
peaze'), gourd-like *macócqwer* and roots called *openauk* that were deli-
cious 'boiled or sodden'. One of the most marvellous was a herb
known as *uppówoc*, which the Spaniards had found in the West Indies
and called *tobacco*. 'The leaves thereof being dried and brought into
powder', Harriot wrote,

> they use to take the fume or smoke thereof by sucking it through pipes
> made of claie into their stomacke and heade; from whence it purgeth
> superfluous fleame & other grosse humors, openeth all the pores &
> passages of the body; by which meanes the use thereof, not only
> preserveth the body from obstructions; but also if any be, so that they
> have not beene of too long continuance, in short time breaketh them:
> wherby their bodies are notably preserved in health, & know not many
> grievous diseases wherewithal wee in England are oftentimes afflicted
> . . . We ourselves during the time we were there used to suck it after
> their maner, as also since our returne, & have found manie rare and
> wonderful experiments of the virtues thereof; of which the relation
> woulde require a volume by it selfe.

No wonder his countrymen wanted this stuff. Harriot's message was
clear: the marvels of the New World were not just a source of wonder
and a challenge to classical inventories, but a resource to be colonized
and exploited.

That was ultimately how the inhabitants of these lands were also
regarded. Harriot says that the colonists were as wonderful to the
natives as these people were to them:

> Most thinges they sawe with us, as Mathematicall instruments, sea
> compasses, the vertue of the loadstone in drawing iron, a perspective
> glasse whereby was shewed manie strange sightes, burning glasses,
> wildefire woorkes, gunnes, bookes, writing and reading, spring clocks

* The French priest André Thevet published an account of the colony of France
Antarctique in Rio de Janeiro in 1557 which, while also describing new plants including
peanuts and tobacco, contained a great deal of exaggeration and invention. His rival
Jean de Léry called it a tissue of lies.

that seeme to goe of themselves, and manie other thinges that wee had, were so straunge unto them, and so farre exceeded their capacities to comprehend the reason and meanes how they should be made and done, that they thought they were rather the works of gods then of men, or at the leastwise they had bin given and taught us of the gods.

Many writers played up the 'wondrous' aspect of the new races and cultures they encountered, exoticizing and exaggerating their strangeness much as early medieval adventurers had turned the people of faraway lands into dog-headed races and monsters with heads in their chests. Conditioned by the accounts of Pliny, Marco Polo and John Mandeville, Europeans tended to *expect* strangeness and marvels in distant lands: their outlook almost demanded that they find pygmies, cannibals and Amazons, not mere men like them. By withholding the purpose of foreign customs (or perhaps not bothering to discover it in the first place), travel writers sought to emphasize the apparent 'oddness' of how other cultures behaved, feeding the European curiosity for the rare and peculiar. Eventually this tendency palled, however, as readers began to suspect they were being fed a diet of pure imagination. 'We [are] already overstocked with Books of Travels', says Jonathan Swift's Gulliver. 'That nothing could now pass which was not extraordinary; wherein I doubted, some Authors less consulted Truth than their own Vanity or Interest, or the Diversion of ignorant Readers.'

The commercial potential of exotic foreign flora such as tea, bananas and pineapples provided a strong imperative to make their study a systematic enterprise. The Dutch East India Company established a colony at the Cape of Africa in 1652, founding what became Cape Town in South Africa, and there the company ran a botanical garden from the 1680s at which the indigenous plants were assessed for their medicinal virtues. Andreas Cleyer, a Dutch soldier with medical training, studied the medicinal plants of Ceylon and Indonesia, such as colocynth (bitter apple, a purgative), when he became senior physician to the Dutch East India Company in Jakarta. The company helped to finance the establishment of, and provided specimens to, the Amsterdam Botanical Garden in 1682 – an early example of corporate sponsorship of a public enterprise.

Ever since Columbus's voyage, the virtuosi became intent on

taming, ordering and cataloguing the marvels of the New World. In this abidingly curious age, natural histories of exotic flora and fauna sold well. One of the first, *Historia general y natural de las Indias* (1535–49) by Gonzalo Fernández de Oviedo y Valdés, had many editions in many languages, and displayed unusual care in the quality of its information. According to historian Paula Findlen, Oviedo was 'a worthy precursor to Francis Bacon in the development of important ideas about the relationship between the quality of good testimony and the reliability of knowledge'. The great Bolognese collector Ulisse Aldrovandi tried unsuccessfully to win funding from several rulers for a journey to the Indies that would enable him to expand his catalogues.

The Spanish king Philip II was more generous, sending his physician Francisco Hernández to New Spain (Mexico) in 1571 to write its natural history. He was less than contented, however, when Hernández failed to send back any material for several years, and then returned an undigested mountain of notes and illustrations loosely arranged into sixteen volumes. Philip asked his new physician Nardo Antonio Recchi to put this mass of information into some kind of order, but although Recchi made many copies of the documents and spent several more years compiling them, still they seemed no closer to publication. The manuscripts eventually passed to Recchi's nephew Marco Antonio Petilio, and as rumour spread of the riches they contained, many collectors and virtuosi were eager to get their hands on them, including Aldrovandi and Giambattista Della Porta.

In 1611 Federico Cesi, who was constantly looking for specimens of New World plants to cultivate in his gardens, obtained the rights to publish Hernández's work under the auspices of the Accademia dei Lincei. But this project too ran into difficulties. Access to the manuscripts was piecemeal, as Cesi and his colleagues either bought pages from Petilio or were forced to run back and forth to his house to double-check the material. The great quantity of illustrations raised problems of cost, and Cesi was forced to commission what were now copies of copies, in a woodblock style that was somewhat inferior to the quality of illustrations by then being produced elsewhere. An initial volume was printed in 1628 but Cesi's death two years later added another obstacle. His fellow Lynx Francesco Stelluti persisted, however, and in 1651 (shortly before Stelluti's own death) the full *Rerum medicarum Novae Hispaniae Thesaurus (Treasury of Medical Matters of New*

A Mexican civet and toucan from Francisco Hernández's *Mexican Treasury.*

Spain or the History of Mexican Plants, Animals and Minerals, commonly known as the *Mexican Treasury*) came off the presses. It was no miscellany of curiosities, but was carefully annotated, classified and cross-referenced: a work of truly scientific natural history. All the same, even the best efforts of Recchi and the Lincei had failed fully to tame the chaos of Hernández's notes, or to expunge all remnants of a penchant for unlikely wonders in the manner of medieval and early Renaissance bestiaries: among the civets, toucans and bison were a two-headed serpent and a bizarre dragon.

The *Mexican Treasury* highlighted the problematic status of illustration in natural histories. Quite aside from the third-hand nature of the images in this case, the woodcuts show how hard it still was for artists to resist stylization, extrapolation and anthropomorphization. Many of these creatures sport human-like grins or grimaces, while a 'Mexican dog' has evolved into an utterly implausible, hyper-muscled mixture of bull and pig with a dog-like head. Dürer's notorious 'armoured rhinoceros' of 1515 would not have looked out of place in this company.

The Lincei acknowledged these problems. Classical writers such as Pliny and Galen had warned against illustrations on the grounds that they could be no substitute for the real thing and might, in their compromised yet persuasive verisimilitude, be merely misleading. In his own contributions to the *Treasury,* the Lynx Johann Faber admitted that 'painters often make big mistakes' and quoted Pliny's *Natural History* on the matter: 'Painting is indeed deceptive, and with its many colours, is particularly so in the copying of nature, where it also falls short as a result of the varying ability of copyists.'

But how could unfamiliar things be conveyed if not with images?

Opportunities for direct contact with the novelties and marvels in faraway places were sorely limited – even if the objects themselves were brought back, they were generally shut away in private collections. Besides, how else but pictorially could one present things without names? Like it or not – and Cesi too fretted over the limitations and dangers of illustration – the spread of *Wunderkammern* and the desire to collect and catalogue all of nature meant that pictures were indispensable. More often they were encouraged and celebrated rather than treated with caution. The German botanist Leonhart Fuchs wrote in 1542 that 'Nature was fashioned in such a way that everything may be grasped by us in a picture.' Aldrovandi considered illustration vital to his encyclopaedic project; as historian David Freedberg says, the hundreds of his manuscripts that survive 'testify to his omnivorous recording of every imaginable kind of fact about natural history and to his attempts to record them by visual means'.

The draughtsman was in that case as important as the naturalist; indeed, it was not always obvious that there was a distinction between them. Artists employed by princes and nobles to catalogue their collections in visual form, such as the Flemish illustrator Joris Hoefnagel, who served in the courts of both Duke Albrecht V of Bavaria in Munich and Rudolf II in Prague, often rendered their images with both high artistic skill and impressive scientific precision. Hoefnagel's celebrated four-volume work *The Four Elements* (1575–82) depicted a vast array of living creatures – especially insects, previously thought too small to be worthy of close consideration. Hoefnagel assigned each creature to one of the four classical elements, and in the manner of an emblem book they were accompanied by texts taken mostly from biblical or classical sources. Harriot's *Briefe and True Report* was meant to be accompanied by illustrations by the artist John White, who came on the expeditions, but funding problems meant that few of White's engravings were included. Some of them were instead adapted by the entymologist Thomas Moffett for *Theatrum insectorum* (*Theatre of Insects*) (1634), or used in other accounts of New World exploration.

The new philosophers of Europe who lacked the means or the desire for travel combed these compendia avidly for information about distant lands, and were eager to acquire their own sources. In 1661 a committee at the Royal Society drew up a list of key questions to ask merchants and sailors, while the physician and FRS John Woodward wrote a pamphlet in 1696 explaining to travellers how properly to observe nature and collect specimens. Robert Hooke cultivated the company of an explorer named Captain Knox, who had sailed to Africa, Mauritius and India. Hooke would entertain Knox with food and beverages at Jonathan's Coffee House in Change Alley, the site of the original London Stock Exchange, and over bowls of chocolate Knox told him about strange fishes, cinnamon trees, and a herb from India called *gange* or *bangue*. Hooke duly related these tales to the Royal Society, saying of *gange* that it 'doth, in a short Time, quite take away the Memory and Understanding', so that the patient 'is very merry, and laughs, and sings, and speaks Words without any Coherence, not knowing what he saith or doth'. At length the *gange* taker falls asleep, but wakes wonderfully refreshed and free of any previous ailment of the stomach or head, so that this herb might, Hooke concluded, be a valuable medicine. (A hypochondriac, he was always on the lookout for those.)

Boyle too welcomed travellers such as navy officers to his (sister's) house in Pall Mall, where they would find themselves grilled relentlessly about anything at all: mining in Hungary, the tides of the South Seas, the topography of Africa.* One Italian traveller claimed to have seen in the African desert 'a Creature, bodied like an ox, head like a pike fish, taile like a peacock'. Boyle also scoured published accounts of travels, keeping careful notes of any snippets that appealed to his magpie mind, such as this by the Frenchman Henri de Feynes in *Voyage par terre depuis Paris jusques la Chine* (1630):

He relates that in the Persian Gulfe there is in an Island which he names not a place calld Barrin, very famous for a pearle Fishery & exceeding beneficiall to the Persian King, & that the Pearles there, which are the properly calld Orientall ones are found in certaine oysters of the breadth of a large plate, & that they have this peculiar & Ingenious way in their fishing, that after the Divers have brought up the oysters in their hand they chaw them warily & by that meanes find what pearles they have in them.

* Boyle's encouragement of visitors eventually got the better of him. John Evelyn wrote that he 'often Entertain'd those who came to visite him, Ever with something rare or new', but in his later years he was forced to announce his visiting hours on a board outside his front door.

Quite what conclusions Boyle expected to draw from such accounts is not at all clear – they are the literary equivalent of the marvels of the cabinet collector. It was indeed, Boyle insisted, pure curiosity and not commercial interest, which led him to accept a directorship with the East India Company in 1669 (although this did not prevent him from making some canny investments).

The useful medicines brought back from the New World don't entirely account for the almost universally rosy picture that reports gave of the opportunities it presented. Partly this upbeat view was sheer salesmanship: explorers hoped to encourage colonization, which would make it easier to establish lucrative export businesses. But partly it reflected the optimism of discovering an unanticipated richness in the world. It's no coincidence that several of the utopian fictions of this age, including those by Thomas More, Francis Bacon and Tommaso Campanella, locate their idealized states in or towards the Americas.

Yet the traveller's passion for new sensations and experiences drew the same condemnation from conservative theologians that curiosity had always attracted. Henri de Castela, a priest of Toulouse, worried that pilgrims would be susceptible to 'vain and vicious curiosity' as they passed through unfamiliar lands, and as a result he advised in his guide to pilgrimage in 1604 that one should try to avoid intercourse with anyone en route, even if this meant feigning deafness, muteness and blindness: a perfect metaphor for the attitude that opponents of curiosity felt the devout should adopt towards the world in general.

Such isolation was unthinkable for that inveterate and sophisticated traveller Michel de Montaigne, who praised 'an honest curiosity to inquire into all things', and displayed it abundantly on his travels throughout Europe in the 1580s. He loved nothing better than to discover local customs: how, for example, the Swiss threw their plates into a basket in the middle of the table after eating their meat, or how their beds were so high that one had to mount steps to reach them. He enjoyed collecting strange stories, like the one in which a Frenchman named Germain had been a girl until, jumping over some obstacle on his path, he found that 'virile instruments' had popped out between his legs. All the same, Montaigne advised scepticism as the default response to what one hears, even to the extent of doubting that we can be sure about anything: 'Nothing certain can be established about one thing by another, both the judging and the judged being in continual change and motion.' This assertion of the grave limitations on human understanding troubled seventeenth-century rationalists such as René Descartes and Blaise Pascal, who could not accept what to them seemed to be a defeatist attitude to all philosophy. Yet Montaigne was sympathetic to the view of many experimental philosophers that the aim of enquiry and scepticism was to make wonders cease, to disperse dumb amazement with rational curiosity: 'Let this universal and natural reason drive out of us the error and astonishment that novelty brings us.' The traveller could expect new and extraordinary sights and experiences, but in the end all could be brought within the scope and compass of a rational system.

Collecting the world

Collectors of curios mostly welcomed the influx of new and wondrous objects from afar. The Sun King, Louis XIV of France, was delighted as exotic animals joined his menagerie at Versailles, including ostriches and an elephant. It wasn't obvious where the accumulation of curiosities ended and scientific classification began. Illustrations of the specimen collection of Frederik Ruysch, first professor of botany at the Amsterdam Botanical Garden, show them arranged in ostentations juxtapositions with the elaborate blurring of art and nature characteristic of the cabinet. Joris Hoefnagel's illustrations, meanwhile, have been accused of endorsing the spirit of the cabinets by asserting the superiority of the rare and unusual over the familiar.

But this profusion of novelty challenged the idea that cabinets of curiosities could represent a microcosm of the world, for it began to seem that there was simply *too much in the world* to make that possible. According to writer Patrick Mauries, the abundance of the Americas played its part in creating 'a fragmented vision of multiple worlds' at the end of the sixteenth century. From this time, he says,

> it was no longer possible to embrace creation in all its diversity at a single glance. It was this tension between the desire to exhaust to the full every aspect of the real world, and to contain it within a finite space, and the increasingly clear impossibility of such an undertaking that lies at the heart of the cult of curiosities.

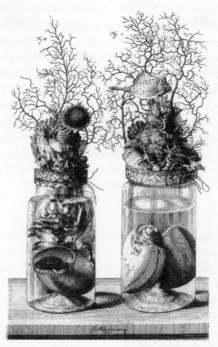

In the *Thesaurus Animalium Primus* (1710) of the Dutch botanist Frederik Ruysch, nature is presented in a highly constructed and extravagant manner.

As a result, the obsession for collecting began to lose some of its mystical agenda and instead became a more nakedly materialistic quest for the rare, exotic and costly. The cabinets and collections no longer aspired to be comprehensive catalogues of creation, rather taking on the aspect of attention-grabbing exhibitions. In the mid-seventeenth century, some private collections began to resemble public museums. One of the first collectors to spot this commercial potential of his hoard was the English botanist John Tradescant, who was gardener for several English noblemen before entering the service of Charles I in 1630 as Keeper of His Majesty's Gardens, Vines and Silkworms. Tradescant gathered much of his collection during travels abroad in search of new botanical specimens for his patrons: in 1625, for example, the first Duke of Buckingham George Villiers instructed him to scour the New World 'for all manner of rare beasts, fowls and birds, shells, furs and stones'. Tradescant displayed his stock of coins, plants and ethnographic objects at his house in Lambeth, South London, known as the Ark, and opened it to the public for a door fee. A visitor in 1634 confessed he was 'almost persuaded a Man might in one daye behold and collecte into one place more Curiosities than hee should see if hee spent all his life in Travell'. The Ark became one of the essential sights for visitors to London, and a regular stream of sailors and adventurers offered new specimens. Among the sober botanical and zoological specimens were fantastical items that showed the legacy of the cabinets of curiosities, such as a griffin's head, 'a natural Dragon, above two inches long', and 'Two feathers of the Phoenix tayle'. A German traveller gave an extensive account of what might be found at Lambeth:

> In the museum of Mr John Tradescant are the following things: first in the courtyard there lie two ribs of a whale, also a very ingenious little boat of bark; then in the garden all kinds of foreign plants, which are to be found in a special little book which Mr Tradescant has had printed about them. In the museum itself we saw a salamander, a chameleon, a pelican, a remora, a lanhado from Africa, a white partridge, a goose which has grown in Scotland on a tree, a flying squirrel, another squirrel like a fish, all kinds of bright colored birds from India, a number of things changed into stone, amongst others a piece of human flesh on a bone, gourds, olives, a piece of wood, an ape's head, a cheese, etc; all kinds of shells, the hand of a mermaid, the hand of a mummy, a very natural wax hand under glass, all kinds of precious stones, coins, a picture wrought in feathers, a small piece of wood from the cross of Christ, pictures in perspective of Henry IV and Louis XIII of France, who are shown, as in nature, on a polished steel mirror when this is held against the middle of the picture, a little box in which a landscape is seen in perspective, pictures from the church of S. Sophia in Constantinople copied by a Jew into a book, two cups of rinocerode, a cup of an E. Indian alcedo which is a kind of unicorn, many Turkish and other foreign shoes and boots, a sea parrot, a toad-fish, an elk's hoof with three claws, a bat as large as a pigeon, a human bone weighing 42 lbs., Indian arrows such as are used by the executioners in the West Indies – when a man is condemned to death, they lay open his back with them and he dies of it, an instrument used by the Jews in

circumcision, some very light wood from Africa, the robe of the King of Virginia, a few goblets of agate, a girdle such as the Turks wear in Jerusalem, the passion of Christ carved very daintily on a plumstone, a large magnet stone, a S. Francis in wax under glass, as also a S. Jerome, the Pater Noster of Pope Gregory XV, pipes from the East and West Indies, a stone found in the West Indies in the water, whereon are graven Jesus, Mary and Joseph, a beautiful present from the Duke of Buckingham, which was of gold and diamonds affixed to a feather by which the four elements were signified, Isidor's MS of De natura hominis, a scourge with which Charles V [the Holy Roman Emperor] is said to have scourged himself, a hat band of snake bones.

In other words, this was the traditional parade of wonders and curiosities, but now identified explicitly as a 'museum'.

After Tradescant's death in 1638, his royal appointment passed to his son, also called John, recently returned from a mission in Virginia 'to gather up all raritye of flowers, plants, shells, &c.' Around 1652 the younger Tradescant sought the help of his friend, the antiquarian Elias Ashmole, and the physician Thomas Wharton to catalogue the collection. The resulting volume, *Musaeum Tradescantianum*, was published under Tradescant's name in 1656. Ashmole not only advised on the project but also funded it, an act of apparent generosity larded with a strong dose of self-interest. For Ashmole was an avid collector himself, and for such a man the Tradescant museum was an irresistible prize.

Ashmole's connections with the Royal Society are a reminder of how diverse a range of tastes and attitudes were represented in their ranks. A staunch Royalist (he was appointed by Charles II to catalogue his coin and gem collection) and a lawyer by training, he had no real knowledge of science but was the typical amateur dilettante, with an agenda often quite at odds with the new breed of experimentalists. His second marriage in 1649 to a rich widow twenty years his senior was calculated to allow him the private means to indulge his antiquarian and scholarly pursuits 'without being forced to take paines for a livelyhood in the world'. All the same, once he was appointed Windsor Herald and Comptroller of the Excise, he was ensured an independent income that kept him wealthy after the death of his wife in 1668 and enabled him to continue adding to his collections of coins, prints, books and curios. Ashmole reminds us that the figure of the

Renaissance courtier was alive and well in the late seventeenth century: he took delight in the fine points of royal ceremony, protocol and heraldry, as befitted his position as Windsor Herald. And he was captivated by Hermetic and occult ideas, acting as a patron to the astrologer George Wharton (no relation to Thomas) and taking particular interest in the works of John Dee. According to the antiquarian Anthony Wood, Ashmole was 'the greatest virtuoso and curioso that ever was known or read of in England before his time'.

Inevitably these enthusiasms sat uneasily with some of the experimental philosophers. They did not so much mind Ashmole's interest in alchemy, and indeed his collection of alchemical texts, *Theatrum Chymicum Britannicum* (1652), won the admiration of Seth Ward, John Wilkins and Thomas Browne, and helped to secure Ashmole election as an FRS in 1661.* But other of his passions were considered suspect, even superstitious. John Evelyn was scornful of Ashmole's interest in astrology, and when Robert Hooke visited him in 1677 he recorded that he saw 'Dees and Kethes and many other Books and manuscripts about, chymistry, conjurations, magick, &c.' A classic product of what Frances Yates has called the 'Rosicrucian Enlightenment', Ashmole made several petitions to be admitted to that shady fraternity.

In the course of his work with Tradescant junior, Ashmole somehow persuaded him to agree that the entire contents of the Ark should pass to him on the event of Tradescant's death. Ashmole's legal expertise no doubt helped ensure that this contract was irrevocable, although Tradescant, regretting his decision, tried his best to undo the rash agreement. When he died in 1662, his widow Hester continued to contest the case, but Ashmole was relentless, even buying the house adjacent to the Ark and constructing a door that allowed him to enter the Tradescant premises at will. By the mid-1670s he had ownership of virtually the whole collection, leaving Hester with precious little prospect of financial security from it. She was found drowned in a pond in her Lambeth home in 1678 – most probably a suicide, although that scarcely exculpates Ashmole.

Ashmole's next act of seeming generosity also had a self-serving aspect. In 1675 he donated the collection to the University of Oxford,

* He was mentioned as a candidate for election at the first official meeting of the society in November 1660.

stipulating that it was to be housed in a building that would bear his name: the Ashmolean. Like the Ark, it admitted the public for a fee, and in effect it became the first public museum in Great Britain, opening its doors in Broad Street, next to the Bodleian Library, in 1683.* Now the cabinets of curiosities were truly no longer a private microcosm for the rich and noble, but wonders for general consumption – to the dismay of the German scholar Zacharias Conrad von Uffenbach when he visited the Ashmolean in 1710, to whom the presence of 'ordinary folk' in such a place was a rude shock.

The second English public museum to benefit from the generosity of a collector of foreign rarities was grander still. In 1753 the death of the 93-year-old Irish physician Hans Sloane meant that his impressive collection of over 71,000 plants, exotic materials, ethnographic objects and other curiosities, as well as 50,000 printed books, could (on payment of £20,000 to Sloane's heirs) pass to the king, George II, for 'the use and improvement of physic, and other arts and science, and benefit of mankind'. By an Act of Parliament that year, the Sloane collection supplied the basis for what became the British Museum.

By the time of his death Sloane had been collecting for nigh on seven decades. He began in earnest when he travelled to Jamaica in 1687 as physician to the Duke of Albemarle, who had become the island's governor. There Sloane collected about 800 botanical specimens and, on his return to England, 'shew'd them very freely to all lovers of such Curiosities'. He brought back animal specimens too, pickled in rum, brandy or brine, although not the live alligators or snakes with which he set sail from the West Indies – these died, or escaped and were shot, during the voyage.

Sloane set up a medical practice in Bloomsbury, London, in 1689, and made the most of his overseas experiences by selling quinine and importing chocolate for its alleged digestive benefits. 'Chocolate is here us'd by all People, at all times', he wrote from the New World:

> The common use of this, by all People in several Countries in *America*, proves sufficiently its being a wholesome Food. The drinking of it actually warm, may make it the more Stomachic, for we know by

---

* This building is now the Oxford Museum for the History of Science – the Ashmolean was moved to grander premises in the nineteenth century.

Anatomical preparations, that the tone of the fibres are strengthened by dipping the Stomach in hot water, and that hot Liquors will dissolve what cold will leave unaffected.

Although in the West Indies chocolate was taken with honey and pepper, this was a little strong for European tastes. Sloane recommended that it be drunk with milk, and as such it joined coffee and tea as a fashionable, curative beverage in the drinking houses of the capital. Sweetened, it hid the bitterness of quinine as a tonic for children. John Locke recommended it for gout, and if he followed Sloane's advice of mixing it with opium, there's no doubt that it would have brought relief from that painful condition.

Sloane cultivated a friendship with Robert Boyle in the last years of the great virtuoso's life, and gave him new substances from his travels for the scientist to study. Sloane's specimens also helped to furnish the encyclopaedic *History of Plants* (1686) by the botanist and FRS John Ray. This tome was part of a series of Baconian natural histories funded by the Royal Society, including the *History of Birds* (1676) and *History of Fishes* (1686) by Francis Willoughby. Ray helped to complete both works after Willoughby's death in 1672, but the latter in particular was so expensive to produce and so indifferently received that it left the society unable to fund Isaac Newton's *Principia* a year later (see page 204).* These huge compendia sought to impose a systematic taxonomy on the new flora and fauna amassed from the New World and elsewhere by collectors like Sloane. But he complained that the experimental philosophers were apt to be rather careless with his collection, being 'so very curious, as to desire to carry part of them home with them privately, and injure what they left'. Such abuses contributed to a delicate tension between private collectors and scientists, as the collectors avidly snapped up and squirrelled away exotica that the philosophers longed to investigate.

As his business expanded, Sloane found clients among the cream of the British aristocracy, including Queen Anne, George I and George II. He also became eminent in the scientific community, being

---

* For a time the unsold copies of *History of Fishes* were used by the impoverished Royal Society as a makeshift currency for paying its members' expenses: Edmond Halley was obliged to accept a stack of volumes in payment for astronomical work.

elected president of the Royal College of Physicians in 1719 and then succeeding Isaac Newton as president of the Royal Society in 1727. His collection grew with his fame, partly by his buying up or inheriting* other private collections, partly from gifts from his clients, but also because he was able to fund his own expeditions to the West Indies. He eventually ran out of space in his Bloomsbury residence, even after he bought the house next door too, and so he moved everything to a grand manor house in Chelsea, near the royal Physic Garden, where he is now commemorated in the name of Sloane Square.

The Royal Society built up its own collection. As Sprat attested in his *History*, the society aimed to acquire 'a General Collection of all the Effects of Arts, and the Common, or Monstrous Works of Nature'. The idea of such a 'repository' was first mooted in 1663, in which year John Wilkins donated his own collection. The following year various members were asked to seek out items in a more systematic fashion: the merchant and FRS Thomas Povey, who had good foreign contacts, was requested 'to procure for the society a collection of all sorts of curious woods, minerals and petrified substances; which he promised to do'. In February 1666 Robert Hooke reported to Robert Boyle that 'I am now making a collection of natural rarities, and hope, within a short time, to get as good as any have yet made in the world, through the bounty of some of the noble-minded persons of the Royal Society.' That same month the trove was suddenly much expanded when the Royal Society bought up for the bargain price of £100 the collection of the traveller and virtuoso Robert Hubert, who had exhibited it to the public near St Paul's Cathedral. This coup inspired Sprat to claim in 1667 (with considerable exaggeration) that the society had 'already drawn together into one Room, the greatest part of all the several kinds of things, that are scatter'd throughout the *Universe*'.

The collection was constantly swelled by a stream of donations from well-wishers and benefactors; announcing its 'museum' in 1666, the Royal Society promised donors that their items would be assured

---

* Collectors often bequeathed their objects to like-minded enthusiasts so that they would not be dispersed among heirs more interested in the monetary than the antiquarian value. In this way Sloane acquired the large collection of William Courten in 1702.

better care in this institutional repository than in a private cabinet. These objects included an armadillo, a crocodile, petrified wood, metal ores, strange bones (one said to be a giant's thigh bone, but in fact from an elephant) and stones (one shaped like 'the secret parts of a woman'), a dodo's leg, luminescent rocks (see page 325), 'bezoar stones' formed in the digestive tracts of animals, and the kind of marvellous miscellany that typified the virtuoso's hoard: a herb allegedly grown in the stomach of a thrush, and a 'tooth taken out of a testicle'. They fuelled numerous meandering discussions among the Fellows. Anything was apt to turn up at Gresham College: the Duke of Norfolk gave an Egyptian mummy, the merchant Thomas Crisp an oddly twisted elephant's trunk. A chicken with four legs arrived from Surrey. Some specimens, such as a four-foot cucumber, were necessarily ephemeral, not to mention unenlightening.

The botanist Nehemiah Grew, who was given the task of cataloguing the collection, lamented the obsession of collectors and virtuosi with unusual objects. He wanted the collection to grow into something more comprehensive – which is to say, more Baconian – saying that this 'Inventory of Nature' needed 'not only Things strange and rare, but the most known and common amongst us'. But well-meaning donors foiled that aspiration, for they seemed to feel that only exotica were worthy gifts; if they sent an egg, it would be a 'monstrous' double-shelled one. Grew had the thankless task of trying to cover over this curious miscellany with a patina of scientific authority, calmly explaining that the 'Cross of wood, growing in the form of Saint Andrews Cross' was not marvellous but grafted. He drew up complex classification schemes which in retrospect seem more like codifications of the arbitrary and incomplete than reflections of any organization in nature. A shell might be described, in a dizzying proliferation of branching categories, as 'conick, with the *Turban*: long, with the Right Lip broad, expanded and turned out, armed with fingers on both lips'.

As well as suffering the whims of its benefactors, the collection was dogged by the lack of a good home and adequate curatorial attention. After the Great Fire it was moved for several years from Gresham College to Arundel House on the Strand. As the Royal Society fell on hard times in the late seventeenth century, there was no prospect of employing staff to look after and display the items. Inevitably the responsibilities fell to the desperately over-committed Robert Hooke, who could barely discharge them. When Uffenbach visited the collection in 1710, he was shocked at the state of it, saying:

> Hardly a thing is to be recognized, so wretched do they all look . . . the finest instruments and other articles [are] not only in no sort of order or tidiness but covered with dust, filth and coal-smoke, and many of them broken and utterly ruined.

From time to time efforts were made to do something about that. Very shortly after Uffenbach's visit, the Royal Society moved to new premises in Crane Court, off Fleet Street, and the collection was given a purpose-built gallery, and in 1712 a committee was appointed to curate it. But its fortunes continued to fluctuate.

Reports and specimens continued to arrive from far-off lands to fuel debate among the Fellows. They discussed the merits of foreign medicinal herbs and treatments, such as those of the physician for the Dutch East India Company, Willem Ten Rhijne, in Java. In 1683 the Royal Society elected to publish Ten Rhijne's treatise on Asian medical practices, including the first European account of acupuncture. But in an age when medicines and medical treatment could be at least as fierce as the conditions they were supposed to treat, it was hard to know how reports of their effects should be assessed. Vomiting, for example, was often considered a good sign, a purgative cure rather than a symptom of poisoning. And lacking either the notion of or the resources for what we would now call clinical trials (it was not obvious then why a hundred patients should furnish any more reliable a test than one individual), the virtuosi tended to rely on self-experimentation. Hooke took, among other things, wormwood tincture, 'steel in wine', ammonium chloride, antimony, mercury and opium, all with results that probably depended as much on his mood as his medicine.

Speaking in tongues

The discovery of the variety of human culture invigorated interest in an old aspiration: to facilitate human discourse through a 'universal language'. The biblical story of Babel seemed to confirm that such a language might exist, and it was widely believed that humans possessed an innate capacity to speak the original 'Adamic' tongue of Eden, a language of divine origin. John Dee believed that angels speak in this language, and it was what his wily collaborator Edward Kelley claimed to be able to hear and understand during scrying sessions with Dee's crystal ball. It was the language with which Adam in Paradise had given all things their 'true' names.

Divine naming conferred Gnostic knowledge of all things: to name was to know, and thus to gain power over the thing named. There is surely an element of this notion in the way Columbus gave names to the islands he first encountered in the West Indies: San Salvador, Isla de Santa María de Concepción, Fernandina, Isabella, Isla Juana, each of them claimed for his sponsors and his religion by this simple act of labelling. To mystics, gaining insight and power from words alone must have seemed an attractive short cut. The thirteenth-century Majorcan philosopher and theologian Ramon Lull, who learnt foreign languages such as Arabic to enlist new converts to Christianity, believed that all knowledge could be recorded in a system of symbolic logic from which truths could be deduced. The 'Lullian Art' did not exactly amount to a language but was more akin to a kind of logical algebra for encoding knowledge, and as such is sometimes considered to represent the beginnings of information science. For Lull, the primary use for his system was to enable all theological questions to be settled absolutely – in favour of Christianity, naturally – through a quasi-mechanical process of logical manipulations. With this in mind, he described primitive 'calculating machines', made from concentric discs with symbols around their circumference that could be turned like rotary slide rules. Lull's system, described in his *Ars magna* (1305), was the inspiration behind the Jesuit philosopher Athanasius Kircher's *Ars magna sciendi* (1669), in which he tried to extend the logical algebra to encompass all knowledge. The Greek inscription on the book's title page gives Kircher's philosophy in a nutshell: 'Nothing is more beautiful than to know the All.' His approach is hard to fathom now. He implies that the divine mechanism of the universe will be revealed if we can only discern the proper way to characterize and categorize qualities and relationships, in disciplines ranging from physics and medicine to jurisprudence and theology. This was in some ways reminiscent of the optimistic belief of the collector (and Kircher was one of the most avid: his Museo Kircheriano in Rome was another proto-museum) that the system of the world would become transparent once one had an itinerary of its constituents that was complete and properly organized.

There is perhaps a hint of the mystical power of words in Francis Bacon's insistence that 'there arises from a bad and unapt formation of words a wonderful obstruction to the mind'. This expresses the belief of Baconian encyclopaedists not merely that poorly chosen terminology is a hindrance to clear communication but that a judicious system of naming things will capture the true relations between them. As a result, words would be more than convenient labels: they might supply a kind of algebra encoding and elucidating the deep patterns of nature. As John Wilkins put it in his *Essay Towards a Real Character and a Philosophical Language*, published by the Royal Society in 1668,* 'we should, by learning the *Character* and the *Names* of things, be instructed likewise in their *Natures*'.

Thus, while efforts to construct a form of scientific Esperanto – of which Wilkins' was the most elaborate and ambitious – were partly intended to lubricate the international exchange of knowledge among scientists, travellers, merchants and diplomats, they were also motivated by the Judaeo-Christian belief that a universal language could actually generate knowledge rather than merely communicate it. The experimental philosophy gave fresh impetus to this quest, because it strove to encompass the full variety of *things* in the world, and to work backwards from that riotous profusion to general organizational principles. As historian Sidonie Clauss puts it, 'The central thesis of Wilkins' linguistic philosophy was the implicit correlation between the development of a new outlook on the physical world and a new attitude towards the methods we use to describe our thoughts about it.' This meant that it was not enough merely to agree on new, universal words to replace our parochial ones; the grammar and syntax of the language had to be based on the order of nature herself.

The theological implications of this idea were acknowledged by Descartes, who felt that, while such a universal language might be

---

* It was in fact first published in 1666, but the entire print run was destroyed in the Great Fire, and poor Wilkins had to redraft it from scratch.

possible, actually implementing it would demand a return to a pre-lapsarian world:

> Do not hope ever to see such a language in use, for that presupposes
> great changes in the order of things, and it would require the entire
> world to be an earthly paradise, something not to be proposed except
> in the world of romance.

It isn't clear that Wilkins knew of Descartes' discouraging view. But he learnt about the obstacles the hard way.

The Cartesian Marin Mersenne also believed in a universal philo-sophical language, but his vision took a rather striking form. Because he considered music to be a quantitative affair that was reducible to physics, physiology and mathematical reason, he imagined scientists conversing in song or even by playing lutes to one another: a charming image that one can't help but wish to see put into practice. This *harmonie universelle* betrays an adherence to the old Lullian tradition that knowledge ultimately consists in symbolic or arithmetic rather than semantic form. Kircher, who corresponded with Mersenne, also believed in a form of universal music, which he described in *Musurgia universalis* (1650). Here he quotes the legendary Hermes Trismegistus as saying 'Music is nothing else but knowing the ordering of all things.' This idea that music is somehow a key that unlocks a universal order is evident also in the works of both Robert Fludd and Johannes Kepler – another indication that these diverse individuals, at one time locked in dispute, were in some ways closer in their views than they liked to acknowledge. Nor were their convictions so far towards the fringes of the new philosophy in the latter part of the seventeenth century; Samuel Pepys bought a copy of Kircher's *Musurgia* in 1668 and professed it to be 'a book I am mighty glad of, expecting to find great satisfaction in it'.

It is significant that attempts to devise a universal language in the seventeenth century coincided with the demise of Latin as the lingua franca of the educated Western word. For as it became normal to publish books and treatises in the vernacular, their potential readership diminished. Often philosophers might be forced to wait impatiently for translations of renowned new works into their own tongue, while authors might provide Latin translations if they wished for their writings to be read abroad. There was also a feeling among men such as Wilkins that the religious and civil strife of the early part of the century was at least in part a result of failures of communication. A new universal language might therefore contribute to social and political stability.

In exposing Europeans to quite different methods of writing, such as Chinese, as well as to cultures that had no written forms at all, travel and exploration highlighted the contingent nature of the familiar Roman alphabet. Wilkins was particularly struck by the way that Chinese characters forego the European method of constructing words from an alphabet and instead use a symbolic means of representation to denote *things*. He did not in fact properly understand how Chinese writing worked, but concluded anyway that the multiplicity of letters is a nuisance – their variety, he wrote, is 'an appendix to the Curse of Babel'.

Partly for this reason, and to avoid the frustrating idiosyncrasies of declensions and conjugations, Wilkins sought to replace or simplify the rules of syntax by attaching systematic short prefixes to simple words. He stipulated several desirable qualities in a universal language:

1. The words should be brief, ideally no more than two or three syllables.
2. The words should be '*plain* and *facil*'.
3. They should be readily distinguishable from one another.
4. They should be euphonious, 'of a pleasant and graceful sound'.
5. They should be methodical.

Unfortunately, Wilkins' syntactic system did not favour his third precept. He devised a taxonomy of concepts in which words for particulars were built up from those for more basic generalities by the addition of single letters. Thus from Zi (beasts) comes Zit (dog-like beasts) and finally Zita (dogs themselves). (The words here are transliterations of his new symbolic system.) This creates a welter of words which, having only small differences in spelling and meaning, were rather hard to tell apart. The distinctions might be clear on paper, but it seems that the human brain does not encode semantic information in so concise and methodical a manner.

Wilkins believed that just 3,000 or so words should suffice to convey most concepts, which did not seem to impose too heavy a burden on the learning process. (It's commonly held that knowledge of about

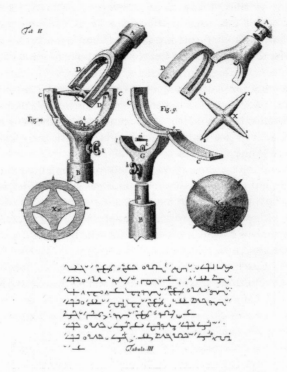

Robert Hooke described the mechanism for his spring-balance watch using John Wilkins' 'universal language'.

this many Chinese characters is sufficient to enable one to read most of what is in the newspapers.) He believed that it should take only a few months to master the language.* But he admitted that some things could not readily be included in it, listing for example titles of honour and office, garments, games, drinks and meats, and tools for trades – a ragbag of concepts that, precisely because it seems not to be comprehensive, makes one rather worried about the limitations.

* Descartes was even more optimistic about the effort needed to acquire a well-ordered universal language, estimating that 'it would not be too much to expect men of common intelligence to master the skills of composition in the new language within less than six hours, provided the aid of [a] dictionary'.

Wilkins specified a system for converting ordinary text into his invented set of characters – basically a kind of encryption, which served rather to undermine his desire for transparency. Apparently it was not hard to use Wilkins' method to convert an ordinary sentence into his new alphabet, but reversing the process was so difficult that, of the few examples of its use, even fewer have been deemed worth the effort of decoding. On the other hand, this made Wilkins' system rather effective for *concealing* information, and it is ironic that Robert Hooke used it in that regard to record his jealously guarded balance-spring watch mechanism in 1675.

It isn't entirely surprising, then, that Wilkins' mighty effort was to little avail, and very few copies of his book were sold. If he was disappointed, he was probably not surprised: he admitted in his introduction that many new and useful inventions had been rejected before and that 'there is reason enough to expect the like Fate for the design here proposed'. In a further irony, one of the obstacles to its wide acceptance was that the book was written in English. Some scholars in continental Europe, including Leibniz, called for the Royal Society to furnish a Latin translation – a daunting task for such a big tome, which John Ray gamely undertook, although his results were never published.

Classifying the world

The real significance of Wilkins' *Essay* lies not so much with the language and notation itself as with the way that it prompted an attempt to provide a 'methodical enumeration' of the natural world – to deduce the organization of all things, supplying the framework on which the syntactic structures were hung. The book contained tables of relationships, clear descendants of those demanded by Francis Bacon for the systematization of 'natural histories', in which the natural world was subjected to rigorous classification of the sort that was soon to become central to subjects such as botany, zoology and chemistry. Ray and Willoughby were delegated by the Royal Society to lend their assistance, and it was hoped (but never realized) that Wilkins' system would provide Nehemiah Grew with the framework for classifying and cataloguing the society's collection.

In this respect, the perception that the scientific enterprise needs a universal language can be seen as a rather too literal expression of a genuine requirement for an agreement about terminology. Science cannot be an international affair, the virtuosi began to appreciate, until we can be sure that we are all talking about the same thing. John Wilkins called for the standardization of units and quantities such as measures of time, distance and weight – a dawning recognition that science needs quantitative consistency. It could not have been easy for the Renaissance navigator, for example, to cope with the fact that the English mile, defined only in 1592 as 1,760 yards, was different from the Scots mile (abolished in 1685), the Irish mile, the nautical mile and the immense German *Meile* (about 7.5 km), not to mention the Portuguese *milha* and the Russian *milya*. Standardization remained a long way off in Wilkins' time (it is still being refined and debated today), but at least the principle was acknowledged.

This need became ever more pressing as the sheer multiplicity of the world was increased by travel and exploration. The greater the diversity of animals, plants, minerals and all natural forms and forces, the more urgent it was to assign them names to avoid confusion. It was from the late seventeenth century that scientists became particularly concerned to systematize technical names, most famously in the taxonomic scheme of species classification devised in the 1730s by the Swedish botanist Carl Linnaeus. Linnaeus's interactions with specimen collectors helped to convince him that a rational and organized system of classification was needed. (He relied on these contacts because, being averse to hot places, he was reluctant to travel to see these organisms in their natural habitats. Later he was able to dispatch his own students to these distant lands.) His treatise on the classification of plants, *Systema Naturae* (1735), introduced the twofold Latinate system of labelling them by species and genus – a scheme that established rules for all discoveries in natural history yet to be made. Soon thereafter Linnaeus met the Dutch banker George Clifford III, a director of the Dutch East India Company, who ran a botanical garden in Heemstede in North Holland of which Linnaeus became the curator. With Clifford's encouragement and funding, Linnaeus travelled to England in 1736 to see the Chelsea Physic Garden and the fine collection of Hans Sloane, whose *Natural History of Jamaica* he had recently read.

The language of science

According to Galileo, nature already had its own natural language, which is mathematics. One need only contrast this view with the approach of Linnaeus to realize that Galileo's perspective (which many scientists share today) is a parochial one that fails to acknowledge the true scope of nature and the limitations of geometry and mathematics to describe it – an issue I consider more fully in the next chapter. Descartes, another deductive logician, was wary of paying too much attention to nature's profusion. He feared that, as historian Paula Findlen says, 'it cluttered the mind with more than it could ever possibly hope to know and with no end in sight to what else there was to be known'. This tension between particulars and generalities is the central dilemma of curiosity.

Yet the wide applicability of mathematical methods to the sciences of astronomy and mechanics fostered the view that some kind of 'universal mathematics' might advance several fields of enquiry. In his essay 'A General Scheme or Idea of the present state of Natural Philosophy', probably written around 1666, Robert Hooke claimed to have created a 'philosophical algebra' which could turn systematized natural histories into axioms. This idea was clearly inspired by Bacon's new organon: as Hooke wrote, 'Of this Engine, no Man except the incomparable *Verulam*, hath had any Thoughts . . . but there is yet somewhat more to be added.'* His philosophical algebra, he says, can enable 'the Discovery of the more internal Texture and Constitution, as also of the Motion, Energy, and operating Principle of Concret Bodies, together with the Method and Course of Nature's proceeding in them'. By calling it an algebra, Hooke gives the impression that it is something like a symbolic language, and like Wilkins' 'universal character' it seems to have been a scheme that represents natural histories in such a way as to reveal their deep content and order. Within this dream that natural philosophy might be conducted by the manipulation of symbols we can see the remnant of the old

---

* Besides Bacon's attempt to devise a 'discovery engine', he was explicitly concerned about the problem of a universal language. In the 'Deficiences of Knowledge, to be Supplied by Posterity' at the end of the *Novum Organum*, he lists 'A Philosophical Grammar; or an account of the various properties of different languages, in order to form one perfect pattern of speech.'

Aristotelian view that science can be as deductive and inexorable as geometry.

Hooke does seem to have formulated his 'algebra' in some form, since he claimed that in the development of the spring-balance watch 'I did from an Art of Invention, or mechanical *Algebra* (which I was then Master of) find out and perfect this contrivance, both as to the Theory and Experimental verification thereof.' But sadly, he never wrote down how his system worked (at least, no such description survives), and so we can only guess at that. It may have been something close to Bacon's laborious subdivision and categorization of the natural world, accompanied by a long list of questions that help to turn natural histories into principles. Hooke displays something of this encyclopaedic tendency in his 'General Scheme' essay, listing twenty-nine classes of 'natural history' and supplying, for purposes of illustration, ninety-three questions about the atmosphere and the weather. Of course, this is no more reliably comprehensive than Bacon's new organon – with a little effort, one could no doubt come up with ninety-three more, and they will all be dependent on your imagination and on what you already know and believe.

However, philosopher of science Mary Hesse thinks that Hooke's philosophical algebra was more akin to what today is often identified as the scientific method: a systematic enumeration of hypotheses which are then tested against observation. Hooke seems to take such an approach in a published lecture on comets, in which he describes the various possible laws of gravity and what their consequences might be for cometary motion: an inverse relationship of gravitational force with distance, an inverse square law (the right one), inverse cube, and so on. Such a system again relies on being able to pose a problem in mathematical form; but in any case it seems clear that Hooke seldom practised what he preached, relying on intuition and informed guess-work more than on any systematic method. To seek a historical 'invention of the scientific method' is a fool's quest; what we see instead in the early modern period is the evolution of a scientific sensibility, which was in the end far more useful.

7 Cosmic Disharmonies

Before their eyes in sudden view appear
The secrets of the hoary Deep
  John Milton, *Paradise Lost* (1667)

Let us therefore not try to discover more of the heavenly and
immaterial world than God has revealed to us. [His] laws are
within the grasp of the human mind; God wanted us to
recognize them by creating us after his own image so that we
could share in his own thoughts.
  Johannes Kepler, Letter to Hewart von Hohenburg (1599)

'The proposition that the sun is in the centre of the world and does
not move from its place is absurd and false philosophically and formally
heretical, because it is expressly contrary to Holy Scripture.' This
notorious pronouncement of the Holy Office of Rome in 1633 seems
to epitomise the Church's opposition, at the peak of the Counter-
Reformation, to investigations of nature that claimed to have discerned
its divine plan. In that respect, it could be interpreted as the last great
assault on the wickedness of curiosity.

But the story is not that simple. For a start, Galileo's advocacy of a
heliocentric (sun-centred) universe was supporting an idea that had
been in print for over sixty years, to which Rome had responded with
little more than grudging tolerance punctuated by occasional and rather
tokenistic censure. It is quite conceivable that Galileo's discourse on
the Copernican theory would have been given a comparably mild
reception from the Church had it not been for a particular conjunction

of circumstances and personalities. Moreover, the arrangement of the heavenly bodies was a topic that had been debated for centuries; astronomy was a standard component of the scholastic curriculum, and not at all the product of the new philosophy's endorsement of curiosity. It was, you might say, part of the old science, not the new.

The part played by Galileo in the unfolding narrative of humankind's exploration of the cosmos is significant but complicated, and certainly not the one that popular tradition tends to insist on. The same is true for seventeenth-century astronomy as a whole. Some science historians have made the 'new astronomy', based on the Copernican model and pursued with telescopes and precise measuring instruments rather than the naked eye, a cornerstone of the cherished Scientific Revolution. One suspects that this is because astronomy – more specifically celestial mechanics, the motions of the heavenly bodies – epitomizes the kind of quantitative, mathematical science from which many historians and scientists have felt most comfortable constructing such a revolution. Perhaps too it was helpful that the astronomy of this period so obviously offers up the heroes – most notably Galileo and Newton – that Whiggish history requires.

But it is misleading to present the changes in astronomy as a monolithic progression, much less one impelled by any new curiosity about the cosmos. For one thing, the impact of the telescope was to reveal not so much where the other worlds are as what they look like – the topic of the next chapter. Despite its implications for Galileo himself, the Copernican model provided only the backdrop to, and not the driver of, the new astronomy of the seventeenth century. And the book that eventually explained how this system was held together in a web of cosmic interaction – Newton's *Principia* (1687) – was not the jewel in the crown of the new philosophy but something of an anomaly, at the same time more old-fashioned and more advanced than anything Newton's contemporaries were writing. Very much, in fact, like Newton himself.

All the same, the *Principia* marks a shift in emphasis in contemplating the stars. Since antiquity, one of the major motivations for astronomy was not pure curiosity but the demands of astrology. The better the movements of the planets were known, the more accurately one could use them to predict the future. In antiquity, says the historian of ancient science Geoffrey Lloyd, 'the study of the heavens was not undertaken for its own sake: rather it was driven by a desire to acquire, somehow, advance knowledge of what was in store for the King or state'. Copernicus's sun-centred theory was received into that same context: many welcomed his new model because it offered a simpler and more precise basis for astrological forecasts. Indeed, Copernicus's pupil and publicist, the mathematician Rheticus (Georg Joachim), presented his master's work as a major advance in astrology, and said that the renowned Italian humanist scholar Pico Della Mirandola would never have written his influential criticism of astrology if he had had available a system as accurate as that of Copernicus. There was some unintentional irony in this, for astrologers were constantly having to explain their failed predictions on the basis that theirs was something of an inexact science: their usual exonerating mantra was that the stars incline but do not compel. All the same, there remained deep unease about the conflicts between the various available tabulations of planetary positions, and this was one of the reasons for the detailed observations made in the seventeenth century by the likes of Tycho Brahe and Johannes Kepler. As the historian of science John North puts it:

> Copernicus lived among scholars and clerics for many of whom the study of astronomy was to be justified first and foremost in terms of its potential application to the art of prognostication, whether of human fortunes, or of such natural phenomena as the weather.

Astronomy was thus very much what we would now call an applied rather than a pure science: it was conducted not for its own sake, but to serve the 'technology' of astrology. Moreover, it was partly for that reason and partly because of the desire to respect the Scriptures that astronomers until the seventeenth century were concerned only to find a scheme that rationalized what they saw, rather than trying to deduce how the cosmos 'worked'. So long as a mathematical description of the heavens allowed for good predictions, astrologers did not overly concern themselves with the question of whether this was truly how matters were arranged up there. That's what made Galileo different.

And to the extent that astronomers did think in terms of a particular physical arrangement of planetary orbits, the issue was merely that of which design God had chosen, rather than how it operated. That's to say, the prevailing view (even for Copernicans) was the

Aristotelian one in which the planets, sun and moon had been fixed to transparent crystal spheres during the Creation; the only question was how the spheres were nested. And that is what made Newton different (although not Newton alone). He began to ask how forces between the celestial bodies could make their organization stable and self-sustaining, instead of assuming that they were all merely pasted in place by God on some great rotating armature. In this regard, then, the great seventeenth-century astronomers might indeed be said to have wrested the status of the heavens from a fait accompli that encodes divine mysteries to a mechanism whose workings are ripe for curious investigation.

Loops and cycles

In the Middle Ages the movements of the seven known celestial bodies – the sun, moon, and five planets Mercury, Venus, Mars, Jupiter and Saturn – were rationalized by the geocentric theory of Ptolemy, in which Earth's central cosmic location reflected simultaneously a solipsistic projection of the rigid hierarchies of medieval culture and a pious statement of humility. God made the universe with man at its centre, but this meant that our terrestrial sphere was but the lower heaven, a zone of impermanence and corruption.

This imagery meant that the Ptolemaic system remained a theological and poetic truth even for some who recognized the scientific virtue of the Copernican alternative. John Milton seems to have been one such. He eulogized Galileo, whom he visited in the 1630s during the astronomer's years of confinement at his villa in Arcetri, near Florence. And yet the cosmos of *Paradise Lost* is resoundingly Earth-centric because our world was required to be the unmoving stage of Satan's drama, with the heavens above and hell below.

Whatever its theological resonance, this model of the universe forced astronomers to accept strange looping orbits for the planets, which seemed occasionally (but predictably) to reverse their journeys across the backdrop of the stars. Ptolemy explained this oddity as the result of epicyclic paths: the planets were considered to progress around Earth on circular orbits called deferents, but at the same time to describe smaller circles (epicycles) superimposed on these. As the planetary motions became more accurately known, more epicycles had to be

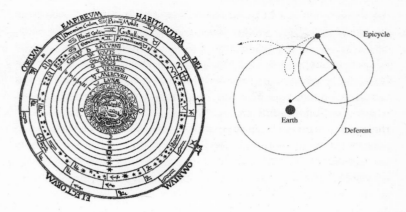

The Earth-centred Ptolemaic cosmology (left), and the planetary motions defined by epicycles (right).

added to fit the observations. So long as the paths were constructed from combinations of perfect circles, it could be argued that they maintained the 'perfection' expected of Aristotle's heavenly sphere. But perhaps equally responsible for the longevity of Ptolemy's rather cumbersome model was the fact that no one felt too concerned by its complexity because it was after all merely a mathematical description, a convenient way of fitting the measurements. That was more or less how the heliocentric model was presented too in the preface of Copernicus's book by its publisher, the Lutheran minister Andreas Osiander, in an attempt to avoid religious censorship. The attraction of the heliocentric theory was that it made the heavens simpler to conceptualize: out went all the epicycles, to be replaced by perfectly circular orbits around the sun.* What mattered primarily was that the description worked in a predictive sense: it enabled astronomers and astrologers (who were generally the same people) to forecast where the planets would be at any moment, and when conjunctions – coincidences of the bodies in the same region of the sky – would occur.

---

* Actually, even that is not quite true, although it is usually how Copernicus's theory is presented. The fact is that, because the planetary orbits are *not* perfect circles (see page 196), Copernicus still needed to include some small epicycles – thirty-four, in fact – to fit the data. But this was still a more economical solution than that of Ptolemy.

This ability made mathematical astronomy a kind of magic – for who else but wizards could predict future events so accurately?

It is largely for this reason that Nicolaus Copernicus's heliocentric theory, published in his *De revolutionibus orbium coelestium* (*On the Revolution of the Heavenly Spheres*) (1543) when the author was on his deathbed, excited relatively little controversy. It's true that Martin Luther called it foolish to try 'to reverse the entire science of astronomy', and his colleague Philip Melanchthon said the idea suffered for 'want of honesty and decency'. But Copernicus's book was not banned by the Roman Church until 1616, and most clerics regarded it as little more than a regrettable folly. They gave barely a thought to the question of how the heavens were 'really' arranged. With the Reformation shaking the foundations of their faith, they had more pressing affairs on their minds.

The matter could not be so easily ignored once Galileo forced the issue by insisting that the mathematics encoded a truth rather than a convenience. More to the point, he was prepared to flaunt that notion under the noses of the bishops, and to defy their commands to desist. He took the battle to Rome itself, armed with little more than some rather flimsy new evidence and an unshakeable conviction in his own intellectual superiority. In retrospect, it's no surprise that it ended badly.

Fair tests

It isn't incidental that the greatest astronomer of the late Renaissance was the son of a music theorist – for music had been allied with the architecture of the heavens since the times of Plato and Pythagoras. Besides his knowledge of the mathematics of music – he proposed a forerunner of the equal-tempered scale – Vincenzo Galilei was a distinguished lute player and a textiles merchant of modest means in Pisa. His son became a professor of mathematics at Pisa in 1589, and three years later took up the same position at Padua. By 1610 his renown won him the patronage of his former pupil Cosimo II de' Medici in Florence. Galileo was surely angling for that when he dedicated his booklet *Sidereus nuncius* (*The Celestial Messenger*)* to the duke

---

* *Nuncius* can mean either 'message' or 'messenger', but the latter is preferred by most translators as it reflects something of Galileo's self-image and personality – half

in March, and it worked – by July he was the prince's court philosopher and mathematician. *Sidereus nuncius* described telescopic observations of a rugged, pockmarked moon and announced the discovery of four moons around Jupiter. Both implicitly called into question the geocentric model and its insistence on the Aristotelian perfection of the celestial bodies – a challenge that was deepened by Galileo's observations of sunspots in 1611.

Already Galileo was propagating controversial and even dangerous ideas. But when Robert Bellarmine, the principal of the Jesuit College in Rome, consulted its leading mathematician Christopher Clavius about Galileo's claims, Clavius endorsed them, reporting that he too had personally seen the moons of Jupiter that Galileo described. All the same, Galileo travelled to the holy capital in 1615 to secure official approval from the Church for what he considered to be his 'divinely inspired' new view of the heavens.

Galileo is often portrayed as the first modern scientist, on the grounds that he tested hypotheses experimentally. Regardless of whether this is really how science is done today, it is not a concept that can comfortably be imposed on the seventeenth century, when there was a rather different understanding of experiment, hypothesis, and the relationship between the two. We've already seen how the purpose of an experiment was frequently to demonstrate the truth of a theory, not to test it. Besides, it might be considered sufficient for an experiment to show roughly what it was predicted to show, without worrying about the precision of the concordance. On the other hand, small deviations of a theory from experiment might be used to justify its wholesale rejection. When Galileo's assertions about falling bodies were tested in 1612 by Aristotelian professors by dropping two objects of different mass from the Tower of Pisa, the fact that they didn't quite reach the ground at the same instant was held to be evidence of Galileo's error. (Galileo correctly pointed out that Aristotle predicted a huge difference in the timings of descent, which the experimenters wanted to hide behind the tiny difference observed – and which in any event Galileo had rightly ascribed to air resistance.)

When Galileo did measure theory against experiment, it seems possible, and hardly reprehensible by the standards of the day, that he

---

egoist, half divine emissary.

was happy to tidy up the observations to fit the expectation. Some historians of science have doubtless gone too far in discounting entirely, on the grounds of limitations to seventeenth-century timing techniques, his empirical deduction of the mathematical relationship between distance travelled and time taken for a ball rolling down the beautifully crafted wooden inclined plane that he used in experiments at Padua. Yet those pre-stopwatch methods can hardly have been accurate enough to supply incontestable evidence for Galileo's (correct) suggestion that the distance is related to the square of the time. It is hard to check, however, because Galileo did not follow the modern protocol of reporting his data. As was typical of experimentalists at the time, he described the apparatus in some detail but then said simply that he had conducted many tests and they had all agreed exactly with his expectations. This emphasizes the role of such experiments, which were supposed to supply verification rather than raw data.

Moreover, one has to interpret with care Galileo's statement about the authority of experiment, which at face value sounds very much like the empirical scepticism that informs Karl Popper's famous 'falsification' principle in the scientific method: theories are not proved but only disproved. 'One single experiment or conclusive proof to the contrary', says Galileo's alter ego Salviati in his treatise on mechanics *Discourses Concerning Two New Sciences* (1638), 'would be enough to batter to the ground . . . a great many probable arguments.' But this skates over the difficult issue of what counts as a genuine falsification. Scientists are faced every day with experiments that 'don't work' – which is to say, which don't give the expected result. Sometimes this is because the theory informing the expectation is wrong. But the problem might be instead with the experiment, in which case it is often only because of the researcher's conviction that the theory *must* be right that he or she discovers the procedural flaw. Alternatively, such a flaw might never come to light; yet rather than abandon the theory, the researcher might conclude that some rogue factor has invalidated or misdirected the experiment – a particularly common situation in the life sciences, where it is very hard to identify or keep track of everything that is going on. In other words, it can be wrong to discard an idea the moment an experiment seems to 'disprove' it, at least before thoroughly examining and repeating the measurement.

In fact, according to a thesis advanced (independently, and in different contexts) by philosophers Pierre Duhem and Willard Van Orman Quine, experiments are logically unable to falsify *any* hypothesis, because the problem is under-determined: one never has enough information to be confident in a negative conclusion. If an experiment gives a result that doesn't agree with a prior hypothesis, this doesn't necessarily mean that the hypothesis is wrong. It could instead be that the theoretical understanding of the apparatus is incomplete. Because experience permits scientists to be surer about some theories and assumptions than others, science can nonetheless function as an effective predictive and problem-solving method. But it is as well to remember that this 'scientific method' is founded on empiricism rather than logical rigour.*

Galileo unwittingly alludes to this problem in a much-quoted passage in *The Assayer*, his response to the theory of comets advanced by the Jesuit mathematician Orazio Grassi. Here he satirizes the absurdity of taking ancient wisdom for granted with reference to the claim that the Babylonians cooked their eggs by whirling them in a sling:

> If an effect, which has succeeded with others at another time, does not take place with us, it necessarily follows that in our experiment there is something lacking which was the cause of the success of the former attempt; and, if we lack but one thing, that one thing alone is the true cause; now, we have no lack of eggs, nor of slings, nor of stout fellows to whirl them, and yet they will not cook, and indeed, if they be hot they will cool the more quickly; and, since nothing is wanting to us save to be Babylonians, it follows that the fact of being Babylonians and not the attrition of the air is the cause of the eggs becoming hard-boiled.

The principle here is sound: to identify the cause of a phenomenon experimentally, one must find what uniquely must be omitted from the procedure in order for the effect to be suppressed. But in practice that can be an extremely difficult thing to identify; indeed, this inductive method is formally impossible, since one can never be sure that

---

* It has been claimed that such ambiguity about experimental proof made the Catholic Church's resistance to Galileo's 'demonstrations' of heliocentrism a more strictly scientific position than Galileo's. But that argument, ignoring the real reasons for Rome's opposition, is not much more than sophistry.

some other alteration might not also have the effect. Perhaps you do not have to be a Babylonian, but just to whirl the sling in Babylon, or to have it done by a man with a darker skin, or to do it at precisely the speed used by the Babylonians . . .

This is not to suggest that Galileo's experimental approach was flawed.* It is simply to point out that it left open many ambiguities about how to conduct a fair and exhaustive test, and what counted as proof – so that, in establishing which ideas should be accepted into natural philosophy, it was understandable that factors such as prior convictions, deductive reasoning, and authority should still have been afforded important roles. In other words, while Galileo was on the right track, it is anachronistic to regard with contemptuous impatience anyone who doubted his methods or conclusions.

It is with all this in mind that one should approach the issue of what happened to Galileo in Rome.

The trial

In the late nineteenth century, two American educators sought to use the 'Galileo Affair' to drive a wedge between science and religion. The chemist John William Draper and the diplomat and historian Andrew Dickson White, a co-founder of Cornell University, both wrote books asserting that Galileo was throughout his career condemned and vilified by the Church in a campaign that ended in his torture for supposed heresy. This, they said, was symptomatic of the religious opposition to science and curiosity through the ages.

Even if this thesis has been somewhat softened over the intervening years, it still colours the popular picture of Galileo's 'persecution'. But the story that Draper and White told was gravely distorted and in some respects simply wrong. The Church's attitude towards Galileo's Copernicanism and his new astronomical observations have little bearing on its (highly ambivalent) views on curiosity as such. The real reasons for the trial of Galileo are complicated and in some respects

* His arguments in *The Assayer*, however, certainly were. Grassi asserted that comets are superlunary, being further away than the moon. Although it offers a fine polemic in support of his approach to science, *The Assayer* makes out that Grassi is wrong and that comets are insubstantial 'toy planets' that can be dissolved in a few days as they follow their heavenly course.

still imperfectly understood. But it is clear that Galileo's case was, as Arthur Koestler put it in his 1959 book *The Sleepwalkers*, 'an isolated, and in fact quite untypical, episode in the history of the relations between science and theology'. Because the merest hint of an apologia for the Church is apt to render some scientists apoplectic (and let's not forget that Galileo was not officially 'pardoned' by the Vatican until 1992), one is obliged to state that enforcing him to recant on heliocentrism in any circumstance, let alone when he was an old and infirm man, was deplorable, ugly and anti-intellectual. Moreover, Koestler went too far in laying the blame fully on Galileo's arrogance and vanity and downplaying the culpability of Rome. But that does not alter the fact that the true story is far more nuanced than the one that has traditionally been told.

For one thing, Galileo was initially feted by Rome when he travelled to the Jesuit College in 1611 to seek Clavius's support for the discoveries described in *Sidereus nuncius*. He was granted an audience with the Pope, Paul V, who showed him unusual respect. The principal of the college, Bellarmine, looked favourably on this famous mathematician and was himself not absolutely opposed to the Copernican model. But he felt that, because it would constitute such a profound challenge to prevailing interpretations of Scripture, it should not be countenanced without compelling evidence to support it. In 1615 he wrote that:

> If there were a true demonstration that the Sun is at the centre of the world and our Earth in the third sphere, and that the Sun does not revolve around the Earth but the Earth around the Sun, then we would have to use great care in explaining those passages of Scripture that seem contrary . . . But I cannot assume that there is such a demonstration unless someone shows me one.

Galileo was equally concerned that the Copernican universe should not be seen to conflict with the Bible: as he wrote, 'two truths can never contradict each other'. But he believed that, as Copernicus *was* demonstrably correct, the Church authorities were obliged to revise their interpretations. 'Scripture cannot err', wrote Galileo, 'but its interpreters can, especially when they would always base themselves on the literal meaning of the words.' In fact, one reason why Galileo was so concerned to press his case is that he feared the Church would lose authority if it insisted on an interpretation that diverged from what (in his view) experience showed. A committed, if perhaps not scrupulous, churchgoer, he did not want to challenge religion but to preserve it. He argued that the Bible employs figures of speech that are not to be taken literally but are merely ways to make the text accessible to ordinary people. Just as God does not really walk and talk like a man, so the sun did not really stand still (but just appeared to do so) so that Joshua and the Israelites might defeat their enemies in battle. In response, he was firmly reminded that he was a philosopher and not a theologian.

To defend his case in Rome, he was asked to prove that Earth moved. He attempted to do so with reference to the tides: he thought that the daily rise and fall of the oceans was caused by rotation of Earth on its axis. But this, the central pillar of Galileo's claim that Copernicanism was supported by hard evidence, is quite wrong. The tides are due to the revolution of the moon around Earth: they are the oceans' response to the moon's gravity (augmented by that of the sun). But Galileo would never have accepted an explanation in terms of such an 'occult' attractive force – when Johannes Kepler suggested as much, he dismissed it. (This explanation was not established until Isaac Newton laid it out in his *Principia*.) In any event, some formidable philosophers of science, including Karl Popper and Pierre Duhem, have pointed out that Bellarmine and the Church authorities were technically correct to assert that heliocentrism was still just a hypothesis for which there was no direct evidence, so that Galileo had no strong basis for claiming that the Copernican model was more than an abstract convenience.

Yet his bullish insistence finally forced Pope Paul V to take action. In 1616 he commanded Bellarmine to summon Galileo and tell him that he must either renounce the idea or at least cease to teach, defend or discuss it. Galileo went to Rome as ordered, but his mood was hardly contrite: he made it clear to anyone who would listen that he felt he had proof of Earth's movement. The Florentine ambassador in Rome wrote to Cosimo II in some alarm, saying that his court mathematician was pursuing his case 'with little prudence or self-control . . . He is deceiving himself, and will get into trouble with those who support him.'

All the same, Galileo acceded to the Pope's demands, and Bellarmine, who was keen to ensure that his friend was not deemed guilty of wrongdoing, issued Galileo with a certificate absolving him from all charges of heresy. Yet it was at this time that Copernicus's *De revolutionibus* was for the first time placed on the Index of forbidden books: it was not declared heretical, but was 'suspended until corrected'. If this decision cannot be attributed entirely to the trouble Galileo was causing, nonetheless there seems little question that the two were connected.

Bellarmine died in 1621, but three years later Galileo felt he had an even more powerful ally in Rome. Cardinal Maffeo Barberini, a long-time friend and admirer, was elected pope, taking the title Urban VIII. Galileo travelled to Rome to congratulate him, and was again warmly received. He now felt confident enough to disregard the injunction of 1616 by preparing a book that presented his case for Copernicanism. *Dialogue Concerning the Two Chief World Systems* was ostensibly an open-minded exploration of the merits of the Ptolemaic and Copernican models as debated by three protagonists. But Galileo's sympathies were glaringly obvious. His own mouthpiece was a philosopher named Filippo Salviati – a real-life astronomer and friend of Galileo (and a fellow member of the Accademia dei Lincei), who had died in 1614. The supposedly neutral party, albeit with an evident Copernican bias, is another old friend, a Venetian noble named Giovanfrancesco Sagredo, while the geocentric case is made by an Aristotelian named Simplicio. It's true that there was a sixth-century Greek philosopher named Simplicius who wrote on Aristotle; but nevertheless it was obvious enough to any reader that this man was a buffoon whose name connotes a simpleton.

Salviati explains that telescopic observations of the moon, as described in *Sidereus nuncius*, showed Earth to be no different from the other celestial bodies. But the key 'proof' is again Galileo's erroneous tidal theory. Indeed, he had wanted to call the book *A Treatise on the Tides*, but Urban VIII disapproved because this seemed to imply that the tides furnished a clinching physical argument for heliocentrism. So Salviati's arguments were in fact either circumstantial or wrong, even if his conclusions were correct.

The problem with the *Dialogue* went beyond its lack of balance. Simplicio was not only a rather hopeless advocate for the geocentric model; he also appeared to represent a lampoon on the pope himself. He objects that the tides could not offer conclusive proof of Earth's movement because God could have arranged them any way he pleased. This sophistic position, which in effect rules out any logical explanation for any phenomenon at all, was one that had been explicitly voiced by Urban VIII. Salviati proceeds to demolish it with the kind of biting sarcasm that was Galileo's stock-in-trade:

> What an admirable and angelic doctrine, and well in accord with another one, also Divine, which, while it grants to us the right to argue about the constitution of the universe (perhaps in order that the working of the human mind should not be curtailed or made lazy) adds that we cannot discover the work of His hands. Let us, then, exercise these activities permitted to us and ordained by God, that we may recognize and thereby so much the more admire His greatness, however much less fit we may find ourselves to penetrate the profound depths of His infinite wisdom.

This book was funded by Federico Cesi, the leader of the Accademia dei Lincei, who had consistently and effectively helped the pugnacious astronomer get his works past the ecclesiastical censors. Cesi and Cassiano Dal Pozzo had managed this skilfully with *The Assayer*, making it so congenial to the Church authorities that Urban VIII allegedly had it read to him at table so that he might laugh at its witticisms. It seemed at first as though the case was going well with the *Dialogue* too. Cesi's influence and diplomacy carried the deliberations along smoothly, and by 1629 the manuscript was ready for the presses, while Galileo was still being treated with honour and respect in Rome. The worst he could expect, it seemed, was to be required to make some small changes before publication.

The *Dialogue* might have slipped quietly through the press had disaster not struck. In August 1630 Cesi died, and his finances were found to be in disarray, depriving Galileo not only of a supremely skilled advocate but also of funds for the project. Now the story gets tangled. The Vatican granted permission, after a fashion, for the book's publication in 1631, but this was contingent on certain revisions and on a complicated procedure of vetting in Rome and in Florence, where the printing was in the end due to happen. The Church authorities seem to have expected that Galileo would take the standard line that heliocentrism was nothing more than a useful mathematical device.

But when the book finally appeared in 1632, its author stated in the preface that:

> I have taken the Copernican side in the discourse and, proceeding as with a pure mathematical hypothesis, I have tried to show, by every means, that it is more satisfactory than supposing the earth motionless – not, indeed, absolutely, but when compared to the arguments of [the Aristotelians].

Moreover, he was found to have omitted some of the Pope's arguments against Copernicanism that he was expected to include.

It was no wonder, then, that Urban VIII prohibited distribution of the book in the summer, and that a commission appointed to assess it concluded that Galileo should come back to Rome to explain himself. But he was ill, and delayed setting out until early 1633. He was comfortably accommodated in the Holy City during his trial.

Galileo did not help his case by arguing, contrary to what was evident to any reader, that 'I neither maintained nor defended in that book the opinion that the Earth moves and the Sun is stationary. Rather I proved the contrary of the Copernican opinion and showed how weak and inconclusive the arguments of Copernicus were.' Told that he was only making matters worse with such a denial, he feigned surprise at what he found on rereading his book:

> Not having seen it for so long, I found it almost a new book by another author. I freely confess that it appeared to me in several places to be written in such a way that a reader, ignorant of my intention, would have reason to believe that the arguments for the wrong side, which I intended to confute, were so expressed that they meant to carry conviction rather than be easily refuted.

He confessed that 'my error has been one of vainglorious ambition, pure ignorance and inadvertence'.

Finally Galileo admitted that 'I do not hold this opinion of Copernicus, and I have not held it after being ordered by injunction to abandon it.' This rather abject and pathetic excuse and recantation is in itself a damning indictment of the bullying tactics of the Church, which held in the background a real threat of torture against the now rather frail and elderly man. He was humiliated further by being made to kneel while the judgement and punishment were read. He was initially sentenced to imprisonment, but this was commuted to the command that he return to his villa at Arcetri and never leave that place without official permission. The likelihood that this frightened, exhausted old man muttered '*Eppur si muove*' as he rose seems as slender as that he thumbed his nose at his inquisitors – this popular story is probably an attempt by Galileo's later hagiographers to preserve his heroic integrity.

Stephen Jay Gould says that Galileo was 'a victim of bad luck and bad judgement, not an inevitable sacrificial lamb in an eternal war between science and religion'. Although one might add that bad faith played a part too, this seems a fair assessment. And while Koestler's suggestion that Galileo 'made no contribution to theoretical astronomy' is unduly harsh, it is fair to say that he added nothing of real substance to the Copernican theory except his influential support. But as we shall see, he was undoubtedly central in opening up the heavens as a space in which curiosity was not limited to a mathematical language. There is no reason to suppose that his trial did much to suppress other astronomers' eagerness for observational exploration of the stars. It did, however, recommend caution about the conclusions one might draw from them.

The emperor's astronomers

That Copernicus's theory was very much an option for those who preferred mathematical neatness, rather than an irrefutable corollary of the observed planetary motions, is clear from the fact that the most famous astronomer of the late sixteenth century, the Dane Tycho Brahe, did not believe it. Tycho devised an ingenious hybrid scheme in which Earth sits at the centre of the universe while the other planets revolve around the sun, which in turn circulates Earth. Tycho sought to preserve Earth's privileged position partly because of the Scriptural arguments for it. But he also believed there were sound scientific reasons to do so, most notably that the positions of the stars relative to Earth do not change as they 'should' if Earth is in motion. (Actually they do change, but because they are so distant, this effect could not be discerned within the accuracy of measurements in Tycho's time.)

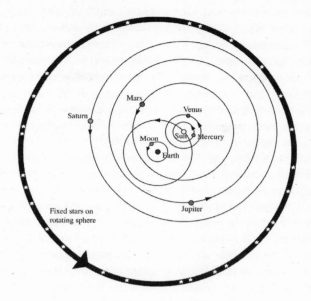

In Tycho Brahe's universe, the other planets all revolve around the sun, which itself (along with the moon) orbits the Earth.

Tycho acknowledged the mathematical advantages of the Copernican system, but felt no need to attribute any physical reality to that picture.

Tycho made his name when, in 1573, he reported a new star in the constellation of Cassiopeia that he had seen the previous year at his observatory in Denmark. This was so disquieting a challenge to the supposedly unchanging state of the Aristotelian heavens that even Tycho could not at first believe his eyes. 'Tycho's star' is now recognized as an exploding star or supernova, caused when the star exhausts its fuel and collapses on itself to trigger a massive nuclear explosion that makes it shine briefly with extraordinary brightness. Tycho became so famous that he was invited to dine with Rudolf II in Regensburg during the emperor's coronation in 1575. On that occasion he cast Rudolf's horoscope, for Tycho was a committed astrologer. Fearful that his celebrated astronomer would be induced to leave Denmark for the glittering court of Prague, Frederick II of Denmark and Sweden gave Tycho an island, Hven, off the Zealand coast, with a pension generous enough to finance a personal observatory that Tycho christened Uraniborg, 'City of the Heavens'. It was a Neoplatonic temple, designed according to principles of geometry and symmetry, and from this privileged position Tycho began to map the stars and planets with precision unmatched anywhere else in Europe (and without the aid of a telescope).

When Frederick died in 1588, Tycho lost his royal patronage. The king's son Christian was just eleven years old, and the palace advisers turned him against the pampered astronomer, who had in truth become a somewhat high-handed landlord to the peasants of Hven. By 1596 Tycho concluded that there was no future for him in Denmark, and he left for Rostock on the German Baltic coast. Two years later Rudolf II offered him a position in Bohemia, where he was welcomed warmly and given a grand pension, a castle and an estate at Benátky, north-east of Prague. He renewed his mission to record the celestial movements, while fulfilling his obligation to provide astrological advice and horoscopes for Rudolf. His forecasts were not always reassuring. He predicted that the emperor would be assassinated, just as the Polish-born Henry III of France had been in 1589, giving the paranoid Rudolf even more reason to retreat from the world.

While still in Germany in 1598 during the interim between Uraniborg and Prague, Tycho received a copy of a book with a long Latin title that is generally shortened to *Mysterium cosmographicum* (*The Cosmographic Mystery*), written by a young German astronomer named Johannes Kepler, a teacher at the Protestant school in Graz, Austria. Tycho was impressed by what he read, even though the cosmography described in the book was Copernican. He was not the only admirer: a copy reached Galileo in Padua, who considered it speculative but was sufficiently stimulated to express to Kepler his delight at having such 'a comrade in the pursuit of truth'. Galileo confessed to the German astronomer that he too held Copernican views but was wary of admitting it openly.

Born in a town in Württemberg in 1571, Kepler was raised in a harsh environment. His father was, in Kepler's own words, 'an immoral, rough, and quarrelsome soldier' who was often away fighting wars as a mercenary, leaving Johannes in the care of his bad-tempered mother Katherine. Notwithstanding her son's poor health and eyesight, she put him to work in his grandfather's inn. But Kepler's precocious intellect won him escape from these impoverished conditions, first through a scholarship to a good Swabian school and then a place at Tübingen University, where he studied theology and philosophy. The offer of a position at Graz directed him towards mathematics and astronomy rather than the religious life, although he remained deeply devout.

In *Mysterium cosmographicum* Kepler explained that the heliocentric orbits of the six planets (Mercury, Venus, Earth, Mars, Jupiter and Saturn) will fit on to the surfaces of their classical crystalline spheres if these just touch the inner faces and outer corners of the so-called Platonic solids (polyhedra whose faces are identical regular polygons) nested one inside the other. There are exactly five of these polyhedra: the tetrahedron, cube, octahedron, dodecahedron and icosahedron. Kepler had discovered in 1595 that, with a particular order of nesting that places the octahedron innermost and cube outermost, the Copernican system offers a close fit to this scheme. 'How intense was my pleasure from this discovery can never be expressed in words', he said.

To Kepler this cosmic design made perfect sense, for as a Neoplatonist he considered that God had created the universe according to geometric principles.* He believed that he had cracked the code that brought divine harmony to the cosmos. According to the historian of astronomy Owen Gingerich, 'never in history has a book so wrong been so seminal in directing the future course of science'. For however fanciful Kepler's cosmological scheme was, it depended on – and argued for – the validity of the Copernican system.

Tycho had his own views about that, but he knew a talented and original thinker when he saw one. Ever keen to procure bright young assistants, he wrote a letter to Kepler praising his work, while taking care to assert his own authority with some delicately phrased criticisms (those he expressed privately to others were a little more barbed). He made it clear that Kepler would be welcome if the young man ever chose to join Tycho's team.

---

* Needless to say (or perhaps not), Kepler's cosmographic model is sheer numerology, not least because there are more planets to accommodate in the solar system than these six. But this sort of mystical thinking retains an appeal today for tidy minds who like to find teleology in the cosmos, among them the heir to the British throne. A simple back-of-the-envelope calculation shows how fatuous this sort of geometric cosmography is.

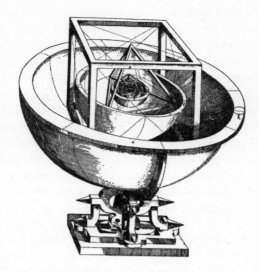

Kepler believed that the planetary orbits could be fitted on to the surface of spheres that just touch the corners of nested regular polyhedra, making the universe a geometrical device.

Kepler characterized his book as a *prodromus*, an introduction, for he intended it to be only the first of a series of astronomical treatises. But his plans were disrupted towards the end of 1598 when Archduke Ferdinand of Austria, the staunch Counter-Reformationist who would later crush the Winter King and reclaim the crown of Bohemia, responded to Protestant agitation by declaring that all Lutherans were to be dismissed from their positions and expelled. Kepler decided it was an apt moment to take up Tycho's offer.

Collaboration served the interests of both parties. Tycho wanted to harness Kepler's sharp mind to help verify the Tychonic model of the cosmos, of which Kepler was privately dismissive. Kepler, meanwhile, hoped to use Tycho's excellent observational data to improve the fit of the measured planetary orbits to his geometric scheme, which he recognized to be far from exact. When Kepler said of his new employer at Benátky that 'he has abundant wealth, only, like most rich men, he does not know how to make proper use of his riches', it was not the reaction of a poor schoolteacher arriving at the castle of a wealthy nobleman: these 'riches' were the Dane's data. Kepler's meaning (and his schemes) are laid bare when he adds: 'Therefore, one must take pains to wring his treasures from him, to get from him by begging the decision to publish all his observations without reservation.'

It's no wonder the relationship was prickly. Despite his assurance to Kepler that 'you will come not so much as guest but as very welcome friend and highly desirable participant and companion in our observations of the heavens', Tycho treated Kepler as a retainer rather than a scientific equal. On one occasion Kepler stormed from Benátky after an argument, determined to sever his ties, while Tycho announced that he likewise wished to end their collaboration. They made up, but it was an indication of how things stood.

It did not help that Kepler's payment for his services was uncertain. Tycho did not hurry to arrange for his assistant a stipend from Rudolf, and even then this was not always forthcoming. Yet Kepler planned to stay only for as long as it took to get his hands on Tycho's precious data. The Dane guarded it jealously, but Kepler hoped that infirmity would force him to relent. 'Now old age steals upon him', he wrote, 'weakening his intellect and other faculties or, after a few years, will so weaken them that it will be difficult for him to accomplish everything alone.'

But Tycho did not have even a few years. Kepler had been at Benátky for barely a year and a half when Tycho caught a bladder infection and died. He had just announced that Kepler would be his key assistant in compiling a vast catalogue of star positions and planetary motions, dedicated to Rudolf and known after their eventual publication in 1627 as the Rudolfine Tables. This was what Kepler needed, and his appointment came not a moment too soon. When Tycho died, Kepler seized control of the data with unabashed opportunism, confessing that 'I quickly took advantage of the absence, or lack of circumspection, of the heirs, by taking the observations under my care, or perhaps usurping them.' Tycho's family tried to win back the rights to the data (and thus to profits accruing from their publication), but Kepler's position became unassailable once he was appointed Tycho's successor as imperial mathematician. The post carried Tycho's astrological duties, which Kepler dispatched with little enthusiasm – he called astrology 'the foolish little daughter of astronomy'*– but apparently with diligence, being possessed with what he himself admitted was 'a childish or fateful desire to please princes'.

The Tables ended up taking even longer to complete than Kepler had imagined, and he published them only three years before his own death in 1630. Tycho had clearly hoped that Kepler would use the observations to vindicate his model – he is said to have repeatedly muttered to his assistant, as he lay on his deathbed, 'let me not seem to have lived in vain'. But Kepler used them instead to refine his picture of nested polyhedra. At first he devoted his efforts to the task that Tycho had set him: to calculate the trajectory of Mars, whose observed path seemed to deviate particularly markedly from perfect circularity. He laboured on this issue for four years before concluding that indeed Mars orbited not in a circle but an ellipse. By analogy (and nothing more at that stage), Kepler concluded that *all* planetary orbits adopt this shape to a greater or lesser degree. This was the principle now known as Kepler's first law of planetary motion, which he announced in his 1609 book *Astronomia nova* (*The New Astronomy*). When Galileo heard of this idea from Cesi in 1612, he did not believe it – the 'imperfection' implied by ellipses was too much even for him to accept.

Hidden forces

If you judge by textbook diagrams, it's easy to misconstrue Kepler's first law: these pictures tend to depict highly elongated ellipses, either to emphasize the point (as on page 198) or to represent the orbits as if seen obliquely from beyond the solar system. In fact the eccentricity (deviation from circularity) of the planets' orbits is generally very small: they are very nearly circular, which is why their true shape did not become apparent sooner. But it was one thing to show that

---

* This does not in itself mean that Kepler was sceptical about astrological influences, but merely about men's ability to discern and predict them. In this, he joins the ranks of those often alleged to be critics of occult sciences such as astrology and alchemy when in truth their doubts concerned practice and not principle. When Kepler assessed his own character, it was in astrological terms: 'In my case, Saturn and the Sun work together in the sextile aspect . . . Therefore my body is dry and knotty, and not tall.' Thus does history resist efforts to make Kepler respectably 'modern'.

elliptical orbits best fitted the data, quite another to explain why they have this form. Kepler's explanation was vague and literally convoluted (perhaps this is why Galileo complained in 1614 that his writing was 'so obscure that apparently the author did not know what he was talking about') but it involved the important assumption that the planets were impelled along their orbits by a motive force emanating from the sun. This is easily misunderstood too. Kepler was not quite saying, as we do now, that the movements are due to a force *attracting* the planets to the sun. Rather, he conceived of a solar influence that hurries the planets along their ellipses; he called it the *anima motrix* or motive soul of the sun. He imagined that this force weakened with increasing distance from its origin.

Here Kepler was inspired by the work of the English scientist William Gilbert, court physician to Queen Elizabeth. In his classic study of magnetism *De magnete*, Gilbert argued that Earth is a great magnet which emanates lines of force that bind the moon in orbit. Kepler accepted this idea but recognized that, like two magnets, Earth and the moon must thus exert *mutual* attraction – a crucial aspect of Newton's later theory of gravity. However, Kepler didn't see the motion of the planets around the sun as being precisely equivalent to that of the moon around Earth. Rather, lines of magnetic force radiating as a kind of immaterial flow from the sun were, he said, deformed into a kind of vortex by the rotation of the sun, and this vortex 'carries the planets with it, drawing them round in a circle'. These paths were distorted into ellipses by some kind of interactions between the sun's magnetism and the magnetic poles of the planets.

This is all a touch confusing, even if Galileo was a little harsh. But the key point is that Kepler was making the sun not just the centre of the planetary motions but its cause. This picture was consistent with Kepler's mystical view of cosmic harmony based on the Neoplatonic notion of correspondences: the sun, representing God, became the physical source (and not just the geometric centre) of the celestial order. And the hypothesis of a (magnetic) gravitational force challenged the Aristotelian view that the planetary orbits were deter-mined by the motions of crystal spheres in a circulating ethereal fluid. However, if the elliptical orbits meant that the Aristotelian ether was no longer characterized by the 'perfection' of circular motion, it surely meant too that the universe could not be modelled as spheres inscribed

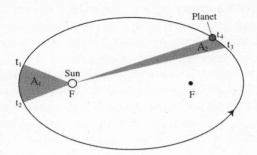

Kepler's second law of planetary motion states that planets sweep out equal areas of arc in equal times. Here A_1 and A_2 are equal, and so are the timespans $t_1 \rightarrow t_2$ and $t_3 \rightarrow t_4$. The two points F are called the focal points of the ellipse; the sun sits at one of them.

within Platonic solids, as Kepler had insisted. Naturally, he did not wish to see it that way, and maintained the fundamental validity of his geometric system until his death.

The other vital aspect of this nascent gravitational theory was that the force exerted by the sun gets weaker with increasing distance, which Kepler needed to assume in order fully to account for the orbit of Mars. It implies that the planets move more or less slowly as their elliptical paths take them nearer to or farther from the sun. Kepler deduced how these variations in speed were described in mathematical terms. The progress of a planet along an arc of its orbit can be considered to sweep out wedge-shaped segments of the ellipse, with the sun at their apex, and Kepler calculated that the time taken for the planet to sweep out two segments of equal area is the same. This is his second law of planetary motion, also adduced in *Astronomia nova*.*

After Rudolf's abdication of the Bohemian throne in 1611, Kepler could see that his position in the court was tenuous, even though Matthias allowed him to stay as imperial mathematician. He made plans to move back to Austria to become the district mathematician in Linz, although he did not go until after Rudolf's death in 1612, by which point Kepler's wife and son had also both died. To add to his woes, his mother Katherine was accused of witchcraft, and Kepler

---

* This is to put the actual train of Kepler's enquiry back to front. He deduced his second law first, and then found by trial and error that it implied an elliptical orbit.

had a long and exhausting battle to prevent her from being burnt at the stake.

All the same, he managed to work on the great synthetic vision of cosmic harmony described in *Harmonices mundi,* in which astronomy, mathematics, music and theology are united to reveal what Kepler considered to be the geometric logic that God has woven into the fabric of the universe. 'I dare frankly to confess', he wrote,

> that I have stolen the golden vessels of the Egyptians to build a tabernacle for my God far from the bounds of Egypt. If you pardon me, I shall rejoice; if you reproach me, I shall endure. The die is cast, and I am writing this book, to be read either now or by posterity, it matters not. It can wait a century, as God himself has waited six thousand years for a witness.

Here Kepler elaborated on Ptolemy's idea that there is a system of celestial harmony, the music of the spheres, whereby each planet may be assigned a musical tone related to its distance from Earth. The ratios of these tones, Ptolemy asserted, are arithmetically simple, like the ratios deduced by Pythagoras for the frequencies of consonant musical intervals: an octave consists of notes in the frequency ratio 1:2, a fifth is 2:3, a fourth 3:4 and so on. Kepler applied this idea to a Copernican cosmos, assigning tones based not on the planetary distances from the sun but their speeds. Scientists today tend to avert their eyes from this Neoplatonic tract until they reach the final Book V, where we find Kepler's third law of planetary motion: the cubes of the mean distances of planets from the sun are proportional to the squares of their orbital period (the time taken to complete one orbit).

Tycho's marvellous naked-eye data had thus guided Kepler to three profound new insights into the motions of the celestial bodies: they trace out ellipses, not circles; each planet sweeps out equal-area segments in equal times; and there is a mathematical relationship between the planet's mean distance from the sun and its orbital period. Are these three independent facts, or is there some logical connection between them? Might they all *necessarily* follow from the same basic cause? Kepler suspected that there might be some principle that unified his planetary laws, but he imagined that the key lay in his spurious musical analogies. And he was constantly diverted and distracted from the question by financial difficulties and the disruptive effects of the Thirty Years War, which forced him to leave Linz in 1626 and settle in Silesia. It was on a journey to Regensburg in 1630 to collect money he was owed that Kepler caught a fever and died.

It is sometimes said of Kepler that what he shares with modern scientists is a determination not merely to record nature but to find a reason why it is that way. But that view elides rather different ideas about the notion of cause. Physicists today typically locate causal action in the operation of fundamental forces. (For other scientists, such as biologists and chemists, causation often has a much looser and metaphorical interpretation: Darwinian 'selective forces' are not like the physicist's electromagnetic force, for example.) But for Kepler, the reason why there should be an arithmetical relation between two quantities such as a planet's distance from the sun and its orbital period (which are not transparently interdependent) was teleological: this is part of a Grand Design according to which the world is built on principles of harmony and proportion. 'God has established nothing without geometrical beauty', he wrote. Thus the reasoning given for his mathematical laws in Book V of *Harmonices mundi* looks bizarre to modern eyes, replete with notation of musical scales and chords that seem to have nothing to do with the motions of the planets or with causes conceived as forces.* If this is closer to the artist's delight in analogy than the scientist's insistence on mechanism, nonetheless

---

* I say 'seem to' because actually Kepler's notion of simple ratios between parameters describing the orbits of celestial bodies does have some foundation, although not in the manner he described. When several objects circulate in gravitational orbit around a larger one, they can occasionally become locked into so-called 'resonances' in which the orbital periods (or some other feature of the orbits) are in whole-number ratios. This is a consequence of the mutual gravitational attractions of all the bodies, crudely analogous to the way swinging pendulums can become locked in step if they are able to influence one another. The phenomenon was first investigated by the French mathematician Pierre-Simon Laplace in the eighteenth century, and the situation in which three or more orbiting bodies have periods in simple ratios is called a Laplace resonance. This is the case for three of the Galilean moons of Jupiter – Ganymede, Europa and Io – which have periods in the ratio 1:2:4. Pluto and Neptune are also in a 2:3 resonance around the sun.

it was a form of reasoning with which Isaac Newton later showed himself to be very comfortable.

The principles

Kepler had unwittingly already discerned the unifying principle he sought: it is the attractive force between the sun and the planets, which we now call gravity. But it is scarcely surprising he failed to appreciate that this is all that matters. How can an inwardly directed force, like that which pulls falling objects to the ground, sustain the planets in circular or elliptical orbits rather than just pulling them down on to the sun's fiery surface? Until Isaac Newton solved this problem in the 1680s, it was generally believed (even by Newton) that an impulse additional to gravity was required to account for planetary mechanics: one that pushed these bodies along their circulating paths. For Kepler this was the *anima motrix*; but many astronomers ascribed it to the rotating vortices with which René Descartes proposed to fill the cosmos. In Cartesian mechanics there is no empty space: it is packed with particles of various sizes, some of them 'infinitely' small grains formed by collision and attrition of the larger particles. These particles are in constant motion, and their mutual collisions shape their paths into circulating vortices. In the heavens these vortices were considered to form stratified bands, each of them entraining a circulating planet. Thus Descartes' mechanics did not permit the notion of 'force' as an occult interaction through space: all motion was caused by the direct interaction of particles.

Newton was not ready to reject the Cartesian vortex as superfluous to celestial mechanics until he had reformulated the entire science of motion. How and why things move had puzzled philosophers since antiquity. It is clear enough that a heavy crate dragged along the ground is being kept in motion by the force imparted by the dragger, since the motion stops as soon as the force is withdrawn. But a pebble thrown through the air keeps moving of its own accord, even when no longer in contact with the hand of the thrower. How can that be? Aristotle postulated that some kind of force is imparted continually by the medium (air) through which the projectile moves. But in the Middle Ages the preferred view was that the projectile acquires a certain amount of 'motive agency', called impetus, during the throwing action, which is gradually used up as it travels along its trajectory until it is eventually exhausted. In the fourteenth century the French philosopher Jean Buridan proposed that this decline of impetus was not spontaneous but was caused by the influence of gravity (that is, the heaviness of the stone) and air resistance. This implied that, without such countervailing influences, motion would not stop.

Galileo articulated the view most clearly in his law of inertia, which stated that a body moving on a level surface will continue in the same direction at a constant speed unless it is disturbed. In other words, bodies have an intrinsic tendency to remain in motion.* The principle was sometimes expressed in an anthropomorphic way: all bodies *resist* a change in their state of motion, as if reluctant to alter it. This resistance to change was called inertia, a term coined by Kepler.

In his *Principia philosophiae* Descartes restated Galileo's principle of inertia as his first law of motion: 'each thing, as far as is in its power, always remains in the same state; and that consequently, when it is once moved, it always continues to move'. His second law expresses a corollary: 'all movement is, of itself, along straight lines', by which Descartes means that an impulse applied to a body induces movement only in the direction of that impulse. Newton in turn reformulated the laws of mechanics in his own *Principia* (or to give it its full title, *Philosophiae naturalis principia mathematica*, the *Mathematical Principles of Natural Philosophy*). Newton's first law states that bodies remain at rest or moving at constant speed unless acted upon by a force. His second law describes the effect of such a force: it induces a change in velocity (an acceleration) in the direction in which it acts, which is inversely proportional to the body's mass. In modern terms this is expressed as the equation force = mass × acceleration. Newton's third law states that for every action there is an equal and opposite reaction: when two billiard balls collide, each exerts a force on the other that changes its trajectory. These principles supply the groundwork for the

---

* Galileo realized that this implies an equivalence between objects that are stationary and ones in motion at a constant speed: in the absence of any disturbing forces, each remains in that same state. A corollary is that it is impossible, without some outside frame of reference, to distinguish the one case from the other: motion is relative. This supplied the classical basis for Einstein's theory of special relativity; but it was also a supposition rejected by Newton, who believed that space itself provides a fixed reference frame against which absolute motion can be gauged.

book's analysis of planetary motions, in which Newton finally explained Kepler's laws in terms of the force of gravity alone.

Like Kepler, Newton was born into an unhappy family environment. His father died shortly before his birth, and when his mother Hannah Ayscough remarried she felt no compulsion to bring Isaac into the new relationship, leaving him to be brought up by his maternal grandmother Margery. When Newton's hated stepfather Barnabas Smith died, Hannah (now moderately wealthy from her late husband's estate) returned to the family home in Woolsthorpe, Lincolnshire, and at last accepted that her son might live with her and his half-siblings. He was given an education that, although indifferently pursued, sufficed to take him to Trinity College at Cambridge. Here he displayed a precocious talent, particularly in mathematics and experimental optics, which brought him to the attention of the Lucasian Professor of Mathematics Isaac Barrow. When Barrow vacated the post in 1669, he secured the appointment of his protégé, then just twenty-seven years of age, in his place. This gave Newton the status of a leading scholar, and in 1675 he was elected a Fellow of the Royal Society, although he remained at that stage outside the primary circles of its Oxford–London clique.

Newton's reputation drew the young astronomer Edmond Halley to Cambridge in 1684 to seek his advice on the trajectories of comets, one of which (that which now bears his name) Halley had studied closely during its apparition two years earlier. During their meeting, Halley asked Newton his opinion on the form a planet's orbit would take if it were attracted to the sun by a force that varied in strength with the inverse square of the distance. Newton did not hesitate: it would be an ellipse, he said.

Halley told Newton that he, Christopher Wren and Robert Hooke had also been discussing the possibility that the force of gravity had this mathematical form, and he mentioned that Hooke had alleged that this hypothesis could explain Kepler's elliptical orbits. Halley recalled that Hooke had claimed 'upon that principle all the laws of celestial motions were to be demonstrated and that he himself had done it'. Asked to show his proof, Hooke had boastfully said that he was withholding it for a time 'so that others, trying and failing, might know how to value it when he should make it public'.

That was enough to try anyone's patience. Wren had goaded Hooke by doubting that he really had such a proof, whereupon Hooke had promised to reveal it. But two months had passed and nothing was forthcoming, so Halley now decided to put the question to Newton. He responded airily that he himself had actually shown four years earlier that the inverse-square law could account for the movements of the planets. But he could not at that moment lay his hand on the proof. Halley implored him to write it down again.

Newton had a good personal motivation to do so. He had already fallen out with Hooke over his theory of light (see page 347), and they had exchanged some fractious correspondence in 1679–80 over this very subject of celestial mechanics. So Newton had no wish to see his priority lost to the man he had come to regard as his rival. But working his proof into something he deemed fit for the public eye turned out to be a far from cursory exercise. Over the next eighteen months he laboured feverishly on the manuscript, and his hyperactivity during that period was described in melodramatic terms by his secretary and distant relative Humphrey Newton:

> I never knew him take any Recreation or Pastime, either in Riding out to take the Air, Walking, bowling, or any other Exercise whatever, Thinking all Hours lost, that was not spent in his Studyes, to which he kept so close, that he seldom left his Chamber . . . So intent, so serious upon his Studies, that he eat very sparingly, nay, oftimes he has forget to eat at all, so that going into his Chamber, I have found his Mess untouch'd, of which when I have reminded him, would reply, Have I; & then making to the Table, would eat a bit or two standing, for I cannot say, I ever saw Him sit at Table by himself . . . He very rarely went to Bed, till 2 or 3 of the clock, sometimes not till 5 or 6, lying about 4 or 5 hours.

Newton's treatise quickly grew into something even more far-reaching than a proof that an inverse-square law of gravity could explain Kepler's three laws. The *Principia*, published in 1687 at a cost borne by Halley, is the foundation of all classical mechanics. That it ever appeared at all owes much to Halley, and not just financially: he constantly had to cajole, wheedle and flatter Newton into committing his ideas to print. Ever sensitive to the slightest criticism, Newton was so stung when Hooke (inevitably) claimed precedence for ideas in

Book II after it was read to the Royal Society that he wanted to suppress the third book, in which he applied the ideas to comets. Happily, the diplomatic and irrepressible Halley prevailed.

There's no reason to suspect that Newton's initial response to Halley's question in Cambridge was merely an informed guess; it is virtually certain that he had indeed already shown how an inverse-square law of gravity would produce elliptical orbits.* Newton had been considering the problem of how to derive Kepler's empirical laws at least since the late 1670s, and had realized that the planetary orbits were secured by a balance of forces. Circulating objects have a tendency to fly off outwards along a tangent unless prevented, as is clear if you whirl round a weight attached to a thread and then release it. There is said to be a 'centre-fleeing' or centrifugal force on the body. While circulation persists, this force is opposed by a 'centre-seeking' or centripetal force. In the case of the whirling weight this is supplied by the tension in the thread, but for planetary motions it corresponds to the force of gravity. In the 1670s Newton showed that Kepler's third law, when combined with his own theory of centrifugal force, implied that the gravitational force that held the planets in their tracks followed an inverse-square law.

In the *Principia* he showed that all of Kepler's laws could be rationalized by such a force: that there was no need for additional hypotheses such as Cartesian vortices or crystal spheres to orchestrate the heavens. The reasoning is as follows. Planets are the same as falling apples because they *are* falling towards the object whose gravity determines their trajectory. But it is a fall that never ends. Since planets follow curved rather than straight-line trajectories, Newton's first law implies that they must constantly experience a force, which is gravity. According to the second law, this means that they are accelerating. But this doesn't have to imply that they gradually move faster; it simply means that their direction changes, which is equivalent to saying that the velocity increases in a particular direction (while in this case decreasing in another). Gravity constantly pulls the motion away from

---

* It's possible that he had not mislaid the proof at all when Halley called, but merely chose to keep it from his visitor until he'd had the opportunity to check it – for nothing would be worse than to have Hooke find a flaw in it. Whether Hooke himself had truly produced anything comparable, we do not know – but it seems unlikely.

a straight line, but the centrifugal force balances it perfectly so that the planet doesn't spiral into the sun. It is, you could say, the planets' speed that keeps them in a steady orbit. A stationary Earth would indeed fall into the sun like an apple on to the orchard lawn at Woolsthorpe.*

A part of the puzzle is still missing from Newton's formulation, however: why are the planets moving in the first place? Newton assumed that this was simply how God had arranged the universe at the outset. Once moving, the arrangement is perpetually stable.† Newton felt that it was futile, and outside the realm of natural philosophy, to enquire about these first causes of celestial motion, which he said were 'certainly not mechanical': it was 'unphilosophical to seek for any other Origin of the World, or to pretend that it might arise out of a Chaos by the mere Laws of Nature'. In this way Newton prescribed boundaries to his own curiosity, almost implying that it is impious to enquire about such things. 'This most elegant fabric of the Sun, planets and comets could not arise', he wrote, 'save by the wisdom and power of an intelligent and powerful Being.'

The problem of planetary motions does have a purely mechanical solution, however. Pierre-Simon Laplace showed in the late eighteenth

---

* Newton's famous account of his insight into the force of gravity being occasioned by the fall of an apple in the Lincolnshire orchard may have simply been a colourful embellishment. He related it to his associate at the Royal Mint John Conduitt (who married Newton's niece Catherine Barton) in his old age, but never wrote it down. Conduitt's description leaves it ambiguous whether Newton actually *saw* an apple fall or just imagined it: 'whilst he was musing in a garden it came into his thought that the power of gravity (which brought an apple from the tree to the ground) was not limited to a certain distance from the earth but that this power must extend much farther than was usually thought'. Newton's definitive biographer Robert Westfall, while admitting that the apple story 'is too well attested to be thrown out of court', comments that it 'vulgarizes universal gravitation by treating it as a bright idea. A bright idea cannot shape a scientific tradition.'

† Actually it isn't. Newton realized that, when three or more bodies both move under each other's mutual gravitational influence (the planets exert a small gravitational influence on one another), the resulting trajectories are extremely hard to predict. In fact, they may not reach a steady state of motion at all. It is now thought that the orbits of the inner planets (Mercury, Venus, Earth, Mars) may be mildly chaotic, which means that it is impossible with certainty to project them indefinitely into the future from their current positions. And it appears that the planets have already migrated to their current positions from a different arrangement in the distant past.

century that rotating motion would be the natural consequence of a solar system formed by the condensation of clumps of gas and dust – a veritable Chaos, without doubt – under its own gravity. The same is true of entire galaxies, which explains the common spiral form of our own Milky Way galaxy and many others.* Rotation, it seems, is the default state of gravity's propensity to form suns and worlds: no divine stirring is needed. As Laplace is famously said to have boldly proclaimed to Napoleon, his own cosmological theory had no need of the hypothesis of God.

Wishing for the moon

For all its brilliance, Newton's celestial mechanics could not fully describe the motions of the moon about the Earth. This was clearly a phenomenon of the same type as the planetary orbits, involving gravitational attraction, but the moon's trajectory turned out to be very hard to predict. Newton was acutely conscious of this short-coming and keen to correct it. But to do so, he needed more accurate measurements of the moon's position. Those were being collected at the very time – but to Newton's fury, he could not get his hands on them.

England's national power depended on naval supremacy, to which end a royal commission had persuaded Charles II to establish a Royal Observatory that would apply 'the most exact Care and Diligence to rectifying the Tables of the Motions of the Heavens, and the Places of the fixed Stars, so as to find out the so-much desired Longitude at Sea, for perfecting the art of Navigation'. In 1675 Wren and Hooke had been given responsibility for designing it. The observatory, situated south-east of London at Greenwich, was equipped with the finest telescopes and instrumentation, including a ten-foot quadrant devised by Hooke. This was Baconian experimental philosophy par excellence, emphatically in the service of the state and Crown. And as such, it exposes Bacon's coldly utilitarian agenda; there was power in this knowledge, but little in the way of real curiosity was motivating it.

* Galactic rotation does not necessarily create a spiral form, however – elliptical galaxies, which lack spiral arms like those of the Milky Way, also rotate. In galaxies with irregular shapes formed by the merging of smaller galaxies, the issue of rotation is more complex.

The man placed in charge of the Royal Observatory was John Flamsteed, a mere 28-year-old who had impressed the observatory's main patron Sir Jonas Moore, Surveyor General of the Ordnance Office, as a diligent observer. This he was, but not much more: Flamsteed, appointed the first Astronomer Royal, was not by any means an insightful scientist. Perhaps that was the point – he would not waste his time *thinking*. And indeed Flamsteed dedicated himself to the task with obsessive zeal, but with an unfortunate determination to guard his data as jealously as Tycho Brahe had. When later pressed to explain the delay in publishing them, he loftily replied 'They may as well ask why St Paul's is not finished.'

By 1712, when Flamsteed was still refusing to publish the promised *History of the Heavens*, Newton had had enough. Being now the president of the Royal Society, he had sufficient influence to force the release of the data, arguing that since it had been collected at public expense, it should be made publicly available whether Flamsteed liked it or not (he didn't). In his preface to the ensuing book, Halley pulled no punches in his assessment of Flamsteed's *oeuvre*: 'nothing has emerged from the Observatory to justify all the equipment and expense, so that he seemed, so far, only to have worked for himself, or at any rate for a few of his friends'. (Flamsteed retaliated by later recalling and destroying all 300 or so unsold copies of the book.)

This tension between Newton and Flamsteed came to a head over the subject of comets, which was what had sent Halley to seek out Newton in the first place. In the second half of the seventeenth century they were still widely regarded as harbingers of drastic events. It is sometimes implied that the astronomical work of men like Halley and Wren undermined this notion by making comets just another class of celestial object, but in fact the two views were not incompatible – even in the sixteenth century there seemed to be a perfectly rational way to explain how comets might influence worldly affairs (namely, by the hot, dry vapour they were thought to emanate), and Newton himself later forecast that the End of Days would be caused by a comet colliding with Earth. All the same, comets were certainly being brought within the framework of celestial mechanics, regarded as just another type of body moving in mathematically defined trajectories under the gravitational influence of the sun.

What form did that trajectory take? 1680 saw an almost unprecedented phenomenon: the appearance of two prominent comets in
rapid succession, just a month apart. Flamsteed boldly proposed that
the two objects were in fact one and the same, seen moving towards
and then retreating away from the sun. But why should a comet,
drawn sunward by gravity, suddenly reverse its course? Flamsteed put
forward a hypothesis that in truth did not make a great deal of sense,
and Newton wasted no time in saying so (he made no secret of his
low opinion of Flamsteed's competence). The Astronomer Royal
thought that the sun's influence was magnetic – this notion was still
favoured as an explanation for gravity, even by Newton – and he
suggested that the comet might for some reason have reversed its
magnetic orientation as it approached the sun so that attraction turned
to repulsion.

The mechanism may have lacked coherence, but Flamsteed was
right about the identity of the 'two' comets. Newton subsequently
deduced – and explained in the *Principia* – that comets are like planets
on extremely long and eccentric orbits, sweeping around the sun with
clockwork regularity. The 1680 comet was a rare example in being
clearly visible both on its inward and outward journeys. This brought
the long-standing puzzle of comets in line with Newton's general
scheme for celestial mechanics governed by an inverse-square law of
gravity. Characteristically, Newton never acknowledged that his ridicule of Flamsteed had not been entirely fair.

Outsiders in the revolution

The celestial mechanics of Kepler, Galileo and Newton is often
presented as the keystone of the Scientific Revolution. But the care
with which these men observed the heavens and drew interpretations
from them stemmed not so much from a curiosity to discover how
nature behaves as a desire to verify prior hypotheses. For Tycho, this
motivation was his geo-heliocentric hybrid cosmography; for Kepler
it was the harmonic system of nested polyhedra; for Newton, the
inverse-square law of gravity. Besides, Tycho and Kepler wanted quantitative precision to improve their astrological forecasts, while for
Flamsteed, Hooke and Newton (and no less for French astronomers)
it was needed to improve navigation and ensure naval superiority.

After all, it generally takes more than curiosity to insist on details down to the tenth decimal place.

In seeking the laws governing the heavens, the 'new astronomy' thus did not respect the modus operandi of Bacon's experimental philosophy, in which facts are exhaustively compiled without being prejudiced by a prior theory, or hurried along to provide one. We might nonetheless consider the celestial mechanicians 'modern' in demanding congruence between theory and observation: they do not accept the philosophical precedence of theory over experience that characterizes Aristotle. The combination of deductive reasoning and observational verification displayed by Kepler, Galileo and Newton is what commends them so highly to contemporary scientists. And it is unquestionably a virtue, for science needs much more in the way of prior hypothesis and theory than most members of the Royal Society were prepared to admit: there is no way to boil down a mass of raw data into a theory unless we are prepared to take a leap of faith by suggesting (and then testing) some generative mechanism for it. But one cannot argue that the cosmographies of Copernicus and Galileo, nor to some degree Newton himself, were compellingly supported by the data in preference to the alternatives. In all cases the scientific reasoning was influenced by subjective attachment to a particular idea, or what Bacon would have called 'idols of the cave'.

It makes little sense to criticize them on that score, not just because this is human nature but also because modern science often works this way too. While the idealized picture of the scientific method posits researchers testing their theories to breaking point, one frequently finds them searching more diligently for the flaws in rival theories. Science does not suffer as a result, because nature is the ultimate arbiter: if you can't fit the data, it doesn't matter how elegant the theory is.* And while defenders of particular theories might be partisan in their choice of evidence, one can rely on their rivals to point that out. So there emerges a kind of optimal selection of explanations by free-market competition as each agent does their best to expose the weaknesses of competitors.

---

* Whether or not this means that elegant theories must immediately be abandoned if the 'facts' do not fit is, however, a more complicated matter, as indicated earlier. Einstein was not alone in insisting that occasionally one should place more faith in the theory than in the data.

But what the celebrants of Newton's genius often fail to acknowledge (or even to see) is that Newton was able to make his science mathematical precisely because he chose – whether by intuition or good fortune – the right problems, namely celestial and terrestrial mechanics. To describe the motions of objects, the linear and parabolic and elliptical trajectories of bodies moving under the influence of forces, geometrical reasoning is the right tool. Yet there are very few other phenomena in nature that will yield to this approach, at least on the basis of what scientists in the seventeenth century could know. Chemistry will not – that Newton contributed far less than Robert Boyle in this field was not for want of trying, still less because he was an inferior intellect, but because the purely experimental approach of collecting and comparing observations was the more appropriate at that stage. Geometry had nothing to say about chemistry, nor really about biology and medicine, meteorology, fluid flow, mineralogy, heat and cold, botany and zoology. The point is not that Newton showed 'how to do science', but that numerical precision is useful and indeed essential in astronomy in a way that it was not for, say, medicine.

Besides, it is quite misleading to suggest that Newton's success in explaining mechanics came from looking at the problem in a 'modern' way. For a start, we've seen how the use of mathematics was allied to the Neoplatonic vision of geometric, harmonic order in the world: a mystical idea which happened to work well when applied to the paths of planets, and not so well when, say, decoding apocalyptic prophesy in the Book of Daniel. Both Kepler and Newton should thus have found John Dee, Robert Fludd and Athanasius Kircher comfortable bedfellows (indeed, Newton esteemed Dee), and it is their good fortune to have been blessed with both rare mathematical talent and access to excellent data, which enabled them to cast the longer shadows.

And make no mistake: gravity was for Newton an occult quality. It joined magnetism and static electricity as one of the invisible, 'hidden' forces, causing objects to affect one another at a distance, of which Della Porta and his circle were so convinced. These things lay at the heart of natural magic, and in effect Newton was simply mathematizing that tradition, expressing the belief that occult forces and emanations operated on arithmetical principles. As for where these forces came from, Newton did not espouse the modern view that gravity is an inherent property of matter – it was, rather, a manifestation of the active agency of God (even if allotted in proportion to an object's mass). For Leibniz (who would never pass up an opportunity to criticize Newton), this concept of gravity as an invisible and inexplicable force was a throwback, a 'scholastic occult quality'. Newton's universe is often considered mechanistic, but that is true in a different sense from the literalist Cartesian picture in which nothing can happen until particles crowding all of space push against one another. Newton's world was still pervaded by unseen forces, tamed by numbers and differential calculus. And as above, so below: if it worked for the heavens, he argued, it should work for atoms too.

It's no surprise, then, that the early scientists who contributed most to an understanding of the laws of the heavens were not Baconians, although they benefited from the new philosophy's endorsement of curiosity and shared some of its mystical and magical roots. Newton and Galileo were in truth *sui generis*: both prominent members of progressive scientific societies, yet unwilling to accommodate their ways to any system or dogma except their own. To see how different Galileo's concept of experimental science was from that proposed by Bacon and pursued by the Royal Society, we need only contrast Galileo's Book of Nature, which 'stands always open in front of us for the purpose of our learning' with Bacon's image of nature as wilfully obscure and secretive, her secrets needing to be hunted down and even extracted by force. And Bacon himself complained (with some cause then, although it was less true later in the century) that astronomers were fixated on mathematical descriptions rather than mechanical explanations: in astronomy, he said,

> the pains have been chiefly bestowed in mathematical observations and demonstrations; which indeed may show how to account for all these things ingeniously, but not how they actually are in nature: how to represent the apparent motions of the heavenly bodies, and machines of them, made according to particular fancies; but not the real causes and truth of things.

The common view of the Scientific Revolution, in which the canonical texts are Galileo's *Discourses Concerning Two New Sciences*, Newton's *Principia* and Descartes' *Principles of Philosophy*, is therefore misleading not only because it places at centre stage ideas and methods that were unrepresentative of the mainstream but also because, by insisting on making mathematical physics and mechanics the wellspring, it marginalizes and even condemns other studies and disciplines for failing to adopt an approach to which they were not well suited. The historian of science Alistair Crombie claims that science began to emerge in the Renaissance as 'knowledge of stable universal principles such as mathematicians knew', and that 'the mathematical sciences and arts combined by the end of the sixteenth century to create within European intellectual culture an effective context for solving problems'. But that context actually solved very few problems outside the astronomical. The kind of commentary that makes the mathematization of mechanics 'the crucial act in the scientific revolution', or that in the words of historian Toby Huff equates the emergence of modern science with 'the Copernican-cum-Newtonian world view', can only end up displaying impatience with all these other sciences for lagging behind, for failing to get on with the business of becoming mathematical and Newtonian.

The determination of historians like Crombie (and of many contemporary scientists) to make Galileo not just an innovative genius but the exemplar of how science was to be conducted thereafter perhaps reflects the fact that he seems closest to the typical picture of the modern physicist – uninterested in mysticism, insistent on quantification, and guided by mathematical principles towards an understanding of underlying mechanisms. This, they feel, is the quickest way to make a break with what they see as the earlier systems of credulous superstition. According to Crombie, Galileo saw himself as 'a pioneer of a new natural philosophy that excluded all others because it uniquely could both define and solve particular and quantitative physical problems and in doing so relate them to a general system of explanation'. Crombie argues that this new science was thereby distinct both from the empty logical disputes of scholasticism and 'the uncritical curiosity of natural magic'. But it was precisely because of the severe limitations, at that time, of the scope of Galileo's approach that it was not able to establish itself as *the* method of science, and that instead natural magic was compelled to adapt its voracious curiosity to a *more* critical approach – to sift, to test, to quantify and agree on standards and principles of verification – in order to provide the true basis for building a new science. It was not until much later – the mid-nineteenth century – that Galilean and Newtonian mechanics, so powerful for describing the heavens and the trajectories of cannonballs, came to be successfully applied to the microscopic behaviour of matter (where ultimately these ideas were found to be wanting). They have never yielded theories of chemistry or biology, and it is only gradually that we are coming to accept that Galileo's dictum – 'In the book of nature things are written in only one way' – does not necessarily mean that this must be the case in the textbooks of science.

8 The First Men in the Moon

Perhaps a thousand other worlds that lie
Remote from us, and latent in the sky,
Are lightened by his beams, and kindly nurs'd.
　　John Dryden *Eleanora* (1692)

As soon as the art of Flying is Found out, some of their Nation
will make one of the first Colonies, that shall Transplant into that
other World.
　　John Wilkins, *The Discovery of a World in the Moone* (1638)

Would Galileo have looked so diligently at the moon if he were an
Aristotelian? A traditionalist would have believed there was nothing
to see: it was a smooth crystalline sphere, as uncorrupted as the
celestial bodies must be. Of course, it did not *look* so perfect, but
the blotches were assumed to be the reflected blemishes of Earth.

But Galileo did train his new telescope on the moon, and he realized that Aristotle couldn't be right. For the line separating the
illuminated from the shadowed part of the moon was not smooth
but ragged and interrupted. It looked just like a range of mountains
and valleys caught by the early evening sun, the raking light bright-
ening the peaks but failing to penetrate into the recesses. His description of these observations, accompanied by his own masterful
sketches, in his 1610 book *Sidereus nuncius*, made his name famous
throughout Europe.

He wasn't the first to stare at the moon through the lens, nor
the first to discern its topography. Just months earlier Thomas

The rugged surface of the moon is revealed by a telescopic view of the dividing line between light and shadow, as drawn by Galileo (left) and Thomas Harriot (right).

Harriot sketched the rugged boundary between the bright and dark regions of the gibbous moon, but although his skeletal drawing conveyed a clear notion of the heights and depths of craters, it was more schematic and less realistic than Galileo's. And in any event it remained in his notebook – a true Elizabethan virtuoso, Harriot felt no pressing need to publish his observations.*

Galileo always denied that the lunar topography implied an Earth-like body fit for life – to him, the moon was a barren world. Others reached a different conclusion. Tommaso Campanella got hold of a copy of *Sidereus nuncius* while in prison in Naples, and in early 1611 he wrote to Galileo with sumptuous praise, expressing the conviction that other planets must now be deemed likely of hosting inhabitants. In *Paradise Lost*, John Milton's Angel voices the popular view when he says of the moon that:

> her spots thou seest
> As clouds, and clouds may rain, and rain produce
> Fruit in her softened soil, for some to eat.

* Harriot also seems to have observed sunspots before, or at least coevally with, Galileo, for he drew them in late 1610. While Galileo has Salviati assert in *Dialogue Concerning the Two Chief World Systems* that he saw these solar maculae in the late summer of that year, this may well have been one of Galileo's attempts to claim priority in retrospect. A December date seems more probable.

It is possible that Milton may have seen those spots for himself at Arcetri through one of Galileo's own telescopes.

The new cosmogeography of *Sidereus nuncius* was by no means limited to the lunar landscape. Galileo swept his telescope across all the sky, finding that the Milky Way – thought by most astronomers, including Kepler and Tycho Brahe, to be a nebulous celestial substance – was composed of innumerable stars. There are at least ten times more stars in the heavens, he announced, than had been thought. 'To whatever region you direct your spyglass, an immense number of stars immediately offer themselves to view', he wrote. The stars called nebulae, he pointed out, are in fact 'swarms of small stars placed exceedingly close together'. Such overwhelming abundance must have seemed almost numbing, rendering futile any aspiration to know anything about them.

Most remarkably of all, Galileo saw four 'stars' – in fact, moons – orbiting the planet Jupiter. On one day in January 1610 they appeared three on one side of the planet and one on the other, like this: *O***; two days later they had rotated to a position in which all four were on the same side: O****. Galileo cannily named these 'errant stars' after his patron: the Medicean Stars.* They are more commonly known today as the Galilean moons, although their mythological names Io, Europa, Ganymede and Callisto were chosen by the German astronomer Simon Marius, who in 1614 claimed that he had actually seen them first.

Sidereus nuncius thus announced a plurality of new worlds, and Galileo was feted by poets as another Columbus. His new vision of the moon shattered the bland, crystalline cosmos of Aristotle. By showing that the celestial realm differs from the way the ancients had conceived of it, Galileo's little tract opened up more room for the Copernican hypothesis.

* He has with justification been called a 'court strategist', though Cosimo II modestly declined the suggestion of calling them the Cosmica Sidera, dedicated to him alone. After *Sidereus* appeared, Galileo received a letter from the French court requesting that, if he should find another 'fine star' like these, he might consider naming it after 'the Great Star of France, as well as the most brilliant of all the earth' – that is, the French king Henri IV. It is an extraordinary circumstance that this most 'brilliant star' should thus come begging a favour from an astronomer, and an indication of how the little book launched Galileo's international reputation.

There was more concrete support for heliocentrism in his book. For if Jupiter could follow its orbit without losing a grip on its moons, this negated the Ptolemaic objection that Earth would lose its own moon if it moved. As Galileo wrote:

> We have moreover an excellent and splendid argument for taking away the scruples of those who, while tolerating with equanimity the revolution of the planets around the Sun in the Copernican system, are so disturbed by the attendance of one Moon around the Earth . . . that they conclude that this constitution of the universe must be overthrown as impossible. For here we have only one planet revolving around another while both run through a great circle around the Sun: but our vision offers us four stars wandering around Jupiter like the Moon around the Earth while all together with Jupiter traverse a great circle around the Sun in the space of 12 years.

Perhaps most significantly, by making an Earth-like world of the moon *Sidereus nuncius* undermined the special status that geocentrism awarded to the Earth. This was why it galvanized the hitherto rather subdued debate over Copernicanism. Even if the Earth were to be, for the sake of mathematical convenience, shifted very slightly from its position at the centre of the universe to accommodate Copernicus, nonetheless one could continue to regard it as the unique home of God's beings. But now it seemed that ours might not be the only world. That, as we saw earlier, was a suggestion advanced by Giordano Bruno, and although it was not why he was burnt, it was not welcomed by the Church. Yet just ten years after Bruno's execution, the new astronomy seemed to be providing visible evidence in support of his troublesome idea. In his 1624 essay *Devotions upon Emergent Occasions*, John Donne vividly expressed the dizzying effect of this shift in perspective:

> Men that inhere upon Nature only, are so far from thinking, that there is anything singular in this world, as that they will scarce thinke, that this world it self is singular, but that every Planet, and every Starre, is another World like this; They find reason to conceive, not onely a pluralitie in every Species in the world, but a pluralitie of worlds.

Donne followed the latest developments in astronomy closely – he undertook a challenging journey to visit Kepler in Linz in 1619 – and he gives
us a glimpse of how deeply they affected intellectual life in the early
seventeenth century. (Shakespeare, by contrast, was much more conservative in his natural philosophy.) In *De stella nova* (*On the New Star*) (1606),
Kepler delivered another blow to the idea of a perfect and unchanging
cosmos when he described a new star that had appeared in the sky two
years earlier in the constellation of Serpentarius. Tycho Brahe had previously noted another such stellar novelty in 1572: they were both supernovae, which faded from sight some months after their outburst.

The appearance of new stars was a profoundly disquieting image
for Donne, who wrote in 1611:

> And new Philosophy calls all in doubt,
> The Element of fire is quite put out;
> The Sun is lost, and th'earth, and no mans wit
> Can well direct him where to looke for it.
> And freely men confesse that this world's spent,
> When in the Planets, and the Firmament
> They seeke so many new; then see that this
> Is crumbled out againe to his Atomies.
> 'Tis all in peeces, all cohaerence gone;
> All just supply, and all Relation . . .
> And in these Constellations then arise
> New starres, and old doe vanish from our eyes.

The appearance and disappearance of stars, Donne felt, made it seem
'as though heav'n suffered earthquakes, peace or war'. Yet this anthropomorphic imagery in itself shows that the heavens were already coming
to seem like the world, only further away. That is the explicit implication
in a poem of about 1609 that Donne wrote to the Countess of Bedford,
in which he compares the astronomical discoveries to geographical ones:

> We have added to the world Virginia, and sent
> Two new starres lately to the firmament.

The virtuosi seem to have approved of such poetic use of their discoveries. In his *History of the Royal Society* Thomas Sprat recommended

the new philosophy as a source of fresh metaphors for poets and writers who 'had tir'd out the *Sun*, and *Moon*, and *Stars* with their Similitudes, more than they fancy them to be wearied by their daily journeys round the *Hevens*'. Now, he said, science can supply new imagery 'which may relieve their fellow-creatures, that have long born the burden alone, and have long bin vex'd by the imaginations of *Poets*'.

One of the poets happy to appropriate the imagery offered by the new astronomy was Milton. The cosmic vistas of *Paradise Lost* would be inconceivable without it. Milton mentions Galileo in person when he compares Satan's shield to

> the Moon, whose Orb,
> Through Optic Glass the Tuscan Artist views
> At Ev'ning from the top of *Fesole*,
> Or in *Valdarno*,* to descry new Lands,
> Rivers or Mountains in her spotty Globe.

The plurality of worlds in Galileo's Milky Way also made an impression on Milton, who describes these teeming heavens with breathtaking beauty:

> A broad and ample road, whose dust is gold,
> And pavement stars, as stars to thee appear
> Seen in the galaxy, that milky way
> Which nightly as a circling zone thou seest
> Powdered with stars.

Along with this perception of the vastness and multiplicity of interstellar space came the complementary view: our world seen from outside, lost in the infinite night. As Blaise Pascal put it, 'All this visible world is but an imperceptible point in the ample bosom of nature.' Here we are on the verge of being swallowed entirely by the implacable magnitude of the cosmos. It is this shift in perspective that perhaps most of all characterizes the imaginative response to the new

---

* Fiesole is a Tuscan hilltop town near Florence. Valdarno is a region around the valley of the River Arno in Tuscany.

astronomy: the Earth hanging in space, the seventeenth-century analogue of the famous Earthrise image of the Apollo missions. That vision supplies the terrifying drama and majesty of *Paradise Lost*: Satan first sees the Earth as a 'pendent World, in bigness as a star, Of smallest magnitude close by the moon'. As he approaches, he looks down like an astronaut in orbit and sees the entire world at a glance from pole to pole.*

Not everyone was so awestruck by Galileo's universe. In 'News from the New World Discovered in the Moon' (1621), Ben Jonson refers disparagingly to the telescope as 'a trunk . . . no bigger than a flute-case', and jokes that 'the brethren of the Rosie Cross have [set up] their college within a mile of the moon; a castle in the air'. And it is perhaps to be expected of the author of *Bartholomew Fair* that he could not resist wondering if this 'new world in the moon' has any pubs in it.

Instrumental aid

That Galileo used a novel instrument to gather his evidence from a realm inaccessible to everyday experience was central to the controversy he provoked. A reliance on instrumentation is now so commonplace in science that it passes without comment. The modern laboratory is typically filled with shiny equipment that might produce conditions – extreme temperatures and pressures, high magnifications, sensing of invisible radiations – far outside human capability or discernment. This situation implicitly accepts the view that there is a whole world of phenomena occluded from our direct perception: hidden forces and operations, unearthly environments, whole worlds we cannot see or sense. What a leap of faith it once was to imagine such

---

* Milton, who was largely sympathetic to the new philosophy, had to find an accommodation in *Paradise Lost* between the traditional condemnation of Satan's – and Adam's – curiosity and the investigations of scientists such as Galileo. His position seems to be that curiosity is not bad in itself, but only if it is indulged without restraint. In *Paradise Lost*, says English scholar Patrick Brantlinger, 'Adam and Galileo are both motivated by curiosity, but ingenuity becomes a mania only in Satan, who has not been interested in exploring God's universe, since he claims he is as knowing as God . . . What is asked of Adam is not that he have no curiosity, but that he be temperate.'

things! It seemed natural in Galileo's time to regard human sensation as the arbiter of mundane reality – claims to see 'things which are not there' were traditionally the stuff of witchcraft.

Francis Bacon argued otherwise. Sometimes, he wrote, the human faculty of sensation 'gives no information, sometimes it gives false information', and there are many things which escape the senses 'by reason either of the subtlety of the whole body, or the minuteness of the parts, or distance of place, or slowness or else swiftness of motion, or familiarity of the object, or other causes'. So the philosopher needs artificial aids and instruments that can pry where the eye cannot.

Lenses were an old invention. Eyeglasses with crystal lenses to correct for defective vision were used since the thirteenth century,* and were sufficiently widespread by 1300 that Venice passed a law protecting its citizens from spectacle-makers who passed off ordinary glass as crystal. The Italian scholar Petrarch grumbled about having to use eyeglasses in the 1360s, when his eyesight began to fail him.

But while the magnifying power of the convex lens was a commonplace, the key to the telescope was the use of two lenses: a refracting lens to collect and focus the light, and an eyepiece lens that magnified the focused image. It's not an easy arrangement to get right, for the two lenses have to be just the right distance apart so that the image is magnified precisely at the point where it becomes sharply focused.

It was known long before Galileo's day that spectacle lenses could develop greater magnifying power when combined in pairs. Girolamo Fracastoro wrote in *Homocentrica* (1535) that:

> If anyone should look through two spectacle glasses, one being superimposed on the other, he will see everything much larger and nearer . . . Certain lenses are made of so great a density that if anyone should look through them either at the moon or at any one of the stars, he would judge them to be so near that they all but surpass actual towers.

The English astronomer Thomas Digges says in the introduction to his father Leonard's book *Pantometria* (1571) that Digges senior had

---

* It has been suggested that this innovation prolonged the influence and power of older members of the clergy, since it allowed them still to read and study the Scriptures.

been able to see things seven miles away with a suitable arrangement of lenses.

Della Porta's interest in optics led him to consider how visual acuity might be improved by lenses. Mindful of an old story that Ptolemy used lenses to see ships from (impossibly) far off, he was casually dismissive of Galileo's telescope: in the habitual manner of the scientist past his prime, he asserted that he had already explained the basic principles of the device decades ago. 'I have seen the secret use of the eyeglass and it's a load of balls [*coglionaria*]', he wrote bluntly to Federico Cesi in August 1609. 'In any case, it's taken from book 9 of my *De refractione*.' It was not entirely an idle boast, however, for Della Porta accompanied his assertion with a sketch of the instrument he had imagined: two nested tubes, one with a convex lens and one with a concave, which go 'in and out like a trombone'. But the claim that he deserved credit for the invention of a practical telescope is thin. Not only had he never actually made one (and it is not so easy, as Galileo's rivals discovered), but he never recognized the significance of the invention for studying nature. Neither, despite his claims in *De telescopio* (1611), did he ever really figure out how it worked. All the same, when *Sidereus nuncius* was published (a year before Galileo became a member of the Accademia dei Lincei) the academy's secretary Francesco Stelluti wrote to his brother that this newcomer was heading for trouble by asserting his priority with the innovation: 'Giambattista Della Porta wrote about [the telescope] more than thirty years ago in his *Magia naturalis* . . . so poor Galileo will be besmirched.' Eventually, however, Della Porta was gracious enough to concede that he had never even come close to Galileo's achievements. Writing to Johann Faber in Rome, he said:

> I really am delighted that my rather rude and slight invention has been elevated to such use by the talent and resourcefulness of the most learned mathematician Galileo Galilei, who has shown that many planets wander in the heavens, and that so many new stars shine again in the firmament, which had lain hidden for so many centuries.

In any event, Galileo never claimed to have *invented* the telescope. That was allegedly done around 1600 by an illiterate Dutch spectacle-maker named Hans Lippershey in the town of Middelburg in Zealand.

The traditional story of this discovery has all the hallmarks of romantic apocrypha: two children were said to have been playing in his shop when they put a pair of lenses together and found they could see clearly the weathervane on the distant town church steeple. Another story has it that Lippershey learnt this trick from another Middelburg inventor, Zacharias Janssen – although this tale is doubtful too, for it originated with Janssen's son, and besides, Janssen was still a child in 1600 and his later conviction for counterfeiting does not give one confidence in his probity. In any event, the first clear account of the invention occurs only in 1608, when Lippershey petitioned the Dutch government to be given a monopoly on telescope-making 'for the utility of this country alone'. It was not granted, and neither was the similar appeal made only months later by James Metius of Alkmaar. Within a few years there were several other instrument-makers in the Netherlands claiming to know the secret. Galileo was told of the device by his friend Paolo Sarpi in Venice. Because he already kept a workshop for making eyeglasses and other mathematical and surveying devices, he was in a position to experiment. Before long – and apparently without having set eyes on one of the Dutch instruments – he had boosted the power of the telescope to a magnification of around twentyfold, transforming a potentially useful military and navigational device into a powerful scientific instrument. All the same, Galileo never truly understood how telescopes work either. Like many scientists, he was an empiricist with sound intuition, not a technician with theoretical knowledge of the device.

When he wrote *Sidereus nuncius*, Galileo was working in the court of Cosimo II in Florence, who sent the tract to Rudolf II in Prague. Rudolf asked Kepler what he made of it, and Kepler – who had hitherto, like Tycho, relied on naked-eye observations – saw at once how the telescope could transform astronomy. He wrote to Galileo to offer his support and congratulations, prompting the Italian to respond that 'you are the first and almost the only person, who, after a cursory investigation, has given entire credit to my statement'. This congenial relationship did not prevent Galileo from guarding his priority: when Kepler asked to borrow one of his instruments, the crafty Galileo understood that thus equipped he would become a formidable rival, and so he prevaricated.

But by August of 1610 Kepler had got his hands on a telescope anyhow, and was exhilarated by what it could do. 'Oh telescope, instrument of much knowledge, more precious than any sceptre!', he wrote in his optical treatise *Dioptrice* a year later. 'Is not he who holds thee in his hand made king and lord of the works of God?' He immediately set about deducing how it worked. A user of spectacles himself, Kepler had developed a theory of lenses and human vision in his *Paralipomena* (1604), and in *Dioptrice* he reiterated and refined this theory before going on to explain (albeit only in approximate, qualitative terms) how the combination of a convex and a concave lens could produce telescopic images. He also described how a telescope might be made with two convex lenses, although he never made one himself. This sort of instrument is now known as a Keplerian telescope, and has the advantage over the Galilean telescope (with concave and convex lenses) that it has a wider field of view.

Evidence of the eye?

New instruments such as the telescope and the microscope are rightly acknowledged as having made a critical contribution to the new philosophy. Their significance was not just that new things could be seen, but that *mechanical art* was thereby made an essential aspect of what had hitherto been a matter for the judgement of naked experience. Despite the egalitarian suggestions that the new philosophy was something any man could practise with sufficient care and without the need for lofty academic titles, it seemed that in fact you needed to have (and to know how to operate and interpret) the right equipment. That was a new entry requirement into the ranks of the cognoscenti – and specialist equipment did not come cheap. Philosophers became privileged not because of what they knew but because they had a better way of finding out.

Yet why should anyone assume that the intervention of an instrument necessarily corrects defects of sensation, rather than making them worse? It was thought, with good reason, that lenses and prisms alter the light that passes through them: not only is it refracted (meaning in effect that its path is altered), but it was suspected – until Isaac Newton showed otherwise – that the spectral colours seen in sunlight after passing it through a prism were *produced* by the prism and not merely *revealed* by it. Quite aside from all this, the instruments were tricky to master: they were not easy to focus and had a small field of view. Anyone who has used even a modern optical microscope will appreciate that the eye needs to be trained to see clearly: at first you might discern nothing at all. This makes it more understandable that some of Galileo's critics refused to look for themselves through his telescope. The Aristotelian philosopher Cesare Cremonini at Padua (who became Galileo's main model for Simplicio in his *Dialogue Concerning the Two Chief World Systems*) saw no reason to try to see what 'no one but Galileo has seen' – and besides, he said, looking through the eyeglass gave him a headache. Another doubter did claim to have tested Galileo's instrument 'in a thousand ways', and to have found that sometimes it produces double images of single stars – raising doubts about the multiplicity, or even the existence, of the 'Medicean stars' of Jupiter. 'The spyglass', said a sceptical professor at the University of Perugia, 'makes us see things that are not really there.'

Naturally, Galileo denied these suggestions vigorously. In 1611 he wrote (probably with a little exaggeration) that:

> Over a period of two years now, I have tested my instrument (or rather dozens of my instruments) by hundreds and thousands of experiments involving thousands and thousands of objects, near and far, large and small, bright and dark; hence I do not see how it can enter the mind of anyone that I have simple-mindedly remained deceived in my observations.

Despite such protestations, Galileo was forced to confront the unsettling possibility of instrumental artefacts just a few years later. Some months after *Sidereus nuncius* was published, he found that the planet Saturn seemed to be a 'threefold' object: there were what looked like two small circular bodies protruding like ears on opposite sides of the planet. As he wrote to his patron in Florence:

> I discovered another very strange wonder, which I should like to make known to their Highnesses . . . keeping it secret, however, until the time when my work is published . . . the star of Saturn is not a single star, but is a composite of three, which almost touch each other, never change or move relative to each other, and are arranged in a row along the zodiac, the middle one being three

times larger than the lateral ones, and they are situated in this form: ○○○

This made Saturn even more peculiar than Jupiter with its four moons. And it was further evidence that the heavens were far more complicated than tradition allowed, more cluttered than 'perfect'. Other astronomers, attempting to check the claim, said that to them Saturn seemed oval in form, or perhaps rectangular. With characteristic overconfidence, Galileo dismissed this as the blurred result of poorer telescopes.

But when he looked at Saturn again two years later, the two smaller 'stars' had disappeared! He frankly admitted to being perplexed, and was forced into a half-jesting allusion to myth: 'Now what is to be said about such a strange metamorphosis? Perhaps Saturn has devoured his own children?' Or could it be – and this is the nub of the matter – that 'it was an illusion and a fraud with which the glasses have for so long deceived me?' Perhaps, he admitted, those who maintained that 'all the new observations are deceptions which cannot exist in any way' had a point. In the end he wisely concluded only that 'I need not say anything definite upon so strange and unprecedented an event; it is too recent, too unparalleled, and I am restrained by my own inadequacy and fear of error.' Galileo must have been glad of this uncharacteristic humility when, in 1616, he found that Saturn seemed to have taken on yet another shape, apparently adorned with two 'handles', which he depicted without further comment in *The Assayer* in 1623. The reason for these changes was not fully understood until 1656, when Christiaan Huygens argued that Saturn is surrounded by a flat ring of material that we see from different perspectives as the planet orbits the sun: sometimes tilted, sometimes edge-on (when the thin rings seem to vanish).

This dilemma thus forced even Galileo to doubt his own senses – or rather, to doubt that his prized instrument can deliver reliable information. Notice how Galileo puts it: not that he has misinterpreted the data, but that *the telescope deceived him*. Of course, it's easy to read that as the defensiveness of a workman blaming his tools; but the question was a genuine one: *can the instrument be trusted*?

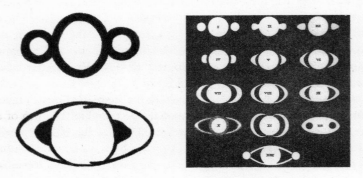

Galileo's drawings of Saturn represent the ring as lobes (top left, 1610) or handles (bottom left, 1616). Christian Huygens realized that these are caused by a ring viewed from different angles (right).

Kepler's dream

Despite its impact, Galileo's suggestion that the moon might be a world like ours was not a new idea. The Greek–Roman historian Plutarch speculated in *De facie in orbe lunae* (*On the Face in the Moon*) that the lunar globe might have mountains, seas and inhabitants, and Cicero in the first century BC cites Xenophanes as a source of this notion. In his satirical *True History*, the second-century Syrian writer Lucian even wrote about a trip to the moon by sailors borne aloft by a mighty whirlwind, where they do battle with the 'Selenites' who live there. But this was a diverting yarn, not a serious proposition about the nature of that world. As the telescope pulled the heavens closer, Galileo's observations gave permission to contemplate such voyages in reality, and to speculate about what and who we might find: sons of Adam, or some other race?

One of the first imaginative leaps on to the lunar surface in the new age of astronomy was made before Galileo put his eye to the lens. It appeared in an essay or 'disputation' conceived by Kepler in 1593 while he was still a student at the University of Tübingen, by means of which he intended to demonstrate the advantages of the Copernican theory. An observer on the moon would see the Earth appear to move just as we do the moon, he said, and from that vantage point the sun and stars would be shifted. This was an invitation to regard our place in the cosmos as contingent and arbitrary: if we were elsewhere, things would look quite different. As such, it challenged not just the structure of the medieval cosmos but the hierarchy of prestige that this encoded. The essay was too much for the young Kepler's professor, the staunchly anti-Copernican Veit Müller, who refused permission for it to be read out.

Kepler returned to this piece of juvenilia sixteen years later in Prague, where he shaped it into a strange semi-fiction posed in the form of a dream (the *Somnium* of the title) he claimed to have had in 1608 about a man who journeys to the moon. The narrative is transparently auto-biographical, which he later had cause to regret. It tells of an Icelandic boy called Duracotus, whose bad-tempered mother Fioxhilde discharges him into the service of a sea captain. He sails to Tycho Brahe's island of Hven, to which the captain has sent him with a message for the great Danish astronomer. Tycho takes the lad as an apprentice, and he stays for five years before returning home. There his mother is as grouchy as ever, but Duracotus discovers to his amazement that she seems already to know all that he learnt from Tycho. This is because she is a 'wise woman', in communication with spirit beings or demons from the lunar land of Levania, 50,000 miles up in the ether. One of these demons explains how they sometimes carry humans with them on their interplanetary journey, hurling them up over mountains and seas with a violent shock. (Kepler correctly senses that these people would need to be protected from the cold and the thinness of the air.) Thus Kepler's solution to the means of getting an observer from the Earth to the moon is rather disappointing in that it invokes no inventive Renaissance technology but is wholly supernatural.

A Levanian demon seizes Duracotus and takes him on this voyage. Once they arrive on the moon, the demon explains that it too has a moon, called Volva, which is presumably the Earth: 'For Levania seems to its inhabitants to remain just as motionless among the moving stars as does our earth to us humans'. Kepler adds in an explanatory note:

> Here is the thesis of the whole *Dream*; that is, an argument in favour of the motion of the earth or rather a refutation of the argument, based on sense perception, against the motion of the earth.

The moon turns out to be inhabited not just by the demons but by other creatures – all of them non-human, for the conditions are too extreme to support humankind. The face of the moon that points towards Volva, called Subvolva, is parched, while the opposite face is dark and covered in ice. The lunar creatures have a lifespan of a single day, yet sound to us now not unlike dinosaurs, being of a 'serpentine nature' – some crawling, some swimming, some winged.

The *Somnium* was largely finished by the end of 1610, but Kepler was forced by his duties in the imperial court to let the project languish, and not until 1621 was the book completed, including a set of footnotes (which grew to three times the length of the main text) that explained the science behind the story. The vicissitudes of the *Somnium* persisted as Kepler was distracted by his mother's witchcraft trial, which had begun in 1615. It seems possible that the manuscript, which was circulated from around 1611, played a part in this affair, for Fioxhilde, obviously modelled on Kepler's own bad-tempered mother, was here seen openly to consort with demons. In one way or another, it was not until 1629 that he was able seriously to commence the process of publication. The *Somnium* was still going through the press when Kepler died in 1630, and its eventual publication in 1634 was overseen by his son Ludwig.

By this time the Counter-Reformation, with its affirmation of Aristotelian dogma, was in full sway, and Kepler was only half joking when he commented that his book might 'provide the fare for us who are being chased off the earth as we travel or migrate to the moon'. John Donne saw a happier solution: in his satirical essay *Conclave Ignatii* (published almost simultaneously in English in 1611 as *Ignatius his Conclave*) he poked fun at the Jesuits (founded by Ignatius Loyola), saying that perhaps they might all be packaged off to the moon instead, where they might 'easily unite and reconcile the *Lunatique Church* to the *Romane Church*'.

Kepler was convinced that Donne's jibe was inspired by the *Somnium*, saying in one of his footnotes that 'I suspect that the author of that impudent satire, the *Conclave of Ignatius*, had got hold of a copy of this little work, for he pricks me by name in the very beginning.' Even though *Conclave Ignatii* was probably completed only around the same time that Kepler's manuscript first began circulating, the link is quite plausible given both how closely Donne followed the new astronomy and how well connected he was. He could have obtained a draft of the *Somnium* from several potential sources – perhaps his close friend, the diplomat Henry Wotton, who kept abreast of such things, perhaps Thomas Harriot, or maybe even James I himself, for Kepler possibly sent the king a draft just as he did his *De stella nova* in 1606, and Donne was an intimate of the royal court. Historian Marjorie Nicolson speculates that Satan's plunge through the chaos of interplanetary space in Milton's *Paradise Lost* might also be indebted to the precipitous voyage described in the *Somnium*.

But if Kepler showed little concern to make his fictional moon journey realistic, he was also willing to contemplate the possibility of a technological means of space travel. His remarks to Galileo on *Sidereus nuncius* in 1610 reveal how those telescopic observations had whetted his imagination and ambition:

> There will certainly be no lack of human pioneers when we have mastered the art of flight . . . Let us create vessels and sails adjusted to the heavenly ether and there will be plenty of people unafraid of the empty wastes . . . In the meantime, we shall prepare, for the brave sky travelers, maps of the celestial bodies – I shall do it for the moon, you, Galileo, for Jupiter.

The new world in the sky

In *The Discovery of a World in the Moone*, John Wilkins also grappled with the implications of Galileo's discovery. He even included in his book a rough sketch of Galileo's lunar landscape, calling him 'the new Embassador of the Gods'. It might seem far-fetched now to infer from the mere observation of the moon's topography that it is an entire world like ours, inhabited by living and perhaps intelligent beings. But the teleological attitude of the seventeenth century made that conclusion almost inescapable. If, as Wilkins argued, the moon is a place fit for habitation, with seas and forests and mountains, why else would divine providence have made it thus if not to put actual beings on it?

Wilkins admits that he had often speculated privately that there might be a world in the moon, 'yet it seem'd such an uncouth Opinion, that I never durst discover it, for fear of being counted singular, and ridiculous'. But he was encouraged to find that not only Plutarch but now both Galileo and Kepler had entertained similar thoughts. Moreover,

John Wilkins' illustration of the Earth as a 'moon' of the moon.

the Copernican hypothesis supported this view, for if the Earth is a planet, 'then why may not another of the Planets be an Earth?'* He maintains that the moon has an atmosphere like ours, and that from that world the Earth would be perceived as a moon, much as Lucian had suggested in his frivolous tale and now Kepler confirmed.

Wilkins also enlists in his support the brilliantly original fifteenth-century cardinal and philosopher Nicholas de Cusa, whose iconoclastic cosmology had already dared to suggest that the Earth moves. He quotes Cusa as saying, nearly two centuries before Giordano Bruno, that 'we conjecture that none of [the stars] are without Inhabitants, but that there are as many particular Worlds and parts of this one Universe, as there are Stars, which are innumerable'. Cusa had speculated that the moon was inhabited 'by Men, and Beasts, and Plants', adding that, if they were men, they must be free from Adam's original sin. He had softened this apparent heresy by assuring his readers that 'perhaps, they had some [sin] of their own, which might make them liable to the same Misery with us, out of

---

* Wilkins extolled the virtues of heliocentrism more explicitly in his follow-up, *A Discourse Concerning a New Planet* (1640), where the 'new planet' is our own, now become one of Plato's cosmic 'wanderers'.

which, it may be, they were deliver'd by the same means as we, the Death of Christ'. But Wilkins agrees with Kepler and Campanella that the inhabitants of another world are unlikely to be men like us: they may be 'of a quite Different Nature from any thing here Below, such as no Imagination can Describe'. Perhaps they are 'betwixt the Nature of Men and Angels'.

While Wilkins expected his book to be scorned ("tis not likely then that this Opinion which I now deliver, shall receive any thing from Men of these Days, especially our Vulgar Wits, but Misbelief and Derision'), his main concern was the reception it would get from the Church. He insisted that his ideas did not contradict Christian orthodoxy. It's true that neither Moses nor St John spoke of any world but this one – but their silence need not imply denial. In any case, Wilkins followed Galileo in arguing that the Bible should not be used to adjudicate on matters of natural philosophy. 'The Negative Authority of Scripture is not prevalent in those things which are not the Fundamentals of Religion', he wrote: things are not ruled out just because the Bible does not mention them, since it is not a book of philosophy but of religion. Wilkins asserts that Moses proclaimed only 'those things which are obvious to the Sense, and being silent of other things, which then could not well be apprehended'. After all, the Bible makes no mention of the planets, but only of the sun and moon, and yet no one doubts their existence. 'If the Holy Ghost had intended to reveal unto us any Natural Secrets, certainly he would never have omitted the mention of the Planets', and many absurdities have followed 'when Men look for the Grounds of Philosophy in the Words of Scripture'.* True enough, though sadly some still do.

There was another doctrinal conflict to address: where does the venerated Aristotle write of other worlds? Wilkins was not one of those who wished to consign the Peripatetic philosopher to the waste heap. But although he takes pains to remain respectful to Aristotle, he says "tis not Ingratitude to speak against him, when he opposeth Truth'. Besides, says Wilkins, Aristotle was not dogmatic on the

---

* Wilkins somewhat compromises this wise position when he approves of Campanella's sentiment that 'the waters above the firmament' mentioned in Genesis refers to seas on other worlds.

matter. While implying the perfection and incorruptibility of the heavens, he had admitted that it is very hard for us to judge the truth when there is so vast a distance separating the stars from us. But Galileo's telescope has changed all that:

> Now its our Advantage that by the help of *Galileus* his Glass, we are advanc'd nearer unto them, and the Heavens are made more Present to us than they were before . . . So that what the Ancient Poets were feign to put in a Fable, our more happy Age hath found out in a Truth.

The new evidence from astronomy should be permitted to correct Aristotle.

Three decades later, Robert Hooke was optimistic that optics would bring the heavens nearer still. ''Tis not unlikely', he wrote in *Micrographia* (1665), 'but that there may be yet invented several other helps for the eye, as much exceeding those already found, as those do the bare eyes, such as by which we may perhaps be able to discover *living Creatures* in the Moon, or other Planets.' Hooke himself studied the moon's surface through a telescope and concluded that the 'vale' then called Mons Olympus (not to be confused with the extinct volcano now so named on Mars)

> seems to be some very fruitful place, that is, to have its surface all covered over with some kinds of vegetable substances . . . So that I am not unapt to think, that the Vale may have Vegetables *analogus* to our Grass, Shrubs, and Trees; and most of these incompassing Hills may be covered with so thin a vegetable Coat, as we may observe the Hills with us to be, such as the short Sheep pasture which covers the Hills of *Salisbury* Plains.

Might we go there to find out for ourselves? This seems incredible, said Wilkins, but then so would a voyage across the wide seas have been deemed impossible to our ancestors. Perhaps the journey might be made with great wings, as Daedalus did – for it was generally believed that the space between the Earth and the moon was filled with air, even if Wilkins acknowledges that the experience of mountaineers shows it to get thinner and colder with altitude. There is, however, a better way. 'I do seriously, and upon good grounds, affirm it possible to make a Flying Chariot', says Wilkins, 'in which a Man may fit, and give such a motion unto it, as shall convey him through the Air.' Maybe it will even carry several men, and their provisions too. Referring to the flying automata attributed to great inventors of the past, he says 'This Engine may be contrived from the same Principles by which *Architas* made a wooden Dove, and *Regiomontanus* a wooden Eagle.'

'Notwithstanding all these seeming impossibilities', he concluded, ''tis likely enough, that there may be a means invented of Journeying to the Moon.' It will be worth the trip, he assured the reader, for one need only 'consider the pleasure and profit of those later Discoveries in *America*, and we must needs conclude this to be inconceivably beyond it'. In other words, the discovery of the New World directly inspired a belief that there were yet more new lands to be found, even if they were not of this planet. Scoffers should remember 'how did the incredulous World gaze at *Columbus* when he promised to discover another part of the Earth'.

Fly me to the moon

By conferring respectability on the idea that the moon is inhabitable and that we might travel there, Kepler and Wilkins stimulated an entire genre of proto-science fiction in the seventeenth century. The first such tale actually predated their visions of space-faring. Although published in the same year as Wilkins' book, it was written around 1628 by the Bishop of Hereford, Francis Godwin, who died in 1633.*
It was called *The Man in the Moone*, and it translates the classical trope of a fantastical voyage into the new context of interplanetary adventure, providing the template for lunar fantasies such as those of Jules Verne, H. G. Wells and Georges Méliès in his iconic 1902 movie *Le voyage dans la lune*.

Godwin's hero, the diminutive Spaniard Domingo Gonsales, is marooned on an island populated by giant geese (*gansas*), which migrate each year to the moon. Gonsales harnesses a carriage to

---

* It has been suggested that Kepler's *Somnium* was the model for Godwin's book – it was, as we've seen, already circulating by the 1610s. But this idea has been challenged by literary scholar William Poole, who considers Robert Burton's *Anatomy of Melancholy* a more likely source of Godwin's astronomical details.

these birds so that they might carry him there. While this sounds like a frivolous device, Godwin seems to have meant it somewhat seriously: he suggests that there might indeed be a constant passage of flying creatures between the Earth and the moon, including locusts, swallows and cuckoos. When Wilkins came across Godwin's book while preparing his own, he was content to take the proposal at face value, saying 'by this means, 'tis easily conceivable, how once a year a man might finish such a Voyage; going along with these Birds at the beginning of Winter, and again returning with them at the Spring'.

Gonsales finds that the moon is a great ocean scattered with islands: the light and dark regions, respectively, that we see from Earth. On the islands dwells a race of lunarians, whose skin is a colour never seen before on our planet. These beings grow to immense stature – a notion mentioned in Kepler's 'official' response to Galileo, *Dissertatio cum nuncio sidereo* (1610), where he states that:

> It surely stands to reason that the inhabitants express the character of their dwelling place, which has much bigger mountains and valleys than our earth has. Consequently, being endowed with very massive bodies, they also construct gigantic projects.

The lunarians live in something like a state of grace, because they banish any of their number 'who are likely to bee of a wicked or imperfect disposition' to the Earth, where the exiles acquire a human-like skin colour by exposure to earthly air. Godwin alludes to the possibility that two 'green children' recorded by the twelfth-century chronicler William of Newburgh who 'fell from the sky' in the village of Woolpit in Sussex may have been such unfortunate outcasts. These children allegedly claimed that they came from the land of 'Martin', a dimly lit place where the sun never rises.*

This is apparently the reason why Godwin's lunar-dwellers worship a god called 'Martinus', an idea that the Catholic Gonsales interprets with some alarm as a reference to Martin Luther.

---

* John Aubrey later interpreted the legend of the Green Children as an account of beings who inhabit a subterranean world inside our hollow Earth, an idea suggested by his friend Edmond Halley.

To call Godwin's book the first science-fiction story is of course little more than an amusing conceit – the genre is a modern one. It is certainly a book written to entertain and beguile, lacking the satirical elements that characterize later fantastic journeys, and it is modelled partly on the picaresque yarn exemplified by *Don Quixote*, which accounts for the Spanish setting. But the book also takes its lead from the travel writing of the age of exploration: from descriptions of great and daring navigations of the world, such as the English adventurer Richard Hakluyt's *Principal Navigations* (1598–1600). Here again, then, this voyage to the moon is considered analogous to other passages to distant lands. The fact that Gonsales continues his adventures on his return to Earth by landing in China implies such continuity, as though the New Worlds of other celestial bodies are after all just other countries with strange inhabitants and customs.

Godwin's story was the model for *A Voyage to Cacklogallinia*, a satirical work of 1727 by the pseudonymous Captain Samuel Brunt. Some have suggested that the true author might be none other than Jonathan Swift, since there are similarities with *Gulliver's Travels* (1726); others have proposed Daniel Defoe. Like Godwin's Gonsales, Brunt travels to exotic foreign parts (the West Indies) before being shipwrecked in a land populated by enormous talking bird-beasts called Cacklogallinians, who fly him to the moon. There they find a race of Selenites, made of stuff somewhat between ordinary tangible matter and an 'aerial Substance', who are in fact the souls of people sleeping on Earth. Once more, Brunt's journey is really little more than an elaborate, imaginative voyage of discovery in which the world discloses its capacity for unexpected new races, phenomena and societies. What the narrator says of his lunar sojourn could as easily have been voiced by any natural historian sailing to new vistas in the Pacific or the East Indies:

> I was allow'd a Week to satisfy my Curiosity, and make my Observations on all the strange things which were there to be seen, which I may justly reckon the most agreeable Part of my whole Life.

The lunar empire

Another inspiration for 'Captain Brunt' was probably the best known of all moon-voyage stories of the seventeenth century, written about twenty years after Godwin's by an author who himself became more fictional than historical. There was probably no Roxane, and surely no immense nose, associated with the real Savinien Cyrano de Bergerac, a French poet and former soldier who enjoyed brief fame after the posthumous publication in 1657 of his books *The States and Empires of the Moon* and *The States and Empires of the Sun*. Those attributes belong instead to the character created by the French poet and playwright Edmond Rostand for his 1897 play that made Cyrano a sentimental hero based only in the sketchiest way on his historical counterpart, about whom rather little is known. Yet it was Rostand's play that revitalized interest in the real Cyrano's curious tales, which were republished in the 1950s in French and English and again claimed as a pioneering example of 'early science fiction'. A further irony is that, unlike Rostand's Cyrano, the seventeenth-century writer had nothing much to do with the town of Bergerac in Gascony: Savinien Cyrano, the son of a lawyer, was born in Paris in 1619 and adopted his noble-sounding name only because his family owned a small estate in Bergerac. Savinien received the typical education of a well-to-do young man, learning the standard humanist syllabus at the Collège de Beauvais, before enlisting as a soldier. He gambled, fought duels, and, weakened by battle wounds, eventually retired from military service aged just twenty or so. In 1641 he returned to studies at the Collège de Lisieux, where he heard Pierre Gassendi lecture. Apparently an intelligent but dissolute dilettante, he began to write essays and letters before embarking on his two great tales of space travel. He never finished the second book, but when the first of them was published in French as *Voyage dans la lune* two years after Cyrano's early death in 1655 by his lifelong friend Henri Le Bret, it was successful enough to justify the subsequent publication of the incomplete sequel *L'histoire des états et empires du soleil*. An English translation was read by Jonathan Swift, and it helped to shape the exploits of Gulliver.

The books are vehicles both to satirize and to debate issues concerned with social customs and natural philosophy. As such, they mark a break with the earlier tradition, exemplified by More's *Utopia*, Campanella's *City of the Sun* and Bacon's *New Atlantis*, of using fictional journeys to present visions of an ideal society. Rather, the lunar civilization allows Cyrano to poke fun at our own by exaggerating our traits or inverting our customs and beliefs to reveal how arbitrary they are. The vision of a perfected society no longer seems tenable: our rules and norms are now made to seem whimsical or foolish, just random choices from an infinite range of options. This relativistic viewpoint was surely prompted by the expanded horizons of both geographical and scientific discovery, by the realization that what one finds in the everyday life of a European is but a fraction of all the things and ways that exist. In Cyrano's new worlds, anything goes, even to the point of blasphemy and heresy.

At the same time, *Voyage dans la lune* offers a glimpse of what message natural philosophy delivered to a bright, educated young man of the mid-seventeenth century. It is a sometimes perplexing mix of the new and the old, of the seemingly occult and the 'modern', the enchanted and the rational. Although the astronomy of Newton and Huygens still lay ahead, Cyrano was familiar with the mechanistic philosophy of Descartes and Gassendi and the experimental bias of Bacon (he was rumoured to have dabbled himself with making ingenious machines), and it seems clear that for him there is continuity of thought between professors of secrets such as Cardano and Campanella and the Cartesian rationalists. His hero, Dyrcona (a near-anagram of the author's name), imagines journeying to the moon after reading Cardano's *De subtilitate*, in which Cardano tells of having been visited by two old men who say they are inhabitants of the moon and then vanish. Mocked by his friends for venturing the idea that 'the moon is a world like ours, which our world serves as a moon', Dyrcona returns to his lodgings to find Cardano's book open on his desk at the corresponding passage. He interprets this as a portent, later concluding that the book was left there by one of those same lunar gentlemen, and he resolves to make a device to travel there. Mindful of the Aristotelian idea that dew evaporates because the sun's heat attracts the vapour, he fastens many bottles of dew to his craft standing in the sunshine, whereupon it begins to rise into the air. But seeing that he is heading off course, he breaks some bottles and falls back to earth. There he discovers that the globe has rotated beneath him and that he is now not in France but in 'New France', or French Canada.

Taken to the viceroy, he explains how he got there, only to have his Copernican viewpoint challenged. Rather than the Earth having moved while he was airborne, the viceroy says, one could equally well assume that he had been dragged over a motionless Earth by a moving sun. And so the two of them set to a friendly debate over helio- versus geocentrism. Revealingly, Dyrcona first bases his argument not on mathematical simplicity but on metaphysical teleology. The sun, being the source of all warmth in the universe, should logically sit at its centre:

> For it would be as ridiculous to believe that this great luminous body revolved round a speck, which is useless to it, as to imagine when we see a roast lark, that the hearth has been revolved about it in order to cook it.

These arguments for what seems a 'proper' arrangement of celestial bodies, invoking hierarchical notions of purpose and design, were also voiced by Copernicus, and it seems that they were at least as persuasive to Cyrano as any argument from parsimony. But the viceroy's response is equally surprising, for he too is already prepared to believe that the Earth moves – but not for scientific reasons:

> I am convinced that the earth revolves, not for the reasons alleged by Copernicus, but because the fires of hell, as the Holy Scriptures tell us, are enclosed in the centre of the earth and the damned, seeking to escape the burning flames, clamber up the vault so as to avoid them and thus cause the earth to revolve, as a dog turns a wheel when it runs enclosed within it.

This point of view is not incompatible with geocentrism, but it offers, even if half-humorously, a very different way of staging the debate than the traditional one of science-driven Copernicanism against the theological Ptolemaic system. Notice too how Cyrano felt the need to find driving forces for the Earth's motion – Dyrcona suggests that it might be caused by 'the rays and influences of the sun', or perhaps from preferential evaporation of vapours from the sunny side of the globe. In any event, the viceroy is won over by Dyrcona's reasoning, admitting that he is now surprised at how Ptolemy's cosmology held sway for so long. That, Dyrcona explains, is because 'the majority of men, who only judge things by their senses, have allowed themselves to be persuaded by their eyes' – a Baconian injunction to abandon the Aristotelian faith in our raw senses.

If Dyrcona's dew-driven ascent seems archaic, that is nothing compared to the way he eventually reaches the moon, in comparison to which Domingo Gonsales's bird-craft seems positively modern. Dyrcona's fall to Canada has left him covered in bruises, and to heal these he rubs himself with beef marrow. He then goes to see what has become of his spaceship, and discovers that a group of soldiers has decided to relaunch it with gunpowder. He leaps on board just as the fuses are lit, and up it goes, a late-Renaissance *Apollo* – but only until the powder has all burnt, whereupon it starts to fall back to the ground. Yet Dyrcona himself finds he is still rising, free from the craft, and he realizes that the moon is sucking him upwards because of the bone marrow rubbed on his body: it was widely believed that the waning moon sucked the marrow out of animals because of some occult attraction.

This seems to be literally a lunatic flight of fancy. Yet, while Cyrano did not necessarily consider it a genuinely efficacious way to rise into the heavens, neither is it obvious that he is merely joking. Rather, like the authors of science-fiction narratives today, he wants only to clear a certain threshold of plausibility so that the reader will not just throw the book down in disgust. Evidently, a reader comfortable with Copernicus and Descartes could also be expected to tolerate the idea that bone marrow is attracted to the moon.

Cyrano also shows us how gravity was popularly conceptualized. He accepts William Gilbert's idea that it is a kind of magnetic force which gets weaker with distance* – at a certain point in his ascent, the moon's gravity takes over from that of the Earth. But although Cyrano recognizes that the moon, being smaller than the Earth, will have 'weaker' gravity, this is interpreted to mean that its force is the same at the lunar surface but extends less far. Arriving at the moon,

---

* In his *Somnium* Kepler assumed a gravitational force whose strength is proportional to the inverse of distance, in contrast to the actual inverse-square relationship.

Dyrcona first encounters a human person who conveniently speaks French, and who introduces himself as the biblical Elijah. His lunar voyage, he explains, was made in an iron chariot, which was pulled into the air by a great magnet hurled in advance. Each time he caught up with the magnet, Elijah would throw it still further out into space and be drawn higher – in effect, a scheme for perpetual motion. Elijah says that the moon was the location of the biblical paradise, and that Adam and Eve had travelled from there to Earth after being banished by God, to escape his wrath.

The true moon-dwellers are gigantic 'man-animals' who walk on all fours, and they consider Dyrcona to be some kind of curious rare beast. He is trained to perform tricks to entertain the crowds, but while doing so he encounters a tall, grey-haired old man who intro-duces himself as one of Cardano's mysterious visitors. He explains that he has also appeared to Agrippa, the fifteenth-century magician Trithemius of Sponheim, Faust, Nostradamus, the Rosicrucians, Campanella and Gassendi – a roll call that reveals the intellectual heritage of Cyrano's tale. This person ultimately identifies himself as the 'demon' with whom Socrates allegedly conversed – a spirit being who can transmigrate between bodies. He explains that his race knows more than humans because they have more senses, for he attests that our sensory limitations prevent us from accessing some forms of knowledge that they take for granted:

> there are perhaps a million things in the universe which you would need a million quite different organs to know. Myself, for example, I know from my senses what attracts the loadstone to the pole, how the tides pull the sea, and what becomes of an animal after its death. Except by an act of faith, you men could no more attain to such lofty conceptions – because you lack the senses proportionate to these marvels – than a blind man could imagine what makes up the beauty of a landscape, the colouring of a picture, or the nuances of the rainbow.

Dyrcona is brought by Socrates' demon – who has now taken on the body of a lunar man-animal – to the king and queen of the moon, where he is introduced to another 'little animal' like him. In a twist that would strike us today as postmodern, this man turns out to be none other than Godwin's Domingo Gonsales – although he is never explicitly named, he is said to be a little Spaniard who travelled to the moon on a device borne by birds. Dyrcona is thought to be the female of this species, so he and Gonsales are put together in the hope that they will mate.

Instead they fall quickly into philosophical discussion. The Spaniard is a free-thinker who left the Earth in despair at being unable to find 'a single country where even the imagination was at liberty'. He was brought before the Inquisition for harbouring the anti-Aristotelian notions that a vacuum may exist in nature and that no substance is heavier than any other – a somewhat confused idea seemingly alluding to Galileo's proposition that all objects fall at the same rate.

Despite his earlier Copernican views, Dyrcona is now portrayed as an Aristotelian traditionalist, apparently because Cyrano needed someone to take that role in order to lampoon it. The lunarians begin to debate among themselves whether these little animals have reason, and having learnt their language (which is evidently not proof enough) Dyrcona can argue the case with them. His views, however, do not impress the judges. 'This Aristotle', they say, 'whose science you make so much of, doubtless made his principles fit his philosophy, instead of fitting his philosophy to basic principles.'

Here Cyrano forsakes any coherence in his characters' positions in order to poke fun at the scholastics' dogmatic rejection of experience in favour of first principles. For, disregarding all that went before, Dyrcona now maintains that on Earth he had begun to suspect the moon was nothing but the classical smooth sphere of Aristotelian cosmology.

> 'But', they all said to me, 'you can see earth here, rivers and seas. What could they all be?'
>
> 'Never mind!' I retorted. 'Aristotle assures us that it is only the moon, and if you had said the opposite in the classes where I did my studies, you would have been hissed.' At this there was a great outburst of laughter.

Yet the lunarians are no better. They consider it atheistic heresy for Dyrcona to pronounce *his* world to be an inhabited place rather than a featureless moon, and so they decide to drown him. In a parody of

Galileo's trial, he is granted his freedom by the king so long as he publicly recants his claim – which he does, saying 'Such is what the priests deem it good for you to believe!'

In these ways *Voyage dans la lune* ventures beyond gentle teasing of philosophical conservatives and enters risky territory. The French aristocracy were unlikely to have appreciated the ribald humour of the lunarians' mark of nobility, which is to wear bronze phalluses. When Dyrcona explains that terrestrial gentlemen carry swords instead, the lunarians are astonished: 'Unhappy country where the symbols of generation are ignominious and those of destruction are honourable!' Far bolder, though, is the attack on Christian belief that unfolds when Dyrcona, debating with Socrates' demon, attempts to defend the idea of the immortality of souls with reference to the Resurrection. 'Who has been putting you to sleep with that fairy story?' the demon demands. Dyrcona is outraged by his atheism, but his interlocutor counters with devastating logic that inverts Pascal's famous wager:

> for if there is [no God], you and I are evenly matched; but if on the contrary there is one, it will be impossible for me to have offended something which I did not believe existed at all, since in order to sin, one must either know it or wish it . . .* How much less should God, in his omnipotence, be angry with us for not acknowledging Him, since it is He who has denied us the means to know Him . . . if a belief in God were so necessary to us, in short, if eternity were at stake, would not God himself have flooded this truth with light, making it as clear to us as the sun, which hides from no one? For to pretend that he wanted to play hide-and-seek with men, as children do, saying 'Peep-bo, here I am!', now masking and now unmasking Himself, disguising Himself from some only to manifest Himself to others, is to create for oneself a god who is either stupid or malicious.

Cyrano knows that such views are too dangerous openly to support, and so he takes care to distance himself from them. Dyrcona calls them 'ridiculous and diabolical opinions', and begins to fear that

---

* Admittedly, the theology here is not so sound, even if common sense commends this view.

his demonic companion may be the Antichrist himself. As though on cue, the two of them are suddenly seized by a great black, hairy demon and taken back to Earth, although Dyrcona suspects his companion is about to be carried further into the fiery hell at the planet's centre. But these pieties fool no one – Cyrano would surely not have put such force into the atheistic argument, only to leave it unaddressed with a hasty drawing of the stage curtains, unless he wanted it to resonate with the reader. His friend Henri Le Bret clearly recognized that, for he censored both the bawdy and the blasphemous passages of the book – the full version did not appear until 1920.

The States and Empires of the Sun was no afterthought, but is prefigured in the moon trip itself, where it is mentioned as a book given (anachronistically, as it turns out) to Dyrcona by one of the young philosophers of the demon's dinner party. Rather marvellously, it is here not actually a book as such, but a kind of audiobook – a machine made of diamond and pearl that contains sound recordings which are played by moving a needle to select the chapter. It was by means of such portable and easily understood devices that the lunarians could acquire great learning at an early age, so that they 'possessed more understanding at sixteen or eighteen than the greybeards do in ours' – a Baconian aid to education that recalls the painted walls of Campanella's City of the Sun.

This self-referential construction continues in the opening of the second book, which relates that Dyrcona has published *The States and Empires of the Moon* to great acclaim and fame – printmakers sell his portraits – but also notoriety. Some are convinced he is a sorcerer, and he is seized by a mob who, discovering Descartes' *Physics* in his saddlebag, believe it to be a work of necromancy, its geometrical drawings of celestial mechanics being interpreted as the pictograms of demon-summoning.

After these escapades, Dyrcona devises another novel way to travel through space to the sun. Here Cyrano picks up again on the debates of the atomists about the existence of a vacuum, which Aristotle denied. His hero makes a box in which a passenger can stand, topped with a hollow crystal icosahedron on to which the sun's rays are focused with concave mirrors. The hot air inside the crystal rises out through a pipe in the top, and air rushes in from beneath to compensate. This sucking in of air pushes the machine upwards. (Never mind that the principle would, if anything, generate a *downward* thrust from the jet of hot air.)

Cyrano was fully aware that the sun is a source of tremendous heat, and so another vague philosophical sleight of hand is needed to sustain the tale. He explains that what actually burns in the sun 'is not fire, but the matter to which it is attached, and the fire of the sun cannot be combined with any other matter'. His journey to the sun supplies a convenient confirmation of Copernicanism: he sees that the Earth really does revolve around it, and that the planets are not stars but shine only with reflected light.

Dyrcona discovers that the sun is divided into many kingdoms, republics and states, each populated by classes of natural objects, such as birds, plants and stones. The birds come from the Earth, for they have a natural yearning to fly sunward. But the main inhabitants of the sun are philosophers, all of them the souls of men who have just died on Earth. All dead souls from Earth come to the sun, but most of them merge with its mass; only the souls of philosophers remain discrete, for they are made of particles that cannot be decomposed. Dyrcona learns all this from an old man he meets on a great solar plain, who turns out to be none other than Tommaso Campanella. He explains that he is hurrying to meet the soul of Descartes, who has just arrived (he died in 1650), and that he holds Descartes' *Physics* in the highest esteem. As they see each other and embrace (uniting magic and mechanism, you might say), the incomplete narrative cuts off.

Space ecstasy

It is not easy to decide what Cyrano was hoping to achieve with his wildly inventive books. Although progressive in spirit, he belonged to no intellectual camp, but seemed eager to embrace a diversity of views about the world. The question remains whether the books of Godwin and Cyrano were intended as mere entertainment – as exotic travel stories, the latter's with a satirical edge – or whether the reader was supposed to regard them as imaginative renditions of a serious idea. Here we have to acknowledge that the concept of an inhabited moon, and perhaps of other planets too, was at the same time both more and less remarkable than it seems today. In the cosmos of Aristotle and Ptolemy, the *purpose* of the celestial spheres was to be sought from a terrestrial perspective: even if the heavens held a perfection denied to the earthly sphere, they were arrayed around us at the centre, and their motions related to events on Earth. This solipsistic view was undermined both by Copernicanism and, in a more parochial sense, by the Renaissance voyages of discovery, both of which forced the possibility that there are other points of view than our own and that in some sense the affairs of the world as we know it are trivial. Indeed, the more other worlds seemed (like Galileo's moon) to resemble ours, the more extraterrestrial life seemed to be inevitable, for why otherwise would God have gone to the trouble of fashioning these bodies with such an abundance of geographical detail? It was common by the mid-seventeenth century to regard the moon as having an atmosphere, water and continents; the rest followed. In this sense, there was nothing much more remarkable about a fiction that imagined a civilization on the moon as one that imagined such a thing on a remote Pacific island.

Any unease, then, was not so much about the scientific plausibility of this idea as about the theological implications, which had caused trouble for Giordano Bruno. Were other worlds Christian, or had they at least experienced their own versions of Christ's promise of salvation?

That question was avoided in Athanasius Kircher's *Itinerarium exstaticum* (*The Ecstatic Journey*) (1656) by making the cosmos lifeless. Kircher recounts how, after listening to a lute concert, he was literally transported: taken by an angel into the heavens, where he could examine the other planets and stars in detail. All other worlds, he insisted, are barren, for in his Ptolemaic universe they are created merely as astrological indicators of events on Earth. Yet they are not the pristine globes of Aristotle. Kircher explains that some of our neighbouring planets are dour, inhospitable places: Mars is a blackened, infernal place of flames and smoke, while Saturn is dark and gloomy. But Venus and Jupiter have on them sweet waters and shining lands of crystal or silver, and they pour down a benign influence on humankind. (Might Christians be baptised in the Venusian waters? he wonders.) For Kircher, the stars are other suns with their own (Copernican!) planetary systems, but equally devoid of life.

The *Ecstatic Journey* deploys the classical trope of the 'dream journey', like that in Cicero's *Dream of Scipio* and Kepler's *Somnium*. Whether Kircher genuinely believed he experienced anything of the sort is unclear; the device is just a vehicle for his revelatory cosmographic speculations on inaccessible realms of experience. Yet he apparently did think that the 'spheres of the heavens' could be reached from his Ptolemaic Earth. In *Turris Babel* (*The Tower of Babel*) (1679) he examined with a literalism somehow both comic and admirable the plausibility of the legend that the tower was built by the ancient king Nimrod, great-grandson of Noah, to storm heaven and revenge himself on God for destroying his forefathers in the Flood. How massive would the tower have to be, Kircher asked, to reach even the lowest sphere, that of the moon, considered then to be around 200,000 miles away (about half the actual value; Kircher gives the distance as precisely 178,682 miles)? He concluded that it would have weighed 3 million tons, and that such a massive object extending into space would have shifted the Earth's centre of gravity and pulled it from its place at the centre of the universe. That mixture of practical mechanics and delirious biblical realism is entirely characteristic of Kircher.

Conversations with a marquise

While Cyrano's journey is an outlandish romp and Kircher's is a mystic vision, the French writer Bernard le Bovier de Fontenelle gave late seventeenth-century readers a speculative account of other worlds that was both scientifically accurate (as far as current knowledge permitted) and supremely witty and urbane. This rare combination reflects Fontenelle's own personality. He was the sort of man whose charm and elegance permitted him to say what he pleased without fear of censure, but whose sharp mind ensured that there was substance beneath the confection. He prompted Montesquieu to remark that one can say many serious and important things while joking.

Fontenelle's *Conversations on the Plurality of Worlds*, first published in 1686, perfectly illustrated his technique. There was plenty here to unsettle the theologian. In a series of conversations over five evenings* between the book's narrator, a philosopher, and a beautiful marquise at whose residence he is staying, the book expounds on the Copernican model in a manner that brings out the shocking implications of our humble, virtually accidental place in the cosmos. Not only might the Earth have ended up by chance in some different part of the solar system – as a moon of Jupiter, say – but Fontenelle dissolves the distinction between the realm of the planets and the sphere of the stars. He repeats the assertion that every star is a sun, each with its own suite of planets, and all of them populated like ours. As John Donne appreciated, this is a bewildering, even frightening idea when you take it seriously. The marquise finds it so when she exclaims that:

> here's a universe so large that I'm lost, I no longer know where I am, I'm nothing . . . This confounds me – troubles me – terrifies me . . . you'd scarcely know how to pick us out in the middle of so many worlds.

Although Fontenelle was aware of the theological complications of other inhabited worlds, one can't say that he affords them much attention, taking only the precaution of presenting the philosopher's assertions as mere speculations to which he (and by implication, the author) holds no binding commitment. His reassurance that no extraterrestrials are human scarcely settles the matter of whether nonetheless they have souls needing salvation. At any event, when *Conversations* was placed on the Index the year after its first publication, Fontenelle displayed little concern; he continued throughout the following decades to bring out new editions to keep abreast of astronomical (and political) developments. During that time his reputation and influence flourished. Adept at cultivating relationships with the scientists of his day, he became secretary of the French Academy of Sciences in 1697, lived to the extraordinary age of ninety-nine, and brought out the last edition of his book in 1742, fifteen years before his death.

The dialogue format for philosophical discourse has of course ancient roots, but Fontenelle gave it fresh expression. The exchange

---

* In later editions Fontenelle added a sixth evening.

between the philosopher and the marquise is light and informal, even novelistic – the philosopher flirts with his gracious host, who tells him with playful firmness to stop such silliness and get on with the science. She is alternately entranced, delighted, horrified and frustrated with what she hears, and while she lacks any training in astronomy or natural philosophy, she is evidently the intellectual equal of the narrator, assimilating in an instant all he tells her. In this respect, Fontenelle does more than provide an educated audience with one of the best popular expositions on science ever written; he challenges the prevailing view that such things are beyond the dainty minds of women, and offers a model of the 'scientific woman' that was to become increasingly apparent in the eighteenth century, as women began both to organize philosophical salons and to participate directly in scientific research. It was for these budding female virtuosi that instrument-makers began to prepare exquisite microscopes (more convenient and cheaper than telescopes), and it was they who the playwright Thomas Wright had in mind when he adapted Molière's *Les femmes savantes* into *The Female Virtuosos* in 1693, graced with incidental music by Purcell.

It would be going too far to make Fontenelle a pioneering feminist – his protagonists observe the prevailing courtly conventions of amorous teasing – but all the same *Conversations* offers a refreshing contrast to the suffocating misogyny and patronization that characterizes most of seventeenth-century culture, whether scientific or not. 'More than the various tracts on the education of girls, Fontenelle's Marquise did much to justify equal instruction for both sexes', says historian Nina Rattner Gelbart. It is surely no coincidence that one of the earliest English translations of the book (published with the travelogue title of *A Discovery of New Worlds*) was made by the female playwright Aphra Behn in 1688 – although Behn was somewhat wary of the content, admitting of Fontenelle that 'one would almost take him to be a Pagan'. There is a glorious indication of changing attitudes in *The Basset Table* (1705) by Susanna Centlivre, acknowledged as the 'second woman of the English stage' after Behn. The play's heroine is a 'Philosophical Girl' named Valeria who, instead of cooing over a dove, dissects it to see if it has a gall bladder. Implored by her lover to elope with him, Valeria exclaims with horror, 'What! and leave my Microscope?'

Fontenelle's philosopher makes perhaps the most accessible and best argued case hitherto for the Copernican cosmos, explaining how the complications of the Ptolemaic system vanish once we are prepared to take the imaginative leap of regarding the motions of the planets from a position different to our own. The marquise is amused, but also persuaded, by this plea to mathematical economy: 'It would seem that your philosophy is a kind of auction, where those who offer to do these things at the least expense triumph over the others.' Better still is the manner in which Fontenelle presents the mechanical philosophy (he considers the circulation of the planets to be explained by Descartes' vortices). Coming from Descartes or Newton, the mechanical explanation of phenomena must seem to the novice to present a cold, bleak vision of the universe (it still does). But the philosopher confesses to it with what seems like almost a regretful apology, making it seem an unfortunate necessity that does nothing to diminish the humour and humanity of his advocacy:

> 'In that case', said the Marquise, 'nature has become very mechanical.'
> 'So mechanical', I replied, 'that I fear we'll soon grow ashamed of it. They want the world to be merely, on a large scale, what a watch is on a small scale, so that everything goes by regular movements based on the organization of its parts.'

This demonstration that it is possible to sugar the pill without compromising the integrity of the message is one that science communicators can still usefully heed.

What is more, the philosopher is non-dogmatic to a degree that sometimes exasperates the marquise. No sooner has she acceded to his arguments than he revises them with an alternative point of view. The moon, he suggests, is a place of continents (the bright regions of the surface) and seas (the darker patches) – but then again, since there are no clouds, perhaps there can be no water after all: 'Those dark places that are taken for seas are perhaps only great cavities.' But maybe the different nature of the lunar air means that vapours can escape from the surface without forming clouds . . . Within his only half-humorous scheme that assigns attributes to planetary inhabitants according to the degree of solar heat they experience, the Venusians are highly vivacious and amorous, while the Mercurians must be boiled to a frenzy, making their planet 'the lunatic asylum of the Universe'. But again, maybe not. Perhaps the planet is composed entirely of saltpetre, whose vapours, says the philosopher, are very cold.

Perhaps what most distinguishes Fontenelle's vision of life on other worlds is his technological optimism about journeying there. He eschews dream voyages and the physical absurdities of Cyrano's spacecraft, saying instead simply that 'The art of flying has only just been born; it will be perfected, and some day we'll go to the Moon.' This, he says, is to expect nothing more than that the great voyages of discovery and conquest of the past two centuries will be continued: 'Did the great seas seem to the Americans any more likely to be crossed?' It is no wonder that so many Americans today still feel that human spaceflight embodies the spirit of Columbus; yet how apt that, even in the seventeenth century, interplanetary travel took as its model the colonial enterprise of trade, sovereignty and profit rather than the expression of pure curiosity.

Meeting the aliens

The first attempt to mount a rigorous scientific case for life on other worlds, without doing harm to Scripture, was made by Christiaan Huygens at the end of his life. He completed his manuscript just months before his death in 1695, and its publication was entrusted to his brother Constantijn, to whom the book was dedicated. Huygens' *Cosmotheoros* finally saw print in Latin and English in 1698, followed shortly thereafter by versions in Dutch, French and German. By this time the number of new worlds had multiplied still further: Huygens himself discovered the largest moon of Saturn, called Titan,* in 1655, and the Italian astronomer Giovanni Domenico Cassini, working at the Paris Observatory, found four more in the 1670s.

---

* Titan is now known to be one of the most extraordinary worlds in the solar system: no bare ball of rock, but a place wreathed in a thick, rich atmosphere of nitrogen and hydrocarbons, with rains, rivers and probably shallow oceans of liquid methane. It is thus a candidate for hosting extraterrestrial microbial life, although this would have to be like nothing on Earth. The probe lander spacecraft sent to this moon by NASA in 2005 was named after Huygens, and was carried there on an orbiting spacecraft named after Cassini.

As Huygens confessed, the possibility of life on other planets could scarcely fail to occur to any person who had followed the new astronomy:

> A Man that is of Copernicus's Opinion, that this Earth of ours is a Planet, carry'd round and enlighten'd by the Sun, like the rest of them, cannot but sometimes have a fancy, that it's not improbable that the rest of the Planets have their Dress and Furniture, nay and their Inhabitants too as well as this Earth of ours: Especially if he considers the later Discoveries made since Copernicus's time of the Attendents [moons] of Jupiter and Saturn, and the Champain and hilly Countrys in the Moon, which are an Argument of a relation and kin between our Earth and them, as well as a proof of the Truth of that System.

He recognized that many before him – he names Nicholas de Cusa, Bruno, Tycho, Kepler – had raised the idea. But he claimed that some of these had done so merely in casual speculation (he was particularly dismissive of Kircher's *Ecstatic Journey*, saying that it contains 'nothing but a company of idle unreasonable stuff'. He, in contrast, now proposed to examine the evidence for life on other worlds with philosophical scrupulousness.

Some, Huygens admitted, will mock the idea just as they mock all scientific speculation as 'the Dreams of a fanciful Head and a distemper'd Brain'. Others will denounce curiosity about such things as improper. But if we had always taken that narrow-minded view, says Huygens, we would never have discovered anything, but would remain at the mercy of a nature we did not understand:

> perhaps they'll say, it does not become us to be so curious and inquisitive in these things which the Supreme Creator seems to have kept for his own knowledge . . . [But] That vigorous Industry, and that piercing Wit were given Men to make advances in the search of Nature, and there's no reason to put any stop to such Enquiries.

As for consistency with Scripture, Huygens repeats the argument that the Bible is not meant to be an exhaustive account of the natural world: it does not mention the 'little Gentlemen round *Jupiter* and

Saturn', for example, the reality of which no one now denies. The more we know about the universe, the more 'we shall worship and reverence that God the Maker of all these things'.

Having established that it is not vain enquiry to ask about the habitability of other worlds, Huygens sets out the arguments in favour of this conclusion. These rest firmly with the Copernican principle: the Earth now being seen to be a planet like the others, it stands to reason that these worlds must be stocked with living creatures, the highest manifestations of God's wisdom and skill. Some rather circular reasoning – things on other worlds must be arranged like those on ours, because otherwise we'd be unusual* – leads Huygens to conclude that the inhabitants of these planets must also include human-like beings with the same kinds of bodies, senses, feelings, even societies, writing, science and music. The corollary, then, is a cosmos thronging with variety, a place of Americas multiplied beyond count, with no end to the study that might be made of them:

> What a wonderful and amazing Scheme have we here of the magnif-
> icent Vastness of the Universe! So many Suns, so many Earths, and
> every one of them stock'd with so many Herbs, Trees and Animals,
> and adorn'd with so many Seas and Mountains! And how must our
> wonder and admiration be encreased when we consider the prodigious
> distance and multitude of the Stars?

Huygens believed, however, that the moon is not a planet like ours but rather, a dry place lacking an atmosphere and probably devoid of life. Yet he concludes with words of astonishing prescience, in the light of current debates about lunar water and its implications for moon bases:

> If we could but once be sure that they had Water, we might come to
> an Agreement, and plant a Colony perhaps there; we might allow it
> then most of our other Privileges, and, with Xenophanes, furnish it
> with Inhabitants, Cities, and Mountains.

---

* The same reasoning still persists today in the search for extraterrestrial life, for example in the idea that life elsewhere must need water because there is no life on Earth without it.

This returns us to Galileo's vision of the moon as an arid, empty ball of dirt. Philosophy's harsh reasoning has squashed the fantasies that made the moon a land of new creatures and civilizations and adventures, stripping it of all wonder and life. Yes, the telescope revealed unanticipated diversity and richness in the macrocosm, but it showed also places where God seemed perplexingly to have wrought and laboured to no obvious end: loci where our curiosity drew a blank. As we shall see, the journey into the microcosm eventually led to the same apparent dead end, where understanding was confounded and purpose denied: into an incomprehensible void, untouched by enchantment.

9 Nature Free and Bound

Reason is not to be much trusted when she wanders far from
Experiments.

Robert Boyle, 'Of Naturall Philosophie' (early 1650s)

In my judgement the use of mechanical history is, of all others,
the most fundamental towards such a natural philosophy as shall
not vanish in the fume of subtile, sublime, or pleasing speculations.

Francis Bacon, *Advancement of Learning*

The frontispiece of Thomas Sprat's *History of the Royal Society* lays out
the society's wares with artful attention to detail. It was designed by
John Beale and John Evelyn* and engraved by the Bohemian artist
Wenceslaus Hollar, who had been plucked from the wreckage of the
Thirty Years War and brought to England by the Earl of Arundel. In
central position is a bust of Charles II, the 'author and patron' of the
society, and dutifully drawing attention to this fact is the first president,
Lord Brouncker. On the other side of the pedestal sits the man who
is hereby credited with the society's founding philosophy, Francis Bacon.
Arrayed around them in the background is the evidence of the society's commitment to experimentation, measurement and practical
application: there are clocks, dividers, some species of rotary instrument. Prominent among these ingenious devices is one that would

* This image was originally intended not for Sprat's book but for a similar eulogy
to the Royal Society that Beale was preparing. When this ran aground, Beale suggested
that the engraving, which had already been begun, might be used instead for Sprat's
work.

Frontispiece of Thomas Sprat's *History of the Royal Society* (1667). On the left is the first president, Lord Brouncker, and on the right is Francis Bacon. The bust is of the patron Charles II. The air-pump is just to the left of the king's head.

have baffled anyone but the cognoscenti. It looks like a globe resting on a stand, attached to a box that is equipped with a crank handle: an impressive machine with an opaque function. As such, it delivers a grave message: this new philosophy depends on technics that are far beyond both the means and the ken of the ordinary person.

The instrument belonged to Robert Boyle, who had recently donated its prototype to the Royal Society. It features here not just as a representative of the specialized equipment needed to pursue the experimental philosophy, but as the exemplar of that. It is an air pump, a mechanism designed to remove the air from inside the glass globe by means of a hand-operated piston pump. It was the most sophisticated item of apparatus known to seventeenth-century science – it has been called the cyclotron of its age – and the *History*'s frontispiece announces that the Royal Society was proud to be associated with it.

The air pump was something quite new in experimental science. For all their craftsmanship, telescopes and microscopes were in the end instruments that helped us to see more clearly what was already there. The elaborate flasks and retorts of the early chemist (the chymist) served the purpose of containing and directing matter, supplying a closed space in which transformations could occur without interference. But the air pump made possible operations that did not and (it was believed) could not happen in nature. It created an artificial environment: a space devoid of air.

Or did it? That was the crux of the matter. Could nature be truly transformed in this way, compelled to adopt such an unnatural state? And if it could, what, if anything, might we learn from it? It is hard to overstate the significance of this question for the experimental philosophy. It was one thing to open up one's eyes and mind to all the dazzling variety of nature, to refuse to deny any question that could be asked about it. But what if the questions did not stop there? What if one could create microcosms that went beyond and outside of nature? Where then are the limits to curiosity?

Comparing the air pump to the machines of modern particle physics has become a cliché in the history of science, but it is more than idle metaphor. The questions and conundrums raised by the air pump were remarkably close to those posed now by the Large Hadron Collider and its ilk. For one thing, only an expert would know both how to operate these instruments and how to interpret the observations they enable. Indeed, only an expert would have any inkling of what questions to ask them. Both devices aim to create anomalies in nature: little parts of the world where the 'normal' rules don't apply. It seems more than coincidence that, just as the air pump sought to eliminate air from its hermetic interior, so the tunnels of the LHC must be evacuated to a degree of emptiness exceeding that of interstellar space. We are accustomed now to the idea that there are places in the universe (in fact, almost all of it) that lack air and are all but free of matter, but the LHC will work only by making a space that is stripped barer than the cosmos.

Most problematically of all, how can we be sure that this synthetic microcosm is governed by rules that bear any relation to those of nature? In the teleological philosophy of the seventeenth century this issue was especially acute. Things happened in nature because God had ordained them, but things happened in the laboratory because the experimenter had done so. Why should one pertain to the other? Was the experimenter, then, a kind of microcosmic God?

Furthermore, when the creation of artificial conditions demands so much cost and expertise, what becomes of science's celebrated reliance on replication? The difficulty of verifying for oneself, with an independent experiment, Boyle's observations with the air pump was scarcely any less than the challenge of reproducing whatever the LHC will reveal. If *this* is where experimental curiosity takes us, can we be sure that what it offers is any less tenuous than the environment it purports to make? Can un-nature really teach us about nature?

Altered states

When Francis Bacon called for the preparation of exhaustive 'histories' of phenomena, he did not just mean we must tabulate everything we find in nature, including its 'errors and monsters'. We must also discover what nature does, he said, 'when by art and the hand of man she is forced out of her natural state, and squeezed and moulded'. This is a routine matter in science today: substances are cooled to absurdly low temperatures or heated beyond endurance, they are subjected to pressures equal to those at the Earth's core, or to magnetic fields strong enough to levitate a frog. Chemists strive to make molecules that are not only non-natural but that test the principles of chemical union to breaking point. Biologists make organisms from genetic combinations that evolution could never engender. The rationale is generally clear enough: extreme conditions and circumstances test our theories to their limits, reveal unexpected new laws and behaviours (such as those dictated by quantum theory), and might produce useful new materials and mechanisms.

There's nothing especially new in that – Palaeolithic humans discovered that natural materials can be usefully altered by heat to create new pigments for cave art, and the synthetic chemical technology of ancient Egypt was a sophisticated craft. But since antiquity, the natural and artificial had been generally considered to occupy separate realms: gold allegedly made by alchemy was thought to be distinct from (and generally inferior to) that which was dug from the ground. Francis Bacon began to challenge this preconception, criticizing 'the fashion to talk as if art were something different from nature, so that things artificial should be separated from things natural, as differing totally in kind'. He felt that, on the contrary, 'things artificial differ from things natural not in form or essence, but only in the efficient', which is to say that, even if they were made by different means, the result was the same. He implied, without quite stating it explicitly, that there was continuity between the world of nature and the ways in which it could be transformed by human agency.

That was by no means obvious, and indeed the prejudice that insists on a distinction is one we have yet to conquer in our cultural conviction (which any microbiologist will disavow) that 'natural' ingredients must be safer than synthetic. The implications were profound. At a time when nature was considered to act through something like an intentional, even anthropomorphic agency, there was no clear reason to suppose that what humans might achieve should be bound by the laws of nature or vice versa. It was a little like asserting that one person had all the capabilities and capacities of another. The assertion of universal laws that governed the operations of nature and humankind was to become a central ingredient of modern science; and while Bacon did not quite formulate or envisage the idea in that manner, his attempt to unite the natural and the artificial started the process.

Others saw no more reason to suppose that truths could be extracted from nature 'forced out of her natural state' and subjected to 'trials and vexations' than to think that a reliable testimony could be obtained from a person under torture. This was one of the main objections to the experimental philosophy raised by Thomas Hobbes, who considered it absurd to think that instruments for 'squeezing and moulding' nature could give us generalizable knowledge. At best, they could offer only observations fitted to the circumstances – and what was the point of that? As we will see, the supposed distorting effects of scientific apparatus supplied critics with one of their strongest reasons to doubt the value of the virtuoso's enthusiasm for experiment.

This was by no means Hobbes's sole reason to cast doubt on experimental study. For one thing, he demanded to know why there is not already variety enough in the circumstances of the world to reveal everything worth knowing: 'Are there not enough [experiments], do you not think, shown by the high heavens and the seas and the broad Earth?' Moreover, he could not understand why the experimental philosophers needed to conduct such an extensive and repetitive programme of research. If you really had to seek recourse in experiment, why wasn't just a single one sufficient to tell you what you needed to know? These objections set Hobbes at loggerheads with the Royal Society, and prevented him from being included in their Fellowship even though he was widely acknowledged to be one of the most eminent natural philosophers in England.* When he argued bitterly with Boyle about the status of instrumental experimentation and manipulation, it was an argument about how to do science, how to obtain knowledge, and indeed about how to be curious.

Falling mercury

Nothing better illustrates the debate about experimentation as a window on the laws of nature than the air pump. It was the flagship of the experimental philosophers, the cherished instrument of their most celebrated representative Robert Boyle. If the air pump could be shown to elucidate the divine mechanism of the world, the entire experimental programme would seem vindicated. Equally, if the air pump failed then all else was thrown into doubt. The globe standing in the shadows behind Brouncker and Bacon therefore endures a tremendous weight of responsibility, both a symbol and a measure of the credibility of the new science.

There was a lot at stake here in another sense too. The supposition that air could be wholly withdrawn from the glass vessel posed a fundamental challenge to scholastic philosophy, which insisted that a vacuum is not possible in nature. The Aristotelian dictum that 'nature abhors a vacuum' bore also on the hypothesis of atomism – for if all things are made of atoms, what can lie between these particles but a void? And if there is no void, how can motion be possible when there is no space for atoms to move into? Here, then, was an example of how curiosity about an apparently obtuse issue – the behaviour of air – opened up deep questions about how the world is constituted.

It is one thing to say that nature abhors a vacuum, another to suppose that she forbids it. This personification of nature seems to

---

* It must be said that there was rather more to his exclusion than that, for the society did manage to embrace a fair diversity of opinion. Hobbes was also considered, with justification, to be a difficult man to deal with: argumentative, dogmatic, and confrontational in a way that ran against the gentlemanly conduct that the Fellows professed to prefer. It's hard not to agree that the Royal Society was wise not to welcome this crabby man into their ranks.

suggest a preference rather than an absolute law: most men abhor filth, but filth remains. Experience with pumps for moving water, for example to prevent the flooding of mines, showed that nature does not lightly tolerate empty space: when air is removed, water rushes in at once to fill the gap, which is after all the very principle of a pump. But was it possible to prevent this, and thereby to force nature to accept a circumstance against her will?

It had been long known that a pump could not raise water above a certain height – typically around thirty-four feet, although that limit was a little lower if the experiment was conducted further above sea level. This presented a problem for mining engineers needing to extract water from deep shafts. The cumbersome solution was to raise up the water in stages of about thirty feet at a time, as illustrated in Georg Agricola's mining manual *De re metallica* (1556). Why this limit? Towards the end of his life, Galileo had been provoked to think about this problem by a letter he received in 1630 from Giovanni Battista Baliani, a Genoese mathematician, pointing out the thirty-four-foot limit of a pump or siphon. Galileo suspected that it stemmed from nature's alleged aversion to a vacuum. If the column of water was any higher, the water's own weight would cause it to break, opening up a void. In his *Discourses Concerning Two New Sciences* Galileo therefore attributed the height restriction on pump action to the 'resistance of the vacuum'.

Between 1640 and 1643 a Mantuan mathematician named Gasparo Berti, working in Rome, devised an experiment that forced the issue of whether a vacuum was involved. He erected a vertical lead pipe thirty-six feet high, capped on top with a glass dome. When the pipe was filled with water and then opened at the bottom, the water ran out so that the level dropped to thirty-four feet, partly emptying the dome. So what was in that space? Berti suspected it was a vacuum; others, including Athanasius Kircher, would not hear of it.

Three months before he died in 1641, Galileo was pleased to receive a visit at Arcetri from the Italian philosopher Evangelista Torricelli, with whom he had corresponded for some time. Torricelli had written with glowing praise for Galileo's *Discourses*, but nevertheless he questioned nature's supposed *horror vacui*. Torricelli suspected that water rises in a pump not to fulfil an Aristotelian duty to prevent a vacuum from forming by the expulsion of air but because it is actively pushed upwards by the pressure of the atmosphere – what he called an 'ocean of air'.

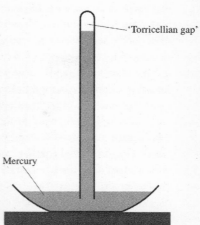

Gasparo Berti invented the water barometer in the early 1640s. It was an inconvenient device: a lead pipe 36 feet high, topped with a glass dome (left). Evangelista Torricelli realised that the same arrangement could be produced on a more manageable scale by using mercury to fill the inverted tube (right). But what was in the empty space at the top, left by the sinking column of mercury?

Torricelli believed that this proposal was vindicated by experiments he conducted in the early 1640s: a variant of Berti's scheme but rendered much less cumbersome by using a much heavier liquid – mercury – which reduced the limiting height of a column from thirty-four feet to less than three. He filled a long glass tube, sealed at the bottom, with mercury, and then inverted it within a bowl of mercury. As with water, the mercury in the tube falls but only for a small distance: its meniscus does not descend to the same level as that of the mercury in the bowl. Torricelli argued that this was because it is held up by the weight of the air pressing down on the mercury in the bowl. If this apparatus is taken up a mountain (not such an easy task with a thirty-six-foot lead pipe), the level of the mercury in the tube falls further.* This, said Torricelli, is because there is less weight

* Blaise Pascal, who made similar experiments and supported Torricelli's view of them, did not bother with so arduous a test, taking his primitive barometer instead to the top of Notre Dame Cathedral and noting a small difference in the fall of mercury. More convincingly, he asked his brother Florin Perier in Clermont to take

of air pressing down at a great height than there is at sea level: as we would now say, the air pressure decreases at high altitudes. In a vacuum, then, the mercury in the tube should fall all the way, equalizing with that in the bowl. With this experiment Torricelli had essentially invented the barometer.

The critical question was not so much why the mercury in the tube fell, but what was left behind in the small space – the 'Torricellian gap' – at the top. For those convinced of nature's abhorrence of a vacuum, the gap was merely air, which could be stretched like rubber. The extent of the fall was then set by the limit to which the air could expand – any further, and a vacuum would appear. The fall of mercury was then arrested not by a push from below but by a pull from above. But Torricelli imagined that the gap was quite void, its extent being determined by the point at which the remaining weight of the mercury column was balanced by the air pressure over the reservoir in the bowl. These were not the only options. René Descartes avoided the notion of a void in his corpuscular, mechanical model of motion and matter by postulating a subtle fluid substance that pervaded all of space. This, rather than air, was what filled the Torricellian gap, he said. Meanwhile, Descartes' rival, the French mathematician Gilles de Roberval, followed Torricelli in postulating a vacuum, but considered this to be actively preventing the further drop of the mercury: the height of the Torricellian gap was then a measure of the degree to which nature's abhorrence of a vacuum could be conquered.

This was the question that Robert Boyle set out to resolve with his air pump. Or rather, one should perhaps say that Boyle wanted to move beyond matters of interpretation, which he said with some justification were all conjectures that could not obviously be resolved by observation. What mattered most was to establish the empirical facts of the matter. In particular, how did Torricelli's apparatus behave when placed in the evacuated chamber of the air pump? One might wish to propose explanations for whatever behaviour emerged, Boyle conceded – but these were secondary and inevitably conjectural,

---

the apparatus up the nearby Puy de Dôme mountain, 4,800 feet above sea level, which Perier did in 1648. The height of mercury was twenty-six inches at the base of the mountain, twenty-three inches at the summit.

whereas the experimental facts, once reliably witnessed, would be incontrovertible and agreed by all.

It is on this apparently humble assertion that the history of curiosity, if not in fact the history of science, pivots. To see why, consider first of all the conventional modern scientific view of Boyle's position.

This maintains that Boyle is quite right: you can quibble all you like about theoretical interpretations, but 'facts is facts'.* An emphasis on empirical observation is the stance of the experimenter who, out of sheer curiosity, asks 'what happens if . . . ?' If we place the Torricellian tube in a chamber and lower the air pressure, what then? Let's say we find the mercury drops lower than it does in the open air. Well then, that is surely the end of the matter as far as experimental fact is concerned: we can write a paper saying 'I placed the device in the chamber, pumped out the air, and the drop increased by five centimetres.' It's an experiment that anyone can do (if I describe my procedure carefully), and can verify for themselves.

But saying 'I did this, and I saw this' is not exactly science. Until we have some inkling of what observations mean, do they not remain mere anecdote, the equivalent of the curios amassed by a collector who shows no interest in organizing them into any scheme or taxonomy? And yet this was indeed what Boyle and his like-minded virtuosi seemed to recommend: Sprat's 'heaping up of facts'. Ultimately, however, their ambition ran deeper, for they shared Bacon's view that it should be possible to boil down the facts into theories and explanations. Boyle did not proscribe interpretation, but he advocated the utmost caution and circumspection in making it. If the mercury drops further inside the air pump, we might reasonably be tempted to say that Roberval's view of the Torricellian gap is problematic (we are gentlemen and so will not simply say it is wrong), for how can the absence of air outside the

---

* Among other things, this naïve view does not acknowledge the shifting meaning of 'fact'. The word derives from the Latin *factum*, meaning deed: something that has been *done* by a person. As such it was originally a legal term, which gradually became applied to what nature 'did' too. In that respect, a 'fact' could be just a lone outcome or observation rather than referring to any generalizable behaviour. It was in this former sense that the French equivalent *fait* was typically used in the seventeenth century.

gap influence nature's degree of abhorrence for a vacuum?* That is fine, so long as we acknowledge that this interpretation is just a suggestion, apparently consistent with the experimental facts but not proved by them.

This is the classic view of how science proceeds. We have a set of alternative hypotheses for how nature works. We do an experiment, and record our observations, and then we ask how consistent they are with each hypothesis. Some of the options seem better to fit the observations than others. So we express our preferences accordingly, and think of another experiment that might help us refine our choices. Then we repeat ad nauseam. All the time, experience is made the arbiter: if a hypothesis doesn't fit the facts, so much the worse for the hypothesis.

I am not about to suggest that this conventional view of the 'scientific method' as a back-and-forth between experiment and theory, predicated on the falsifying authority of nature, is misguided. It is a pretty good description of how scientists mostly work, and how they arrive at an understanding that allows them to make remarkably good predictions about behaviours in new circumstances. But if this is indeed their 'method', then a closer look at what Boyle did to 'squeeze and mould' nature, and what he concluded from it, will reveal some facets that ultimately undermine any attempt so neatly to package up the way scientists exercise their curiosity. To wit:

—this 'scientific method' was not invented in the 'Scientific Revolution'.
—experiment isn't always the arbiter of scientific truth, for reasons both good and bad.
—experiments provide 'facts' only after passing through a cultural filter: they are shaped by the society in which they emerge.
—crucial, decisive experiments are very rare, at best.
—replication of scientific results is not the norm.

If these claims upset you, please be patient.

---

* We'd be wise to maintain this humble attitude, for couldn't Robervalists retort that the degree to which nature abhors a vacuum is proportional to the pressure of the air around it? Why shouldn't they? They could even claim that the result vindicates such a relationship, which they will call Roberval's Law. We would then need to think about how else that putative law might be tested.

Making the machine

There is an image in the 1657 book *Mechanica hydraulico-pneumatica* by Athanasius Kircher's pupil Caspar Schott of the predecessor to Boyle's air pump. This is the device invented by Schott's friend, the German natural philosopher Otto von Guericke, burgomaster of Magdeburg, who stands on the left demonstrating it to a group of nobles. But who is really doing the demonstrating? Guericke is nowhere near the instrument, and neither is any other person. It is apparently being operated by faithful putti, like so many celestial laboratory assistants. It is only rather recently that these cherubs have been given much credit for their labours – today we recognize them as the post-doctoral students and technicians who work day and night to get the equipment working and debugged, and who even these days may find themselves passed over when the Guerickes, supervising from a distance, put their names to the papers and win the awards.

Robert Boyle's assistants were similarly faceless, with the almost

Otto von Guericke's air pump, illustrated in *Mechanica hydraulico-pneumatica* (1657) by Caspar Schott.

sole exception of the man who helped to design and operate his air pump and was his intellectual if not his social equal: Robert Hooke. However, it would be unfair to imply that virtuosi of Boyle's standing were totally hands-off in their experimental work; Boyle, indeed, was quite capable of making himself ill in the zeal with which he pursued his endeavours. Moreover, the experimental philosophy was criticized by some opponents on the very grounds of its reliance on the skills of artisans. Displaying the time-honoured snobbery against manual work, Hobbes asserted that apothecaries, gardeners and other workmen cannot possibly be philosophers, while another critic, Thomas White, wrote that experiments 'belong to Artificiers and Handy-Craft-Men, not Philosophers, whose office 'tis to *make use of* Experiments for Science, not to make them'. All the same, Schott's illustration exemplifies the persistent image of the virtuoso as the effortless worker of wonders whose experiments manipulate divine forces.

Experiments as technically complex as those that Boyle conducted with the air pump would have been impossible without the assistance of practically minded men like Hooke. With some input from the London instrument-maker Ralph Greatorex, Hooke devised and constructed the instrument that Boyle used while living in Oxford in the late 1650s. These experiments were described in Boyle's *New Experiments Physico-Mechanical*. This book included a detailed diagram of the air pump, which offers a splendid example of what science historian Steven Shapin has called Boyle's 'literary technology'. The engraving is, first of all, rather beautiful: it is a carefully composed collage of images, rendered not schematically but in exquisite perspective and chiaroscuro to persuade the reader that this is a real machine. With its swelling tubes and sprouting stopcocks, it resembles nothing so much as a botanical illustration, a dazzling sight as well as a matter of scientific record. And yet – can you imagine trying to build such a device from these instructions? There's certainly a lot of information here, but it is inconceivable that it would be adequate in itself for a replication of the experiment, and indeed not a single air pump was constructed elsewhere in Europe by anyone who lacked prior first-hand experience of Boyle's. The diagram offers a pretence of total disclosure, but with the best will in the world it can be no more than that, because the technology is simply too complicated. It is primarily an exercise in winning trust.

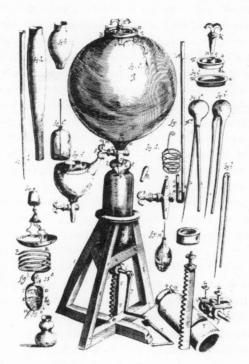

Robert Boyle's air pump, depicted with a visual rhetoric that is as much concerned with persuasion as with explanation.

Hooke's inspiration was the vacuum chamber (to name it anachronistically) made for Guericke: a copper globe that could be evacuated with a pump. In his initial experiments Guericke adapted the pump used by the Magdeburg fire brigade so that it could extract water from the filled chamber. Later he created a device that would pump air out directly. He noted that strong suction developed as the air was removed, and in 1564 he first performed his famous experiment of evacuating two metal hemispheres to which ropes were attached, demonstrating before the Reichstag and Emperor Ferdinand III in Regensburg that two teams of horses could not pull them apart. Popular contemporary accounts of this spectacle – an example of how scientists paraded their marvels before potential benefactors – often imply that Guericke thereby established at a stroke the existence of a vacuum. It should be clear now that the conclusion could be nothing like so clear-cut, for there were several other possible explanations for the cohesion of the spheres.

Boyle felt that Guericke's device could be improved. In particular, using a glass chamber, with an access port through which items could be inserted, would allow the effects of removing the air to be observed. His glass sphere had a cap on the top, sealed with an improvised paste. Guericke subsequently began also to use glass vessels, bell-shaped and obtained from pharmacists. Like Boyle, he showed that candles were extinguished and animals died when the vessel was evacuated. The dramatic depiction of such an experiment by Joseph Wright of Derby in 1768 captures the sense of drama that would have accompanied Boyle's demonstrations at the Royal Society and at his house, albeit with rather too much cosy domesticity to be taken at face value.

Despite Hooke's mechanical virtuosity, the air pump was hard to use. Evacuating the air was laborious – the valve had to be opened and closed on each stroke, and as the amount of air decreased, it was ever harder to force the rest out. The limitations of seventeenth-century materials left the experimenters fighting a constant battle against leakages. Boyle and Hooke totally redesigned the air pump around 1667, donating the old one to the Royal Society, which promptly lost it.

On air

Boyle's *New Experiments Physico-Mechanical* reports forty-three experiments touching on a wide variety of processes. Lacking any a priori reason to judge whether this or that phenomenon might or might not be altered in an evacuated space, Boyle, Hooke and their assistants seem to have looked at everything they could think of, often more or less at random. They tested (probably at the watchmaker Hooke's instigation) the swinging of pendulums, and whether a magnet could be felt by a compass needle when the air was withdrawn (it could). They looked at how sound propagated in the almost airless space: a ticking watch (surely one of Hooke's) went silent, though a struck bell could still be faintly heard, presumably because the sound was conducted through the thread on which the bell hung. They saw liquids bubble and seem to boil without heating as dissolved gas escaped into the partial vacuum. Boyle found that some substances that spontaneously emit light were dimmed without air (see page 336). In short, the evacuated chamber was a new kind of space in which, it seemed, anything might happen.

In cases where Boyle did express a prior hypothesis, it was as likely as not to be totally mistaken. For example, he felt that candles and coals might burn more easily in the absence of air, since there would then be more space into which the 'stifling steams' produced by combustion might expand. Absent any understanding of the role of oxygen in burning, this is not an unreasonable supposition, but of course he found quite the opposite to be the case. Several animals had a hard time of it in Boyle's laboratory: a lark in the chamber began to 'droop and appear sick, and very soon after was taken with as violent and irregular Convulsions as are wont to be observ'd in Poultry, when their heads are wrung off'. A mouse appeared to die, only to revive when the air was let back in. Boyle speculated that the same effects might be expected in humans at high altitudes. He wondered 'whether or no, if a Man were rais'd to the very top of the Atmosphere, he would be able to live many minutes, and would not quickly dye for the want of such Air as we are wont to breathe here below'. He mentioned the report of the Spanish explorer José de Acosta in his 1590 book *Historia Natural y Moral de las Indias*, who found difficulty breathing in the high Andes of Peru and concluded that humans need 'a more gross and temperate air'.

It seems intuitively obvious now that air and other gases should expand to fill the available space, but that is a quite different behaviour from what solids and liquids do, and was not easily understood by Boyle and his contemporaries. Boyle spoke about air as possessing 'spring', which he compared to the way a woollen fleece will shrink on compression and spring back when released. It suggests a continuous elastic medium, in contrast to Descartes' proposal that air is composed of particles in agitated motion. The latter is far closer to the current picture of gases as collections of molecules in motion, although Boyle characteristically did not try to press for any particular mechanistic picture to explain his observations. He did, however, realize that if air expands, its pressure is reduced. The quantitative relationship between these two properties – the pressure is inversely proportional to the volume – is Boyle's law, which does not appear in *New Experiments* but in Boyle's 1662 treatise *A Defence of the Doctrine Touching the Spring and Weight of the Air.*

The air pump belonged to Boyle, but the study of air did not. Hooke doubtless had his own views on the results they obtained together, and many of his experimental plans concerned the nature and behaviour of air. How did it hold up the clouds? How do light and sound pass through it? Why does it support life?* With the virtuoso's impetuous eagerness for first-hand experience, in 1671 Hooke devised and built an air-pump chamber big enough to hold a person, and climbed inside. Fortunately for him it was extremely leaky, allowing the removal of only about a quarter of the air inside – but that was enough for Hooke to report giddiness and pains in the ears.

By weighing air, Boyle calculated that to hold up the level of mercury inside the Torricelli barometer twenty-nine inches above that in the reservoir would require a column 35,000 feet high, if the air's density is uniform. But it was recognized that in fact the air gets less dense with height, on which basis Hooke estimated the total height of the atmosphere. He admitted that this gradation of density might mean that in fact there was no clearly defined edge of the atmosphere at all: the uppermost layer, being highly rarefied, might extend for hundreds of miles, if not perhaps indefinitely, so that it reached the stars and planets. But he suggested that the highest layer below this last one might be twenty-five miles up – a fairly good estimate, as it turns out, of the height of the uppermost stratosphere (about thirty-one miles). Never mind the numbers; the real point is that a relatively simple benchtop experiment was now claiming to be able to say something about the structure of the cosmos.

And what, in the end, *is* air? Boyle's experiments on the extinction of both life and combustion within the air pump suggested not only that both might depend on some ingredient in the air but that it might be the *same* ingredient in both cases. This did not necessarily imply that the vital substance was a *part* of the air – he imagined that the elastic fluid of air was merely the medium in which the life-giving particles were conveyed. Hooke suspected that the combustible

---

* As we've seen, because the air pump's vacuum was imperfect, not all animals placed inside were asphyxiated. Insects survived well, while snakes seemed just to feel uneasy.

ingredient in air is also present in saltpetre, one of the components of gunpowder – an astute intuition, for saltpetre (potassium nitrate) does indeed contain the oxygen that is essential for the powder's detonation, which it will release when heated. It was not, however, until the end of the eighteenth century that oxygen was identified as the chemical substance responsible for combustion and respiration, a discovery that marks the beginnings of modern chemistry. The physician John Mayow may have been largely summarizing the ideas of Boyle and Hooke on air in one of his *Five Medico-Physical Treatises* (1674), in which he postulates a 'nitro-aerial' substance that is consumed as a candle burns in air or as a mouse breathes in a confined space. Mayow also reported that this nitro-aerial gas, released from heated saltpetre, will turn arterial blood red, confirming the connection between respiration and burning. It seems surprising that it took another hundred years for 'nitro-aerial' to become oxygen. The standard tale is that the spurious 'element' phlogiston, supposedly released when a substance burns, confused the issue in the interim. If that's a little too simplistic, it does reflect the challenging character of chemistry, which rarely seemed to offer consistent behaviour: metals burnt in air gain weight (as Boyle showed), whereas wood and other organic materials lose weight. So is combustion a matter of losing or gaining a component of the air?

The philosophical space

The controversial matter of the Torricellian space is broached in Boyle's *New Experiments*. He copied Torricelli's investigations with an inverted, mercury-filled tube, using a narrow tube three feet long and observing that the mercury fell to twenty-nine inches above the level in the dish. Then he inserted the whole apparatus through the opening in the top of the air pump's chamber, sealing it over with putty where the tube stuck out of the top. He pumped out the air and noted that the mercury in the tube fell and fell, but never quite to the same level as that in the dish.* Moreover, if he reversed the operation of the

---

* Boyle reports that it remained about an inch higher, which allows us to calculate the extent to which the chamber could be evacuated: he could reduce the pressure to just three per cent of normal atmospheric pressure, equivalent to its value twenty-three kilometres up in the atmosphere.

pump so as to increase the air pressure in the chamber, the mercury rose above its initial level.

All very nice, and easily explained in Torricelli's terms. The lessening of air pressure in the evacuated chamber reduces the force pressing down on the mercury in the dish, which is what prevents the mercury levels in the tube and the bowl from equalizing. Because, as Boyle acknowledged, the seals of the vessel were not perfect, it could not be evacuated completely, and so the level in the tube remained always a little higher. And if the air pressure was made greater than that of the atmosphere, the mercury was pushed back up.

But what about the gap? Boyle rose loftily above the issue, saying that this experiment simply made no pronouncement on that delicate question. He did not, he professed, 'take upon me to determine so difficult a controversy'. Indeed, he failed to see how this apparatus could allow one to do so – in which case, the experimentalist could do no more than speculate or offer a personal opinion about the matter, without hoping to have that view tested in any way. The air pump, he said, makes 'the controversy about a vacuum rather a metaphysical, than a physiological question; which therefore we shall here no longer debate'. All he would commit to was that the gap was 'almost totally devoid of air'.

This is rather thrilling, for it was (apparently without intending to be) extremely provocative. The urgent question pro or contra Aristotle is whether the Torricellian space maintains a vacuum. And here is Boyle with one of the most elaborate and expensive of devices his age can offer, experimenting on that very issue, and yet saying 'Oh, I can't tell you about *that*.' Worse, in fact: he seems to be saying '*That* isn't a particularly important question after all.' In other words, the experimental philosophy was proposing to change the game altogether – to impose new rules and new standards of proof, and to pronounce out of bounds what seemed before to be the key questions.

This did not prevent Boyle from being considered a 'vacuist', since the antagonists in any polarized debate are never happy until they have placed you in one camp or other. But Thomas Hobbes, at least, was alert to what Boyle seemed to be saying about the nature of experiment. And that was precisely what he deplored about it.

The point of philosophy, Hobbes insisted, was to secure reliable knowledge. And there was only one way to do this: by logical deduction through the power of human reason. That is to say, you start from unimpeachable axioms and work out what must follow from them. In this way, no sane and intelligent person can disagree with your conclusions. That is the Aristotelian method, honoured since antiquity, and illustrated in most exemplary fashion by geometry. According to John Aubrey, the discovery of this method of philosophizing was a revelation to Hobbes when, at the age of forty, he chanced across it in Euclid's *Geometry*:

> He read the Proposition. 'By G—',* sayd he, 'this is impossible.' So he reads the Demonstration of it, which referred him back to such a Proposition: which proposition he read; that referred him back to another which he also read, and sic deinceps [slowly but surely], that at last he was demonstratively convinced of that trueth. This made him in love with Geometry.

Hobbes contested the new philosophy not because he considered experiments to be useless, but because he did not consider them to be philosophy. They simply did not work in the way philosophy was meant to: by starting from solid ground and advancing by deduction. An observational 'fact' was of little value in itself, Hobbes thought, because it was merely something that happened – a kind of anecdote. It was silent about the causes of the phenomenon. And the likes of Boyle made things far worse by insisting that this silence was a virtue, was indeed a crucial part of experimentation – that one should not speak of causes, or do so only in the most circumspect and tentative manner. On what basis then, Hobbes complained, could one suppose that experiment told you anything at all? Experiment was all very well, but only as an amusing diversion that could not aspire to the status of true philosophy. 'Every man that hath spare money', he said, can acquire furnaces, telescopes and engines – but 'they are never the more philosophers for all this'. He added that 'It is laudable, I confess, to bestow money upon curious or useful delights; but that is none of the praisis of a philosopher.'

---

* 'He would now and then sweare an Oathe, by way of Emphasis', Aubrey admits a little reluctantly – his bad language was another reason why some people disapproved of the voluble Hobbes.

That was not just a point of semantic pedantry. For Thomas Hobbes, preserving the integrity of philosophy was much more than a matter of intellectual hygiene. He is remembered now not as a natural but as a political philosopher: his great work *Leviathan* (1651) is famous, some will say notorious, as an attempt to argue from philosophical first principles that absolute monarchy is the only stable form of government. Although begun before the English Civil War, *Leviathan* is in many ways a response to that upheaval, and it asserts that total obedience to a monarch – who is a representation of the collective will and interests of his subjects – is the only means of avoiding social unrest and anarchy. As a staunch Royalist, Hobbes had spent the war in Paris, where he mixed with the Cartesian circle of Mersenne and Gassendi and acted as tutor to the Prince of Wales, the future Charles II.

Hobbes considered his political philosophy to be inseparable from his scientific interests. Indeed, *Leviathan* is predicated on a mechanistic theory of physiology and psychology that takes as its ultimate axiom Galileo's principle of inertia: all bodies remain in uniform motion unless forces act on them. Hobbes attempts to progress in deductive manner from this physical law to a theory of what motivates individual human actions and what this implies for the organization of society. He argues that, in the absence of a supervening power, every man will aim to exploit his neighbour, and in the consequent State of Nature life is 'solitary, poor, nasty, brutish, and short'.

This analogy between the order of nature and the stability of the state was one that Hobbes may have learnt from his mentor Francis Bacon, who stated in *Advancement of Learning* that:

> There is a great affinity and consent between the rules of nature, and the true rules of policy: the one being nothing else but an order in the government of the world, and the other an order in the government of an estate.

Even in ancient Persia, Bacon claimed, the kings took 'the fundamental laws of nature, with the branches and passages of them, as an original and first model, whence to take and describe a copy and imitation for government'. To undermine the practice and authority of philosophy by making it contingent on empiricism is therefore for Hobbes a moral issue that threatens the stability of the social order.

If Hobbes had simply denied all validity to experimental philosophy, he could at least have claimed the intellectual high ground. But he was determined to attack it on all fronts, and so he set out to prove not only that Boyle's results were worth next to nothing but that they were flawed in any case. He missed no opportunity to point out the inadequacies of the air pump. The leakages, he said, were not just points of detail that prevented an ideal experiment; they fundamentally undermined the integrity of the experiment. Through the tiny cracks and gaps in the apparatus, Hobbes sought to pump in the debilitating vapours of scepticism.

It may partly have been Hobbes's technical criticisms of the air pump that stung Boyle and Hooke into trying to improve it by reducing leakage, as they did around 1667. But Boyle felt that his opponent was nonetheless wishing 'to be revenged on an engine that has destroyed several of his opinions'. Yet the leaks were not just imperfections, Hobbes said: they were *inevitable*. In effect he asserted that the composition of air was such as to make it impossible to prevent its ingress – that way, the impossibility of a vacuum did not have to depend on some sort of divine imperative, because it was a *practical* fact. Hobbes said that the air is a mixture of particles in an ethereal, fundamentally fluid and 'infinitely subtle' medium, and that this fluid can therefore find its way around the tightest seal or valve. So while the air pump could sure enough exclude 'straw and feathers', at least some part of the air will always get in. Likewise, this tenuous stuff will fill the Torricellian gap, because it can always slip around the mercury.

This insistence on *defining* air as (in part) a substance that cannot be excluded from an open space exasperated the experimentalists. John Wallis accused Hobbes of defining his terms differently 'from what the words signifie with other Men', while Boyle pointedly explained that by air he meant 'the obvious acceptation of the word, for part of the atmosphere, which we breathe, and wherein we move'. Moreover, Hobbes's philosophical approach effectively imposes a restriction on the questions an experiment can ask. Embedded in his criticism is already an answer to the question 'what is air?', so there is no possibility that the air pump could supply a contradictory view. This did not prevent Hobbes himself from complaining that the experimental approach carried unspoken assumptions (he was and is right about that, even if he is blind to the plank in his own eye).

Hobbes was by no means alone in attacking suggestions that the Torricellian gap was a vacuum. The Jesuit mathematician Franciscus Linus offered a more civil critique, for example asserting that if it was truly empty, not even light could pass through, since this was thought to be conveyed by the fluid aether. But Hobbes was undoubtedly the most intransigent and problematic of Boyle's naysayers. If he can seem now like a grumbling, old-fashioned dogmatist, nevertheless some of his complaints should give us pause. For one thing, they reveal how hard it was, especially in the absence of a substantial body of accepted prior facts (such as atomism, gravity and theories of heat and light), to offer a unique explanation of observed phenomena. More or less any hypothesis might be tailored to meet the 'facts'. So animals die and candles go out when the air pump is evacuated? Why, said Hobbes, that's because removal of some of the air sets up a violent and eventually fatal circulation of the fluid inside. Others supposed that the suction of a vacuum was caused by the establishment of invisibly thin 'threads' between the surfaces involved. And although we like to imagine that science should in the end offer physical causes for phenomena, there is nothing to tell us when those causes should be considered sufficient, and when we must seek causes of causes. The endless regression of causality that today forces us into the near-metaphysics of string theory, M-theory and the multiverse was already foreseen by Boyle when he complained to Hobbes that 'Whatever true cause I told you, you would not then acquiesce to its truth, but would ask me further what was then the cause of this cause, whence it would go on to infinity.' For Boyle, at least, curiosity about *causes* really did then threaten to become a glass-bead game.

What Boyle seemed to perceive was that a 'cause' for Hobbes could not be a basic empirical law of nature – 'this is fundamentally how things are' – but had to observe self-evident principles, like the axioms of geometry. If you try to explain the results of the air pump, as Boyle sometimes did, by invoking the 'spring' or elasticity of air, where did that get you? What is the reason for the air's spring? Unless you can adduce that, said Hobbes, all you have is a 'machine', a thing that produces a certain result but without any knowledge of its inner workings. The real argument here is about *what counts as an explanation.*

At the bottom of Boyle's aversion to an emphasis on causes is not just his nobleman's dislike of the disputation it provokes, but a suspicion that these things are unknowable and therefore mere conjecture. For it is possible, he thinks – and we have as yet no reason to doubt him on this – that effects might not have a *uniquely attributable* cause. The contemporary quest for a so-called Theory of Everything puts its money on that being untrue – on the logical necessity of all that we observe. It is still a big leap of faith. Questions about the multiplicity of causes were raised in Descartes' *Principia*, in which he writes of two clocks which both tell the time reliably but have completely different mechanisms. In contrast to these imponderables, Boyle sought the terra firma of experimental fact.

The historian D. G. James holds the view that actually Hobbes was not so concerned to unearth ultimate causes either – or rather, he was not bothered about the realist view of science which maintains that there is a particular way the world works which we can discover for ourselves. Hobbes, says James, seems sometimes to accept that we may never know the rules that God uses to run the universe – but is unconcerned by that. All he demands is a system that works on its own terms. The aim of philosophy is then not to understand God's work, but to satisfy the demands of human reason: to find causes that, if not necessary, are at least sufficient.

Weighing the evidence

The Fellows of the Royal Society imagined, *pace* Bacon, that some mechanism must exist for turning facts into laws. We saw earlier how Hooke tried to realize Bacon's notion of constructing philosophical tables from which 'axioms and theories' might be extracted. His method apparently involved a formal process of comparing different observations, as if by subtractive mathematics their core principles would be stripped clear of superfluities. In this process, the facts were not merely to be piled up; rather, each should be carefully inspected as though examining a gemstone for flaws, before being placed in the table. Although 'nothing is to be omitted', Hooke wrote, 'yet every thing [is] to pass a *mature deliberation*'. It was the job of the natural philosopher to show 'much *rigour* in admitting [items to the tables], much *strictness* in comparing, and above all, much *slowness* in debating, and *shyness* in determining'. There is thus to be 'a *scrupulous* choice, and a *strict examination*, of the reality, constancy, and certainty of the Particulars that we admit', so that the philosophers were to 'esteem the riches of our Philosophical treasure [not] by *number* only, but chiefly by the *weight*'. This sceptical attitude, whereby 'facts' need to justify their inclusion in the evidence, is the descendant of the practice among the professors of secrets of testing each recipe and claim that they encounter before accepting it. The annals of curiosity now had entry requirements.

Yet how do you know when to trust an empirical 'fact'? To the scholastics, it was purely a question of authority and status: a fact was verified if it could be found in an authoritative text, but otherwise it was mere hearsay. The virtuosi shared the old distrust of lay testimony. To the English writer (and avid collector) Thomas Browne, whose broad interests were strongly shaped by Baconian views, popular culture is a hotbed of superstition and error, which he attacked in his highly successful catalogue of common misconceptions *Pseudodoxia epidemica* (1646). The ordinary people, he complained, are unable to filter their senses through reason, so that for example they consider the Earth to be far bigger than the sun:

> Hopelessly continuing in mistakes, they live and die in their absurdities; passing their days in perverted apprehensions, and conceptions of the World, derogatory unto God, and the wisdom of the Creation. Again, being so illiterate in the point of intellect, and their sense so incorrected, they are farther indisposed ever to attain unto truth.

Francis Bacon insisted that his natural histories be weeded of all 'superstitious stories' and 'old wives' fables'. That was easier said than done; or rather, it was predicated more on a judgement of who was telling the tales than of what tales were told.

But Browne was equally dismissive of the tendency to believe something just because it had been reported by some classical writer. 'There is scarce any tradition or popular error', he said, 'but stands also delivered by some good Author.'* In its stolid march through popular misconceptions from the idea that diamond may be broken by the blood of a goat, to the belief that a badger has legs on one side shorter than on the other or that there were no rainbows before the Great Flood, his book is in itself a veritable cornucopia of curiosities.

While some felt that the most naïve witnesses were potentially the most trustworthy – Montaigne claimed that to obtain a reliable report 'we need a man either very honest, or so simple that he has not the stuff to build up false inventions and give them plausibility' – the new philosophers insisted that only well-informed observers gave valid testimony. The 'eyes of an idiot', said Galileo, are not as good as those of 'a careful and practised anatomist or philosopher'. Boyle emphasized that his experiments were done 'in the presence of an illustrious assembly of virtuosi' – which, for his early studies with the air pump in Oxford, meant Christopher Wren, Seth Ward and John Wallis.

This, however, still required one to take a lot on trust. One heard so many strange reports from around the world, even on good authority, that one ran the risk of seeming credulous by accepting them all. Yet to a gentlemanly club like the Royal Society, it would be terribly bad form to doubt the word of one's social equals. It was partly for this reason that such importance was afforded to replication of experiments and to securing multiple witnesses: not so much because this meant that the report was true (there were plenty of reliably witnessed accounts of the transmutation of metals) but because one could feel absolved from the need to dissent. This convention – which we saw also originated with the academies of secrets – bred scepticism of the tendency in the old natural-magic tradition for experiments to be described as a form of private, first-person revelation, or even for some crucial details to be withheld for the sake of secrecy. As Sprat asserted (doubtless with hyperbole), no claim was ever published in the Royal Society's annals 'till the whole *Company* has been fully satisfi'd of the certainty and constancy, or, on the otherside, of the absolute impossibility of the effect'.

---

* The modern equivalent, equally valid, is that there is no scientific idea so absurd that you cannot find someone with a PhD (indeed, often with a Nobel Prize) to support it.

Yet the issue of transparency remained contentious. Bacon himself had indicated that the scientists of Solomon's House were selective in what they chose to reveal: these men would discuss

> which of the inventions and experiences which we have discovered shall be published, and which not; and take all an oath of secrecy, for the concealing of those which we think fit to keep secret, though some of those we do reveal sometimes to the state, and some not.

Robert Hooke appears something of a recidivist when he suggests that secrecy, as well as guarding against abuses, has a certain cachet:

> For to be open, will be to expose every thing, either to be catched up by some, or to be droll'd upon by others, or sleighted by all. Whereas, on the contrary, a secret is wont to be admired, what ever it be.

This jealous guarding of knowledge was prevalent in the alchemical tradition, and Robert Boyle was not immune to it in his own chemical explorations. He would claim to be holding back information for the public good – lest it should get into the wrong hands – but it is likely that he did so at the insistence of the alchemical adepts who provided it, and who might otherwise have been less forthcoming.* This proprietary attitude to chemical knowledge was not unique to Boyle. Other natural philosophers would barter their alchemical 'secrets', as on the occasion when Oldenburg wrote to Samuel Hartlib in 1659 'I send you here enclosed a Chymical process of vitriol (in acknowledgement of the secret you sent me, which shall not loose [*sic*] the name of a secret for me).'

The society's deliberations on veracity also depended on who was admitted to the group in the first place. There were few tradesmen and practical technologists among its members, and not just because the steep membership fees tended to favour aristocrats and gentlemen. Sprat insinuated that aristocrats were to be preferred

---

* Boyle was acutely aware of the need to cultivate and protect his connections in alchemical circles. When he published some aspects of an alchemical process in the *Philosophical Transactions*, he took the most unusual step of doing so in Latin as well as English, presumably to improve the chances that some adept in Europe would see it and hopefully offer more information.

because their liberty from the demands of commerce gave them greater independence of mind: 'though the *Society* entertains many men of *particular Professions*, yet the farr greater Number are *Gentlemen*, free and unconfin'd'. On several occasions the society debated whether or not its membership should have a broader base – an issue somewhat contingent on whether the aims were, as Sprat had put it, to amass facts and experiments or to develop theories and explanations, a task of which only a privileged few were deemed capable. As William Eamon puts it, 'The problem of what qualifies as knowledge is intimately connected with the problem of who owns it.'

Thomas Hobbes seized on the selectivity and insularity of the Royal Society as another argument against their experimental philosophy. The Greshamites, he said, don't just form their own elite, but have successive elites within their own hierarchy. So much for their supposed 'public' verification of claims. In fact, most of Boyle's work was done in his private laboratory – a word that came into use during the seventeenth century, when it connoted a 'secret' space, with hermetic overtones. In comparing the virtuosi with the idle collectors of curiosities, Hobbes stressed how the latter maintain exclusivity by charging visitors. In his *Dialogus physicus* he says that the Fellows of Gresham

> display new machines, to show their vacuum and trifling wonders, in the way that they behave who deal in exotic animals, which are not to be seen without payment.

Hobbes seems here to have been looking for yet more reasons to undermine the authority of the Royal Society, rather than voicing heart-felt complaints. Nonetheless, he had a point – in fact, more so than even he realized. The implication commonly hinted at by the experi-mental philosophers (both then and now) that they were simply letting nature speak for herself fails to acknowledge that there was a particular process for generating facts from observations. For a start, it tended to require an instrumental technology, the use and interpretation of which depends on expert knowledge. There need be nothing so obviously wrong with that, except that it means that, even if the uninitiated had access to the experiments (which most did not), they still often had to take the results on trust rather than being able to rely on the evidence of their senses. False conclusions caused by instrumental artefacts or malfunctions are still frustratingly routine in science. As far as the air pump was concerned, obtaining reliable and reproducible results was not trivial. The machine was temperamental, and only Hooke could consistently get it to work. Since the Royal Society was so keen to show it off to visitors, that was a constant worry; part of the reason Hooke was taken on as curator of experiments was to ensure that he was always available to put his hands to the pump, as it were. Neither, for all the society's advertised insistence on verification, was it easy to reproduce some of the more elaborate experiments. Only a very few air pumps existed in Boyle's day – Huygens had one, and Guericke, perhaps the English experimenter Henry Power and a few others – and their differences in design and difficulties of operation meant that it was laborious and probably thankless to try to replicate someone else's findings rather than trying to secure a new claim of your own. And if such an attempted replication failed, it wasn't obvious whether this was because the original claim was false, or because you hadn't followed exactly the same procedure, or because (and no one would ever admit this) your own instrument wasn't working well. In fact, in those early days of experimental science it was likely that to know if your instrument *was* working properly you would have to calibrate it against the earlier observations you were supposed to be testing, and so you could hardly claim also to be validating them.

Even when the results had been duly recorded in a notebook, they did not acquire the status of facts. They had first to be processed by the community of virtuosi, something that depended on literary resources and on peer evaluation and consensus. The experimental philosophers gradually refined a literary style designed to persuade by means of carefully selected rhetorical tropes, most especially the adoption of a dispassionate, impersonal tone that contrasted sharply with both the *Sturm und Drang* of the Renaissance magus and the tiresome appeal to authority of the scholastics. And as we've seen, consensus generally meant verification by reliable witnesses – an explicitly social mechanism.

It's no exercise in casual analogy to say that this mechanism for generating experimental 'facts' still operates in science. As historians

Steven Shapin and Simon Schaffer say:

> Any institutionalized method for producing knowledge has its founda-
> tions in social conventions: conventions concerning how the knowledge
> is to be produced, about what may be questioned and what may not,
> about what is normally expected and what counts as an anomaly, about
> what is to be regarded as evidence and proof.

One might say in defence of this modus operandi that it is hard to
think of a better way. There is certainly something pernicious in the
contemporary tendency to make 'expert' almost a term of abuse, as
though no amount of time spent studying a topic can confer any more
authority than the next person. But that isn't the point. Scientists often
react with horror to the suggestion that 'knowledge is a social
construct' because it seems to imply that, say, Boyle's law is an arbitrary product of European culture that might not be expected to apply
in Africa or in the next century. There's surely a problem, however,
if in response to such absurd relativist excesses we propagate an
imaginary notion of how science progresses that takes no account of
its social context or construction. There are too many examples of bad
science become mainstream (eugenics, genetic determinism), or of
good science being ridiculed (prion diseases, 'jumping genes' or trans-
posons) to neglect the social structures that guide the evolution of
science. The fact is that all the attempts to describe what science is
or how it happens – that it is 'organized common sense', or its precise
opposite (anti-intuitive reasoning), or that it is the falsification-testing
of hypotheses, or serendipity sprung within fertile soil, or technology-
led or theory-led or curiosity-led – all these say something true about
the process, but none gives the whole picture. None offers a foolproof
'scientific method' that churns out a steady stream of accumulated
knowledge. The problem is that, because science produces knowledge
that is, for the most part, dependable and precise, we tend to believe
there must be a dependable, precise method for obtaining it. This is
the legacy of Bacon's dream of a 'new organon' that would grind
facts into underlying principles. But the truth is that science works
only because it can break its own rules, make mistakes, follow blind
alleys, attempt too much – and because it draws upon the resources
of the human mind, with its passions and foibles as well as its reason

and invention. This is clearer to see when we look back at the beginnings, because we can afford more indulgence towards the failure of the early pioneers to practise what they preached, and because we can see how they muddled through, and because that muddling through is confirmed by history and does not remain a matter of faith. But unless we are prepared to take this lesson away from the history of science, that history is worth very little.

10 On the Head of a Pin

All philosophy is based on two things only: curiosity and poor eyesight . . . The trouble is, we want to know more than we can see.

Bernard de Fontenelle, *Conversations on the Plurality of Worlds*

So, naturalists observe, a flea
Has smaller fleas that on him prey;
And these have smaller still to bite 'em,
And so proceed *ad infinitum*.

Jonathan Swift, 'On Poetry' (1733)

After dinner on 14 August 1664, Samuel Pepys felt fortified for intellectual enquiry. He went up to his chamber and finished reading Henry Power's recently published *Experimental Philosophy*, which included a substantial section on the microscope. Pronouncing it 'very fine', Pepys was inspired to get out his own microscope and, in the company of his wife, observe for himself – he does not tell us which specimens he chose. But it wasn't easy, they discovered. Pepys confessed to 'great difficulty before we could come to find the manner of seeing any thing'. They succeeded in the end, but Pepys was rather disappointed by the whole experience, saying that he saw 'not so much as I expect when I come to understand it better'.

Pepys was an inveterate dabbler, precisely the kind of well-connected and curious gadfly that the Royal Society attracted in its early days. His acumen both for administration and for cultivating useful relationships, not to mention his extraordinarily energetic persona, elevated him to the position of Chief Secretary to the Admiralty in 1673. Already by the time of his foray into microscopy he was deeply involved in English maritime affairs, sitting on the Navy Board and acquitting himself well during the Anglo-Dutch war of 1665–7. A worldly soul better suited to the affairs of men than the exploration of nature, he showed no particular aptitude for natural philosophy, but tried as best he could and with his trademark enthusiasm to follow the deliberations of the ingenious gentlemen at Gresham College. That kind of positive spirit was really all that was required of any well-bred chap for him to be considered a part of the club, and in 1665 Pepys was elected a Fellow of the Royal Society; between 1684 and 1686 he served as its president.

It is no surprise, then, to find Pepys gamely working his way through Power's book and having a shot at the experiments it described. As Power explained, the philosophers of the Royal Society had been recently investigating all manner of tiny specimens under the microscope's lens and finding in them the most remarkable contrivances. His description of the flea made it sound like a hybrid of a monstrous crustacean and some exotic warrior of the Mohammedan lands:

> It seems as big as a little prawn or shrimp, with a small head, but in it two fair eyes, globular and prominent of the circumference of a spangle . . . He has also a very long neck, jemmar'd like the tail of a Lobstar, which he could nimbly move any way; his head, body, and limbs also, be all of blackish armour-work, shining and polished with jemmar's, most excellently contrived for the nimble motion of all the parts: Nature having armed him thus *Cap a pe* like a Curiazier in warr . . . His neck, body, and limbs are also all beset with hairs and bristles, like so many Turnpikes, as if his armour was palysado'd about with them. At his snout is fixed a Proboscis, or hollow trunk or probe, by which he both punches the skin and sucks the blood through it.

Small wonder that Pepys was keen to see such marvels for himself. To which end, he had been that August to the shop of Richard Reeve in Long Acre, the finest instrument-maker in the country, and there paid the 'great price' of five pounds and ten shillings for a microscope. 'A most curious bauble it is', said Pepys, 'and he says as good, nay, the

A seventeenth-century microscope, with illuminating 'scotoscope' on the left, as depicted in Robert Hooke's *Micrographia* (1665).

best in England, and he makes the best in the world.' With it came a 'scotoscope' – as Pepys describes it, 'a curious curiosity it is to [see] objects in a dark room': a brine-filled globe that focused the light of a lamp on to the microscope stage, illuminating what would otherwise be an unintelligibly dark specimen.

But Pepys was evidently at first frustrated and then a little underwhelmed by what he saw with his expensive bauble. Apparently, what Power testified to with such authority was not something transparent to the casual observer. Skill and understanding were required to manipulate the instrument that mediated between the eye and this new microcosm. And if the operation of the microscope was in itself problematic, this was nothing compared with the question of how to interpret what was seen. The harder they looked, the more the experimental philosophers discovered handiwork on a scale more minute than they could ever have imagined. What astonished and perplexed them was that the world seemed to be so fine-grained, filled with details that the unaided sense of vision could never perceive. To some, this was evidence of God's omnipotence: He could fashion matter with far greater dexterity than humans could ever hope to accomplish, making our finest engineering look crude and clumsy in comparison. To others, the microscope held the promise of revealing at last the tiny corpuscles, machines and mechanisms that many supposed to be the operative agencies in nature's laws and effects. The mechanical philosophy of Descartes seemed about to become more than just a speculative hypothesis. For Robert Hooke, the Royal Society's most accomplished microscopist, this aid to the eye seemed able to expose the *secrets* of nature, revealed now not by Bacon's 'trials and vexations' but merely by an ability to look more closely. He proclaimed that the experimental philosopher could discover

> the subtilty of the composition of Bodies, the structure of their parts, the various texture of their matter, the instruments and manner of their inward motions, and all the other possible appearances of things . . . we may perhaps be inabled to discern all the secret workings of Nature, almost in the same manner as we do those that are the productions of Art, and are manag'd by Wheels, and Engines, and Springs, that were devised by humane Wit.

But this prospect of peering into nature and understanding her mechanisms was accompanied by the suggestion, at once exciting and troubling, that that process of investigation might never end. Inspired in part by the bounty of the microscope, Hooke wrote that 'the limits of Naturall knowledge are as infinite and boundlesse as the quantitys of Natures productions'.

All this tiny detail raised other disturbing questions. Why would a benevolent God have placed the innermost mechanisms of nature beyond the scope of unaided human vision? If He has done so, is it proper that we should spy on his workings? And if we do, can we be sure that human comprehension is capable of interpreting what we see? In short, seventeenth-century microscopy illustrates all the facets of this age of curiosity: the exploration of new worlds and new phenomena, the prospect of making occult secrets apparent, and questions both theological and secular about the wisdom and the utility of the investigation.

Writ small

The microscope and the telescope are often regarded as two of the key enabling devices of the Scientific Revolution. Yet that claim perhaps elides cause and effect, for both instruments seem to have been feasible in principle well before the seventeenth century. We saw earlier that magnifying lenses date at least from the thirteenth century, when the great 'experimental Franciscan' Roger Bacon wrote that with such optical aids 'from an incredible distance we might read the smallest letters and number the smallest particles of dust or sand'. This sounds as though he was already in effect practising telescopy and microscopy. But the crucial difference seems to have been that for several hundred years the lens was regarded simply as a way of improving our ability to see what was already 'there', not to descry new things. It could enhance poor or ordinary eyesight, but no one considered it a window to new worlds. And as both Roger Bacon and Giambattista Della Porta knew to their cost, optical ingenuity was commonly associated with trickery and necromancy – even that arch natural magician Cornelius Agrippa had called optics deceitful. It was an association perpetuated by the use of optics for 'magical' demonstrations, as reported in Caspar Schott's *Magia universalis naturae et artis* (1657). The advances in knowledge occasioned by instrumental optics in the seventeenth century flowed not so much from the invention of new devices as from an acceptance of lenses as aids to scientific study rather than as toys,* conveniences or tools of trickery and deception.

The compound microscope and the telescope (at least that used by Galileo) are basically the same device: two lenses, one convex and one concave, mounted in a tube. The difference is simply in the separation of the lenses. But not all microscopes had two lenses; the most powerful of them had only one. These single-lens or 'simple' microscopes are basically identical to magnifying glasses, but gain their tremendous enlarging power from the small size of the lens, a bead consisting of a sphere formed spontaneously from molten glass. The smallness of the lens meant that the simple microscope was hard to make and to see through, as well as presenting difficulties for adequately illuminating the specimen, and so despite their potentially superior performance they were generally avoided in favour of compound instruments.

Given the ceremony and controversy that greeted the invention of

---

* The prisms bought by Isaac Newton for his famous experiments on light (see page 343) were sold as expensive playthings rather than as scientific instruments. And Pepys's comment on the 'bauble' he bought from Reeve speaks volumes about the attitude towards optical instruments even then.

the telescope, it is surprising that the advent of the microscope happened so quietly. Some give priority to the Dutch lens-maker Hans Janssen and his son Zacharias in the 1590s, others to Hans Lippershey, both of them candidates for the invention of the telescope. In any event, microscopes were available in Europe by the 1620s, although it is not clear whether these, or just the magnifying glass, are what Francis Bacon has in mind when he states that the scientists of Solomon's House are equipped with 'glasses and means to see the small and minute bodies perfectly and distinctly: as the shapes and colours of small flies and worms, grains and flaws in gems which cannot otherwise be seen'.

In 1625 the instrument was given a name, when Johann Faber of the Accademia dei Lincei mentioned the *microscopio* in a letter to Federico Cesi. Faber's friend Galileo had been experimenting with the microscope, and he sent one to Cesi in 1624, asking him to continue these studies and tell him anything interesting he discovered. 'I have observed', Galileo wrote,

> many tiny animals with great admiration, among which the flea is quite horrible, the mosquito and the moth very beautiful, and I have seen with great pleasure how flies and other little animals can walk attached to mirrors upside down. You will have the opportunity of observing a large number of such particulars, and I should be grateful if you would let me know about the more curious. In short, the greatness of nature, and the subtle and unspeakable care with which she works is a source of unending contemplation.

Entomology was one of the earliest uses to which it was put. In 1625 the Lincei published *Melissographia*, a formidable engraving by Johann Friedrich Greuter of three bees, embellished with the fine anatomical details that Francesco Stelluti's microscopic investigations had revealed. This was an accompaniment to Stelluti's text *Apiarium* published in the same year. The *Apiarium* is a bizarre document: a single, huge broadsheet (comprised of four joined sheets) covered with dense, almost unreadable text in a variety of tiny typefaces in which allusion, myth and wordplay are mixed with a treatise on the classification of bees. It reveals how the literary traditions of Renaissance humanism were still active within the nascent scientific enterprise. The *Apiarium*

The *Melissographia* (1625) of Johann Friedrich Greuter, a large engraving published by the Accademia dei Lincei in honour of the Barberini family. The inscription reads 'Observed by Francesco Stelluti, Lincean of Fabriano, by means of a microscope.'

was conceived as an ingratiating offering to the Barberini family of the new pope, Urban VIII (Maffeo Barberini). '[It] has been made all the more to show our devotion to our Patrons, and to exercise our commitment to the observation of nature', Cesi wrote to Galileo. The Swiss physician Theodore Turquet de Mayerne, a Huguenot exile in England and physician to Charles I, seems mindful of this Italian heritage when he refers in the preface of Thomas Moffett's *Theatre of Insects* to the utility of 'lenticular Glasses of crystal' (although whether again this refers to a microscope or a magnifying glass is unclear), saying that 'though you have Lynx his eyes, these [glasses] are necessary in searching after Atoms'.

Athanasius Kircher published drawings of his microscopes in the 1630s, and in 1644 Giovanni Hodierna at Palermo augmented Stelluti's demonstration of the finely wrought nature of insects with his book *L'occhio della mosca* (*The Fly's Eye*). In 1656, the physician and botanist Pierre Borel listed many observations of plants and animals in his *Centuria observationum microcospiarum* (*Century of Microscopic Observations* – meaning one hundred of them). Like Power, however, Hodierna and Borel had to rely on words alone to convey what they had seen. And words can only go so far, not least because they force one to rely on macroscopic analogies – armour, turnpikes, large creatures and so forth. This can be somewhat effective for animals such as insects, which after all have eyes, legs and wings like much bigger creatures. But how to convey less familiar sights?

This is why Robert Hooke transformed the science of microscopy: not because he looked at anything particularly novel, but because he accompanied his descriptions with the most glorious illustrations, generously proportioned. Many of these were probably drawn by his friend Christopher Wren, although Hooke was no mean draughtsman either. His *Micrographia* was published a year after Pepys had struggled with his costly instrument, and it became a best-seller, an essential purchase for any gentleman concerned to keep abreast of the science of the day. Inevitably Pepys snapped up a copy at once, and he stayed up until two in the morning perusing what he proclaimed to be 'the most ingenious book that ever I read in my life'. *Micrographia* is without question one of the jewels of the age of curiosity, a graphic manifesto to the abundance of the world and the rewards of probing it with the tools of human art.

What distinguished Hooke's book, however, was not just the feast it presented to the eyes, but the territory it sought to claim for experimental philosophy in general and for microscopy in particular. Hooke supplied nothing less than a philosophy of the microscopic world. As with the telescope, he wrote, every improvement of the microscope presents 'new Worlds and *Terra Incognita*'s to our view'. It was a matter of perspective: just as the teeming Earth when seen from afar would seem a mere bland speck, so God packs as much detail into the smallest of natural objects that to us seem without form or elaboration. Every grain is a microcosm.

That was the key. Curious experimenters were no longer merely using the lens to see things more accurately – they were able to see *new* things, entirely novel aspects of how nature worked. Today it might seem self-evident that extending our senses beyond their natural capabilities will do this; astronomers look into the X-ray universe, electron microscopes focus their beams more tightly than optical instruments may. But for Hooke and his contemporaries, the idea that there was more to the world than the eye can register was profoundly challenging, and it became the major point of contention about the value of the microscope.

For Leonardo da Vinci the eye is the emissary and arbiter of understanding. It is, he said,

> commander of astronomy; it makes cosmography; it guides and rectifies all the human arts; it conducts men to various regions of the world; it is the prince of mathematics; its sciences are most certain; it has measured the height and size of the stars; it has disclosed the elements and their distributions; it has made predictions of future events by means of the course of the stars; it has generated architecture, perspective, and divine painting.

Nothing separates more clearly than this reverence for the senses (especially the visual) the late-medieval Neoplatonist from the Baconian new philosopher. For Francis Bacon, the eye was no longer the steadfast companion and assistant of the curious mind, but a treacherous ally, never to be trusted at face value. To Robert Hooke, instruments could put that right. Critical of the ancients for their reliance on the 'outward Shape and Figure' of things, he felt that the philosopher should not be satisfied with their superficial appearance, which may be silent or even misleading about their inner workings.

But why would God arrange things this way, placing vital clues about nature out of reach of our quotidian perception? For Hooke this was not so much God's doing as mankind's: the 'mischiefs, and imperfection, mankind has drawn upon itself', he said, stem both from 'innate corruption' – the Fall in Eden – and 'from his breeding and converse with men', the coarsening effect of the world. According to Joseph Glanvill, Adam had been able to see the inner details of matter, as well as the distant stars, without the microscope or the telescope.*

---

* So explicit a view was not universally shared. Henry Power argued that Adam's powers of perception were no different from ours, so that the new world under the microscope was truly being revealed for the first time in human history.

To repair these faults, said Hooke, we must 'rectif[y] the operations of the Sense, the Memory, and Reason'. Comparison with the superior perception of some other creatures (think of the lynx) implies that in humans these operations fall 'far short of the perfection they seem capable of'. In consequence, 'the limits, to which our thoughts are confin'd, are small in respect of the vast extent of Nature it self; some parts of it are *too large* to be comprehended, and some *too little* to be perceived'. This 'disproportion of the Object to the [sensory] Organ', said Hooke, can be rectified if we add '*artificial Organs* to the *natural*'. He speculated that other instruments might some day be devised that improve man's hearing, smell and taste just as the lens improves his vision.

Small is beautiful, and ugly

Micrographia was a project foisted on Hooke by John Wilkins, who in 1663 asked him to relieve the over-committed Wren of the duty to prepare a microscopic study of insects for presentation to the (doubtless indifferent) king in 1663. Some of the book's lavish drawings may first have been made at that time. But Hooke did not restrict himself to the microscopic world. He seized the opportunity to put into print his observations and speculations about a wide range of phenomena, including the principles of capillarity (how water is drawn into narrow tubes and spaces) and his telescopic views of the moon and the stars. He used the book's preface to present an apologia for the new philosophy, alongside an ingratiating dedication to Charles II, who is portrayed spuriously as the author of the Royal Society's mission 'of avoiding *Dogmatizing*, and the espousal of any *Hypothesis* not sufficiently grounded and confirm'd by *Experiments*'.

Nonetheless, there is a calculated arc to the miscellany of his material, which begins with the prosaic peculiarities of human art (printed letters, cloth) when viewed close up, moves through the mineral (glass, sand, charcoal) to the vegetable (mould, sponges, seeds) and culminates with the famous spectacle of the insect world, its victims mostly drugged with brandy (they shrivelled if killed). The masterpiece is the head of a 'grey drone-fly' decapitated and glued on to the sample stage with bejewelled eyes staring opaquely into the lens. Hooke alludes to this infinite regress of gazes by plunging

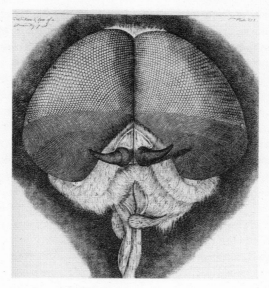

Hooke's microscopic image of a fly's head.

us vertiginously into the multitude of worlds captured inside each of the 14,000 (as he estimated it) facets of the fly's eye, for he claimed rather fancifully to be able to see reflected in each microscopic hemisphere 'a Landscape of those things which lay before my window, one thing of which was a large Tree, whose trunk and top I could plainly discover, as I could also the parts of my window, and my hand and fingers, if I held it between the Window and the object'. What could be more eerie than this vision of the dead fly seeing impassively, and multiplied many thousandfold, the gigantic world and the fateful hand of his nemesis?

Equally unnerving was Hooke's phlegmatic observation of a louse as it sucked out his own blood:

And the Creature was so greedy, that though it could not contain more, yet it continued sucking as fast as ever, and as fast emptying itself behind: the digestion of this Creature must needs be very quick, for though I perceiv'd the blood thicker and blacker when suck'd, yet, when in the guts, it was of a very lovely ruby colour, and that part of it, which was digested into the veins, seem'd white.

As one who delighted in mechanical ingenuity, Hooke was inevitably drawn to the microscope. His duties as curator of the Royal Society obliged him to investigate any objects requested by the Fellows, but he needed little encouragement. He studied everything from a razor's edge to urine, from plant seeds to the printed full stop. And he found that everything he observed was transformed under the lens, where apparent perfection of form was rendered crude and misshapen, making the mundane monstrous. Everything repaid attention, and there was interest even in the most unpromising of materials. That in itself was somewhat overwhelming, for it seemed to deny any hierarchy of significance. Here was the louse, as complex and delicately formed as any magnificent bird of prey. Mould became a magical garden, a vista populated with 'Microscopical Mushroms' and other strange sprouting bodies, and certainly 'not unworthy of our more serious speculation and examination'. Gazing at the thousands of lens-like formations on the fly's eye, Hooke sensed the dizzying, infinite scope of God's power:

> there may be as much curiosity of contrivance in every one of these Pearls, as in the eye of a Whale or Elephant, and the almighty's *Fiat* could as easily cause the existence of the one as the other; and as one day and a thousand years are the same with him, so may one eye and ten thousand.

This is the awesome vista of Milton's *Paradise Lost* through the other end of the telescope, as it were: an endless expanse converging on infinity's inverse.

If that situation presented the natural philosopher with the happy prospect of finding riches wherever he looked, it also eroded any distinctions about where one *should* look – there was, in the meanest of lodgings, in the dust on the shelf and the putrefaction of the ill-tended larder, more than enough fare to occupy the curious mind for days and weeks. And none of this unpromising stuff was quite what it seemed to the raw senses. 'Even in those things which we account vile, rude, and coarse', said Hooke, 'Nature has not been wanting to shew abundance of curiosity and excellent Mechanisme.'

In that levelling of signification, that denial of privilege to any thing or substance, one might easily have detected profanity. Those of a suspicious inclination might even imagine that the microscopists were unduly drawn towards the lowliest of matter, to cinders and decay, to urine and faeces, spittle and sperm. (To investigate the 'gravel' found in urine, Hooke apparently went scooping deposits out of urinals.) To find such things worthy of 'serious examination' could seem not merely perverse but perverted. Meanwhile, the most lofty object might be found under the microscope to be reduced to something shabby and ill-formed; when Hooke investigated samples of miniature writing, wherein the Lord's Prayer, the Apostle's Creed, the Ten Commandments and other biblical extracts were written so small as to be illegible to the naked eye, he 'discover'd of what pitifull *bungling scribbles* and *scrawls* it was compos'd'.

All this exposed how poor a fabricator man is when examined close up. Hooke's characterization of a printed full stop as being 'like a great splatch of *London* dirt' seems more a judgement than a description. The edge of the keenest razor is revealed as a ragged, pitted line, something better suited to harsh butchery than to clean incision. The tip of a needle is not a sharp point but a blunt and pitted cone, while the hairs, bristles and claws of insects are 'many thousands of times sharper'. 'The Productions of art are such rude mis-shapen things', he wrote, 'that when view'd with a Microscope, is little else observable, but their deformity.' John Wilkins commented after reading *Micrographia* that in comparison to the works of nature, 'the most curious works of Art, the most accurate engravings or embossments, seem such rude bungling deformed works as if they had been done with a Mattock or a Trowel'.

God is in the details

This superiority of nature's art over that of humankind was not just an exercise in humility, but became for some of Hooke's collaborators and readers, most notably Robert Boyle, an important theological defence of the experimental philosophy. To look closely at the smallest works of nature, says Hooke, is to discover God's 'greatest Excellencies' – and by implication, to render atheism unthinkable. In commenting on the contrivances of insect feet, he defies the reader to imagine if anyone can 'be so sottish, as to think all those things the productions of chance'.

While it would be uncharitable to deny the genuine piety of these assertions, they were also strategic. The experimental philosophers were acutely aware of their vulnerability to accusations of atheism, not just (or even primarily) because of their impudent curiosity but because their demonstrations of a rule-bound nature within which even marvels and wonders could be accommodated seemed to be leaving ever less room for God's agency. Their response was to find new roles for him.

Robert Boyle laid out this programme of natural theology in *The Christian Virtuoso*, where he argued that God's imprint lies not on the surface of things but in 'the hidden and innermost recesses of them'. To search in those depths was therefore obligatory. The more we know about the world, Boyle claimed, the more it becomes clear that it could not have been put together by 'so incompetent and pitiful a cause as blind chance or the tumultuous jostlings of atomical portions of sense-less matter'. For Leibniz, the microscope beat the telescope for revealing this divine plan: 'Nothing better corroborates the incomparable wisdom of God than the structure of the works of nature, particularly the structure which appears when we study them closely with a microscope.' A clockwork macrocosm shows great wisdom, but cannot compare with the ingenuity and subtlety of the microcosm. The pioneer of plant microscopy Nehemiah Grew expanded on this argument in *Cosmologia sacra: or, A discourse of the Universe as it is the Creature and Kingdom of God* (1701). The microscopists argued that God had put the instrument into their hands precisely so that they might discover the true extent of His ingenious work.

It was in some respects an old argument. Pliny had asserted that small animals better exemplify divine production than large ones: their parts are more refined, more ingenious, and further beyond the reach of mortal art. Thomas Moffett agreed, saying 'I see more of God in the History of Lesser Creatures'. As Henry Power declared in his poetic ode 'In Commendation of the Microscope':

> God is greatest in the least of things,
> And in the smallest print we gather hence
> The World may Best read his omnipotence.

This line of argument was also a useful defence against the accusation that insects were an unworthy subject of scientific enquiry. The early entymologists justified their research with a catalogue of the morally and socially improving lessons to be extracted from their discipline: bees may teach us about architecture and monarchy, ants about industry and democracy, spiders about household management and so on. The French philosopher Pierre Bayle even argued in 1702 that 'there may be more intelligence in invisible animals than in the larger ones'. (With the contemporary notion of 'swarm intelligence', these ideas are back in vogue.)

Enlisting the microworld for the glory of God meant that what one found there had to be regarded not merely as an abundance of intricately designed riches but as an indication of wise planning. God was not idly demonstrating his superior craftsmanship but had fitted everything to its function: as John Ray put it, he displays 'infinite prudence', so that no detail is frivolous. 'Nature is not capricious', proclaimed Bernard de Fontenelle.* In this regard it contrasts with human art, a point made by Cardano in 1560 when he compared the useless virtuosity of writing on grains of rice and flies carved from ivory with the 'celestial' crafting of nature. As Cardano was well aware, this attitude challenges Aristotle, for it asserts that the truth of things cannot be transparently read from surfaces visible to the eye. Here again is the Baconian insistence that nature is much more subtle than our senses or our everyday reasoning can perceive. The particle

---

* This attitude is still with us, although today the economy of nature is seen to result from the exigencies of natural selection, whereby anything superfluous is weeded out as a waste of effort and resources and therefore counter-adaptive. As a general principle that governs the living world this is sound enough, although it is now seen to be risky to push it too far: Darwinian evolution does not force nature to optimize, or to rigorously fit form to function, but only to order things to near-term advantage. For example, although genomes are not quite so full of genetic detritus – 'junk DNA' – as was once supposed, nonetheless it seems clear that our genes are embedded in a considerable amount of baggage and redundant material that has accumulated over the aeons, because there is insufficient selective pressure to force nature to purge its old files. And while biodiversity undoubtedly contributes to the robustness of ecosystems, it seems very likely that the plurality of species goes considerably beyond what might be deemed 'necessary' in that regard. Moreover, intricacy and complexity of form can arise from purely physical processes – think of the geometrical excesses of the snowflake – that operate equally in the living and the inorganic worlds, and which has no *raison d être* beyond the fact that nature's laws happen to permit and indeed to compel it. It is not fanciful to suppose that, when 'nature's economy' is overplayed today, part of the cause lies with a forgotten theological motivation to celebrate the wisdom of the Creator.

accelerators and electron microscopes of today's science attest to the longevity of that conviction.

Hooke and his colleagues were not content to serve up their microscopic marvels solely as indications of God's dexterity and wisdom. As Henry Power explained, by revealing the fine texture and granularity of the world the microscope seemed to offer the hope that what was once occult would become manifest. Most experimental philosophers shared to some degree Descartes' corpuscular view of nature, according to which phenomena could be given a mechanical explanation predicated on the action of tiny particles of matter one upon the other. Even if Newton and others were ultimately to question the Cartesian assertion that this action had to be wholly materialistic, brought into being by direct contact, nonetheless the default position was that nature is grainy at the smallest scales. 'Those effects of Bodies, which have been commonly attributed to *Qualities*, and those confesse'd to be *occult*, are performed by the small *Machines* of Nature', Hooke insisted. Good Baconians were obliged to handle that hypothesis cautiously while it remained beyond the means of experimental probing. Now, however, the microscope seemed potentially to supply such means. The details seen by Hooke, the pits and bumps even in the 'smoothest' surface, seemed to represent the first manifestations of this granular fine structure, and he felt sure that improvements in microscopy would eventually bring the fundamental corpuscles into view. In his *Experimental Philosophy* Power claimed that it might reveal 'the magnetical effluviums of the loadstone, the solary atoms of light (or aetherial globules of the renowned Descartes), the springy particles of air'.

At the same time, the microscope eroded the old Neoplatonic idea of correspondences, according to which kinships and analogies existed between entities that reflected their common mechanism, and which were often revealed by similarities of form: a plant shaped like a particular organ of the body, for example, was thought useful for treating ailments of that part. Under the lens, such resemblances generally vanished, and even if new ones appeared (as they often did), there was no obvious reason to award greater significance to similarities at one scale than at another. Taken to its conclusion, this realization diminished the solipsistic medieval view of the world: it was not a system of signs placed there for our benefit and instruction. As historian Catherine Wilson puts it, 'we are always, in this new world, intruders'.

Pond life

'In every *little particle* of matter', Hooke wrote in *Micrographia*, 'we now behold almost as great a variety of Creatures, as we were able before to reckon up in the whole *Universe* it self.' This was characteristically hyperbolic – Hooke had a tendency to exaggerate both the multiplying power of the microscope and the abundance of the world it revealed – but it reflected Hooke's appreciation of how minute living things could be. Both he and Power described observations of tiny 'eels' found in vinegar, while the Dutch entymologist Jan Swammerdam, whose *General History of Insects* was published four years after *Micrographia*, recorded microscopic observations of water fleas that were barely visible as specks in the water.

Despite this, nothing prepared the experimental philosophers for the letter that the Royal Society received in October 1676 from a linen merchant of Delft named Antony van Leeuwenhoek, in which he claimed to have found that a drop of rain water 'which had stood but a few days in a new tub' teemed with tiny 'animalcules' too small to see unaided. 'As the size of a full grown animalcule in the water is to that of a mite', said Leeuwenhoek, 'so is the size of a honeybee to that of a horse; for the circumference of one of these same little animalcules is not so great as the thickness of a hair on a mite.' Some of these microscopic animals had 'two tiny limbs near the head, and two little fins at the rear of the body'. They had 'different colours, some being whitish and transparent, others with green and very glittering little scales, yet others were green in the middle and white at both ends, and some were grey, like ash'. Leeuwenhoek testified that 'no more pleasant sight has ever come before my eyes'.

This was something new. It was one thing to find that very small animals such as the flea were ingeniously contrived, but quite another to discover animals so tiny that one could not, without the microscope, know of their presence at all. Vinegar eels were gargantuan serpents compared with Leeuwenhoek's animalcules.

In which case, could this Dutchman, a mere cloth seller, be trusted? One wonders how seriously the letter would have been taken had Leeuwenhoek not already established a relationship with the Royal Society in a series of letters that began in 1673, when Henry Oldenburg was forwarded his critique of one of the drawings of a bee's eye in

Micrographia. This communication came via a reliable source, and when Leeuwenhoek began directly to send the Royal Society accounts (in Dutch, which of course Oldenburg could translate) of his observations with the microscope, he was vouched for by Christiaan Huygens' father Constantijn, who said that the Dutchman was 'a person unlearned both in sciences and languages, but of his own nature exceedingly industrious and curious'. It was no great mystery that a linen merchant should be using a microscope, since this enabled him to inspect the quality of his wares. Leeuwenhoek's first original report, on moulds and bee stings, was judged sufficiently sound to be published in the *Philosophical Transactions*.

A steady stream of further results followed: some 200 letters in all, faithfully describing observations which the author offered up for more educated minds to interpret. (Lacking Hooke's graphic skills, Leeuwenhoek employed Dutch artists to sketch what was seen in the lens. Sadly, his friend Jan Vermeer was not among them.) That apparent humility was, however, belied by the man himself, who did not lack ambition or ego, nor hold back from conjectures that others often considered ill-advised. Even in his initial correspondence Leeuwenhoek warned that 'I do not gladly suffer contradiction or censure from others.' To the immense frustration of his collaborators, he guarded his methods jealously; when requested to send out microscopes, he would release only those that were somewhat inferior to the ones used for his crucial observations. Even Constantijn Huygens was eventually driven to distraction by this secrecy, exclaiming after a visit in 1685, 'O what a beast!' Perhaps Leeuwenhoek knew (correctly) that if he made too free with his techniques, he would quickly become redundant as other, more knowledgeable natural philosophers would take over the enquiry. As it was, his unique ability to see new things through the lens brought him visits from royalty and dignitaries across Europe, including Charles II, and the Royal Society got into the habit of sending him requests of things to observe, including blood, milk, bone, hair, saliva, sweat, fat, tears, sponges, fish, muscle fibre and sperm. He was made an FRS in 1680.

Notwithstanding his growing reputation, nor the fact that the observation was testified by eight reputable witnesses including two ministers and a public notary of Delft, Leeuwenhoek's microscopic animalcules stretched credibility. As ever, the proof was in the pudding, and so it fell to Robert Hooke to replicate the experiment.

This was far from easy. Leeuwenhoek used home-made simple (single-lens) microscopes, the greater magnifying power of which could show things invisible to the compound instrument. But these instruments were tricky to make, and Hooke laboured for some months with his compound microscopes without seeing anything comparable.* It is a little surprising, and creditable, that he did not simply give up and conclude that Leeuwenhoek had been misled by optical or psychological artefacts. (Optical effects in the microscope can create the appearance of worm-like bodies, as anyone who has used one will know.) Finally, in November 1677 his persistence was vindicated: a microscope that magnified distances a hundredfold revealed the little swimming things to all the witnesses present:† 'by all who saw them', wrote Hooke, 'they were verily believed to be animals.' He confessed his astonishment:

> I was very much surprised at this so wonderful a spectacle, having never seen any living creature comparable to these for smallness. Nor could I indeed imagine that nature had afforded instances of so exceedingly minute animal proportions.'

These organisms became called protozoa ('first animals'); they still are. But while they are now recognized as single-celled, it was widely believed then that they were anatomically like any other animal, equipped with muscles, tendons, circulating blood and so forth. This seemed to imply a dizzying elasticity of scale: what was seen in the large could simply be reduced wholesale, with no evident prospect of a limit. Hooke had already intimated this persistence of fine detail at smaller scales in his descriptions of snowflakes: 'the bigger they were

* It was only after Hooke had finally succeeded in verifying Leeuwenhoek's claim that he created single-lens microscopes, so that the animalcules might be clearer still. He did this by heating glass rods in a flame and drawing them out into thin strands. When their tips were melted, they retracted into tiny balls, perhaps just a tenth of an inch across, which could be fixed into a small hole in a metal frame. He was improvising, for Leeuwenhoek refused to divulge the secrets of his technique – 'for reasons best known to himself', Hooke grumbled.

† If this was good enough for Hooke, it did not satisfy everyone. In 1686 the Italian anatomist Lorenzo Bellini in Pisa still felt compelled to wonder 'whether this man [Leeuwenhoek] has not assisted his eyes and his microscope with some prepossessions that have made him see things wrongly'.

magnify'd, the more irregularities appear'd in them'. (This property of irregularity in structure over many scales is a feature of objects now called fractals.) To grasp such a regress challenged the capacity of human intuition and reason. As the French physician Nicolas Andry de Boisregard wrote in his pioneering study of microscopic parasites *On the Generation of Worms in Human Bodies* (1700), 'Our imagination loses itself in this thought, it is amazed at such a strange littleness; but to what purpose would it be to deny it? Reason convinces us of the existence of that which we cannot conceive.'

Andry was one of the few prepared initially to credit another of Leeuwenhoek's astonishing announcements, which closely followed his report of animalcules. The Dutchman had begun to study human sperm (his own, rushed decorously but presumably somewhat jarringly from the marital bed) in 1674, this having being on the list of secretions proposed by Oldenburg as subjects for investigation. But it was not until four years later that – his observing skills and perhaps his conceptual sensitivities having been sharpened by the discovery of protozoa in rainwater – Leeuwenhoek reported seeing 'animalcules' with tails in the semen of a man suffering from gonorrhea. He said that he subsequently saw the same thing in the semen of healthy men:

> These animalcules were smaller than the corpuscles which impart a red colour to the blood; so that I judge that a million of them would not equal in size a large grain of sand. Their bodies were rounded but blunt in front and running to a point behind, and furnished with a long tail, about five or six times as long as the body, and very transparent . . . The animalcules moved forward with a snake like motion of the tail, as eels do when swimming in water.

Leeuwenhoek saw the same entities in the sperm of dogs and rabbits; Christiaan Huygens and Hooke independently verified the presence of these 'tadpoles' in human seed. But were they parasites, or might they have something to do with the generative power of semen? The latter notion was supported by the fact that the animalcules were not present in the sperm of young boys or very old men. Andry argued that the shape of these 'spermatic worms', being virtually all head and no body, 'agrees with the Figure of the *Foetus* or human Birth, which when it is little, seems to be no more than a Great

Head upon a long Body, that seems to end in a kind of a Tail'. Thus began the preformationist theories of the eighteenth century, in which every person begins life as one of these homunculi: not exactly a little being with all body parts present, but as a creature programmed to grow that way. But this idea raised some challenging questions. If the swimming things were miniature human beings, where did that leave the role of the female egg? Does each animalcule have a human soul? If so, why are there so very many of them in every drop of semen, and what happens to them all? It was hardly surprising, especially in view of the difficulty of testing any theories of human generation, that some natural philosophers preferred to deny the presence of these animalcules in semen, or otherwise to dismiss them as parasites.

Microscopic demons

That latter interpretation in itself touched on another profound impli-cation of Leeuwenhoek's discoveries: the mechanism by which diseases spread. The association between disease and decay or putrefaction was recognized since ancient times. But does death cause decay or

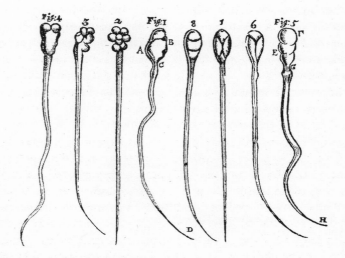

Antony van Leeuwenhoek's images of human and canine sperm.

stem from it? Putrefying substances were known to harbour maggots and other vermin, and were widely thought to generate them according to the doctrine of spontaneous generation.* Yet in this view putrefaction is a life-giving agency: as Paracelsus had written, 'decay is the beginning of all rebirth'. It was from the putrefaction of human sperm that the 'artificial man', the alchemical homunculus, was thought to be created in the Middle Ages.

As for disease, many believed that putrefaction and decay corrupted the air and that this 'bad air' (the Italian *mal aria*) caused illness. Malaria was after all prevalent in regions where water stagnated in the heat of the sun. (Of course, this is in fact because standing water incubates malaria-carrying mosquito larvae.) Girolamo Fracastoro suggested that it was not exactly the air itself that was rendered foul; rather, the putrefaction caused invisible corpuscles to be emitted, and these transmitted the infection. The spread of the Great Plague in 1665, which disrupted the activities of the Royal Society and banished Isaac Newton to fruitful exile in Lincolnshire, was explained by the London apothecary William Boghurst as being caused by such airborne corpuscles.

Evidently, here already are the ingredients of a germ theory of disease. Leeuwenhoek's discovery of invisible animalcules in water could appear to supply the final piece of the puzzle, especially as it quickly became clear that some of these microscopic worms may transmit parasitic diseases. The German mathematician Johann Christoph Sturm seemed to have put the pieces together when in 1687 he suggested that the air is full of animalcules that produce disease when breathed in, unless they are sweated out again. Athanasius Kircher came closer still when he proposed that diseases such as the plague are caused by microscopic 'seeds' of virulent 'worms' that enter the body through the mouth and nostrils. Yet this image of

---

* Robert Hooke's observations of microscopic mites and other creatures already led him to doubt whether spontaneous generation was real. After watching tiny larvae in rainwater develop gradually into gnats, he wondered whether 'all those things that we suppose to be bred from corruption and putrifaction, may be rationally suppos'd to have their origination as natural as these Gnats, who, 'tis very probable, were first dropt into this Water, in the form of Eggs'. Others, such as the FRS Oliver Hill, maintained that the worms themselves were the product of spontaneous generation – an idea that Hooke dismissed with characteristic impatience.

ailments being produced by what one might construe as malicious beings was not so far removed from old superstitions about the role of invisible demons, and the connection seems explicit in Daniel Defoe's *Journal of the Plague Year* (1722), a novelistic account of the Great Plague of 1665. Defoe's narrator says he has heard that if a person with the plague breathes on glass and places it under the microscope, 'there might living Creatures be seen by a Microscope of strange, monstrous and frightful Shapes, such as Dragons, Snakes, Serpents and Devils, horrible to behold'. He admits, however, to grave doubts about the truth of this.

One can say, then, that a germ theory of contagion emerged in the late seventeenth century. However, verifying a direct connection between animalcules and disease was extremely difficult with the tools then available, not least because many contagious diseases are caused by microorganisms such as bacteria and viruses that were still too small for the microscopes of the time. As a result, the idea floundered and fell out of favour until revitalized by Louis Pasteur in the late nineteenth century.

Leeuwenhoek's animulcules confirmed the growing view that what the microscope showed was not just finer details of our own world but veritable new worlds in themselves, populated by inhabitants as insensible of our colossal presence as we generally are of their diminutive existence. This image was invoked in Cyrano de Bergerac's *The States and Empires of the Moon*, in a passage where Dyrcona sits down to dinner with the demon of Socrates and two young philosophers of the moon. One of them expounds on the infinity of worlds:

> Picture the universe, therefore, as a vast organism . . . We, in our turn, are also worlds from the point of view of certain organisms incomparably smaller than ourselves, like certain worms, lice, and mites. They are the earths of others, yet more imperceptible . . . For do you find it hard to believe that a louse should take your body for a world, or that, when one of them travels from one of your ears to the other, his friends should say that he has voyaged to the ends of the earth, or that he has journeyed from pole to pole? Why, doubtless this tiny people take your hair for the forests of their country, your pores full of sweat for springs, your pimples for lakes and ponds, your abscesses for seas, your streaming nose for a flood; and when you comb your

hair backwards and forwards they think this is the ebb and flow of the ocean tides.

Here the microscopic world is interpreted in quasi-Neoplatonic terms as a reduced image of the macrocosm, just as for Leonardo da Vinci the vein-like networks of rivers carry the blood of the earth. Bernard de Fontenelle repeated both this infinitude and its macrocosmic analogy in *Conversations on the Plurality of Worlds*. 'You mustn't think that we see all those who inhabit the Earth', his philosopher warns the marquise:

> there are as many species of invisible animals as visible. We see from the elephant down to the mite; there our sight ends. But beyond the mite an infinite multitude of animals begins for which the mite is an elephant, and which can't be perceived with ordinary eyesight. We've seen with lenses many liquids filled with little animals that one would never have suspected lived there . . . A tree leaf is a little world inhabited by invisible worms, and it seems to them a vast expanse where they learn of mountains and abysses, and where there is no more communication between worms living on one side of the leaf and on the other than there is between us and the Antipodes.

Fontenelle's philosopher finds here good reason to believe that other planets must be inhabited. 'Nature [nature now!] has distributed the animals so liberally that she doesn't even mind that we can only see half of them. Can you believe that after she had pushed her fecundity here to excess, she'd been so sterile toward all other planets as not to produce anything living?'

In *The Anatomy of Plants* (1682), dedicated to Charles II and making much use of microscopical studies, Nehemiah Grew hoped to win over the Royal Society's patron with this geographical metaphor – with the implication that the monarch's realm was thus being increased:

> Your Majesty will find, That there are *Terrae Incognitae* in Philosophy, as well as in Geography. And for so much, as lies here, it comes to pass, I know not how, even in this Inquisitive Age, that I am the first, who has given a Map of the Country.

As with observations of the landscapes of the moon, the age of exploration and empire thrust visions of conquest into this microcosm. In *Micrographia* Hooke eulogises 'the use of Mechanical helps for the Senses',

> both in the surveying of the already visible World, and for the discovery of many others hitherto unknown, and to make us, with the great Conqueror [Alexander], to be affected that we have not yet overcome one World when there are so many others to be discovered.

But what would it take truly to conquer this world, rather than merely to gaze at it in wonder and incomprehension? And beneath that challenge lies the deeper question about following where curiosity leads: *to what end do we wish to know?*

Uncomfortably close

As with the telescope, the microscope did not simply appear and commence to augment our understanding of the world. Novel scientific techniques today have first to establish their veracity and reliability before they are added to the standard arsenal of the experimenter; that was truer than ever at a time when the whole notion of experiment was new and controversial. Between the moment when an innovation becomes technically possible and the time when it is used almost without thought, there lies a period of negotiation in which it vies for acceptance. This was especially the case for optical instruments, with their dubious legacy of association with trickery.

The microscope found easier acceptance than the telescope, but it was not without its sceptics. One of the most vocal, although her influence did not match her social status, was Margaret Cavendish, Duchess of Newcastle. Cavendish was such a remarkable woman – a largely self-taught philosopher who had to endure much derision and condescension within the insufferably chauvinistic and misogynistic culture of natural philosophy – that one longs for her to be putting these self-satisfied men in their place. Sadly, she was on the contrary a stubborn and dogmatic traditionalist, and nearly always wrong, and were it not for her nobility she would probably have been almost entirely ignored.* As it was, she was given the respect due to a duchess, but her writings were greeted somewhat dismissively by most; only the Platonist and vitalist Henry More at Cambridge, whose ideas were close to hers, gave her much credit. Apparently compensating for her shyness by cultivating a flamboyant persona, she was accused of being bold, conceited and crazy: 'Mad Madge' was the epithet favoured by her critics.

Cavendish wrote not only philosophy but also poetry, plays and fiction. Her husband William Cavendish was an amateur scholar who owned telescopes and dabbled in alchemy, and as Royalists exiled in Paris during the English Civil War they had convened the so-called Newcastle Group that included Thomas Hobbes, the diplomat Kenelm Digby and the physician Walter Charleton. Margaret Cavendish was highly sceptical that the experimental philosophy could reveal much of significance about the natural world, and was convinced that the conventional Aristotelian approach of deductive reasoning from prior axioms was a sounder and more reliable route to knowledge.

Her objections to microscopy, described in her confrontational *Observations Upon Experimental Philosophy* (1666), rather typify her mode of criticism, which relies more on assertion of fault than on any demonstration of it, whether logical or empirical. She begins by saying 'I am not able to give a solid judgement of the art of "microscopy", and the several dioptrical instruments belonging thereto, by reason I have neither studied nor practised that art', only then to deliver precisely that:

> Of this I am confident, that this same art, with all its instruments, is not able to discover the interior natural motions of any part or creature of nature; nay, the question is, whether it can represent yet the exterior shapes and motions so exactly, as naturally they are; for art doth more easily alter than inform.

* Her credentials as an early feminist are moreover rather compromised by her suggestion that women, with practical skills well honed by domestic duties such as baking and sewing, should be allotted the tasks of experimental science so that men might be freed up for 'more profitable studies, than in useless experiments'. With stereotyping equal to that of any male writer of the times, she cautioned that women would however make poor alchemists because 'our sex is more apt to waste, than to make gold'.

In other words, Cavendish believed that the microscope lens distorts as it magnifies. The basis for this assertion is never explained, although she argues that if the tip of a needle or the edge of a blade really were as rounded and blunt as they seem in the microscope, they would never be able to pierce and cut as they do. It seems that Cavendish considers mere magnification itself to be a distortion: by making bigger, the lens somehow also swells and distends the specimen:

> artificial glasses . . . can present the natural figure of an object, yet that natural figure may be presented in as monstrous a shape, as it may appear misshapen rather than natural: For example, a louse by the help of a magnifying glass appears like a lobster, where the microscope enlarging and magnifying each part of it, makes them bigger and rounder than naturally they are. The truth is, the more the figure by art is magnified, the more it appears misshapen from the natural, insomuch as each joint will appear as a diseased, swelled and tumid body, ready and ripe for incision.

One can detect a feeling here that magnification is in fact distasteful: a horrid and almost immoral corruption of what it reveals. Cavendish is not merely distrustful of but actually repelled by what the microscope shows. In her satirical novel *The Description of a New World, Called The Blazing-World* (1668) (see page 373), published initially as an appendix to her *Observations*, the empress of the Blazing-World is shown a flea and a louse in the microscope, which 'appear'd so terrible to her sight, that they had almost put her into a swoon'.

This sense of indecency in the microscopic gaze is evident also in Swift's *Gulliver's Travels*. In the land of the gigantic Brobdingnagians, Gulliver is disgusted by their bodies when seen so close up: 'Their skins appeared so coarse and uneven, so variously coloured when I saw them near, with a mole here and there as broad as a trencher, and hairs hanging from it thicker than pack-threads.' Among the common folk he is repelled by the immense lice crawling on their clothes:

> I could distinctly see the limbs of these vermin with my naked eye, much better than those of an European louse through a microscope, and their snouts with which they rooted like swine. They were the

first I had ever beheld, and I should have been curious enough to dissect one of them, if I had proper instruments (which I unluckily left behind me in the ship) although indeed the sight was so nauseous, that it perfectly turned my stomach.

The indecorous aspect of magnification was held to be particularly unsettling for ladies of refinement. For Gulliver it becomes almost a metaphor for the threatened loss of social grace and manner when the personality is placed, as it were, under the microscope:

This made me reflect upon the fair skins of our English ladies, who appear so beautiful to us, only because they are of our own size, and their defects not to be seen but through a magnifying glass, where we find by experiment that the smoothest and whitest of skins look rough and coarse, and ill coloured.

Proportioned between the denizens of Brobdingnag and Lilliput, Gulliver shows us the contingency of the human scale, a mere waymarker on the journey from the infinitely small to the infinitely big. 'Undoubtedly,' he says, 'philosophers are in the right when they tell us, that nothing is great or little otherwise than by comparison.' Modern science has undoubtedly now outstripped our ability to intuit these extremes: we understand but cannot truly imagine the long and complex history that took place in the first second of the Big Bang, and are asked to contemplate infinite multiverses in eleven dimensions or more. That we are insensitive to all bar a trivially thin slice of space and time is surely, however, a blessing for our reason and sanity, as the Scottish poet James Thomson averred in his poem cycle *The Seasons* in the 1720s:

These, concealed
By the kind art of forming Heaven, escape
The grosser eye of man; for, if the worlds
In worlds enclosed should on his senses burst
From cates ambrosial, and the nectared bowl,
He would abhorrent turn, and in dead night,
When silence sleeps o'er all, be stunned with noise.

What do you see?

Cavendish's circular logic that only the eye itself is the 'best optic' – because only the eye can show us things as we really see them – is a desperate attempt to reject any obligation to look at the grotesque microworld. Making an object bigger *by definition* invalidates what you see, she says, precisely because it is bigger:

> those creatures, or parts of creatures, which by art appear bigger than naturally they are, cannot be judged according to their natural figure, since they do not appear in their natural shape; but in an artificial one, that is, in a shape or figure magnified by art, and extended beyond their natural figure; and . . . man cannot judge otherwise of a figure than it appears.

For the same reason, telescopes are to be distrusted when they show us new things in the heavens, simply because (unlike magnified towers twenty miles away) these things can't be there unless we can see them to begin with. For Cavendish, the fact that she cannot or will not conceive of something to be thus is all too ready an argument for why it is not.

She also played the now too-familiar card of making scientists' willingness to revise their ideas into prima facie evidence that they don't know what they are talking about. They are, she said, forever extolling the wonders of their instruments, only soon to be confessing doubts about them and then denouncing them altogether. 'By which it is evident', she wrote,

> that experimental philosophy has but a brittle, inconstant, and uncertain ground. And these artificial instruments, as microscopes, telescopes, and the like, which are now so highly applauded, who knows but they may within a short time have the same fate.

Suprisingly, the year after her *Observations* was published, Margaret Cavendish let it be known that she would like to attend one of the Royal Society's meetings. This request was put before the society in May 1667, and 'after much debate pro and con' according to Pepys's diary entry for the 30th, it was agreed. Among the diversions prepared for the duchess was 'a good microscope', arranged, naturally, by Hooke. At the meeting she comported herself regally, not deigning to engage in criticism but feigning admiration for all she saw. Pepys was disappointed that he did not 'hear her say anything that was worth hearing'. But John Evelyn composed a ballad to mark the occasion, in which he recorded the microscopical demonstration thus:

> But oh a stranger thing, this Dame
> A Glasse they shew'd with an hard name
> I cannot fix upon't
> That made a Louse to looke as big
> As any Sow that's great with pig
> Some Swore an Eliphant.

Evidently, Cavendish's 'admiration' did not mean she was persuaded; she reissued *Observations* and its satirical appendix *The Blazing-World* the following year.

Although Cavendish's conviction that the microscope deformed nature was capricious, there were valid questions about how exactly the images were to be interpreted. It was not just a matter of understanding a structure's function, but of even comprehending it as a structure in the first place. We are predisposed to recognize categories of object in the macroscopic world: to consider leafy things to be plants, faceted things to be minerals and crystals, and so forth. But no one could know which categories, which rules of thumb, applied at scales outside human acuity. The microscopical literature of this time abounds with macroscopic analogies, not just of form but of presumed function. Fungal tendrils are like velvet and carpets, insect cuticle is like military armour, and most famously, the chambers in a thin section of cork were compared by Hooke to the enclosed 'cells' of the monastic hermit – the origin of this word for a compartment of living tissue.*

---

* Hooke considered a cellular structure to be characteristic only of plant tissue, in which the cells are often relatively large and easy to see. A cell theory of all living organisms was proposed only in the nineteenth century by the German physiologist Theodore Schwann.

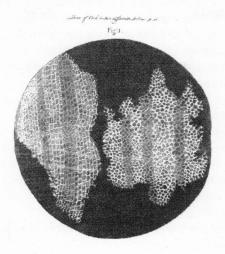

Robert Hooke's drawing of microscopic compartments in a piece of cork, which he called 'cells' – the first instance of this fundamental biological term.

All too often, it was far from clear exactly what manner of three-dimensional object was being seen in the artificially flattened micro-world. As Hooke himself admitted:

> It is exceeding difficult in some Objects, to distinguish between a *prominency* and a *depression*, between a *shadow* and a *black stain*, or a *reflection* and a *whiteness in the colour* . . . The Eyes of a Fly in one kind of light appears [*sic*] almost like a Lattice, drill'd through with an abundance of small holes . . . In the Sunshine they look like a surface cover'd with golden Nails, in another posture, like a Surface cover'd with pyramids; in another with Cones, and in other postures of quite other shapes.

For this reason, Hooke stressed that he did not make an interpretive drawing until he had looked at a specimen from several perspectives. But even that was not always a guarantee of seeing the right thing.

The greatest problem for early microscopy, however, was that it was largely a science ahead of its time. The details of nature's handi-work that the lens revealed were wonderful and unexpected, but there was (particularly in the case of living organisms) no theoretical frame-work for understanding them. Once the astonishment subsided at the discovery that nature goes on revealing order and form as we look ever more closely, the question became whether all these details could be interpreted, and whether they led us to a deeper understanding of the properties of matter. After the first flush of enthusiasm in the 1660s, it started to become clear that, on the whole, they do not.

Certainly, there were a few instances where microscopy genuinely advanced knowledge. Jan Swammerdam dissected insects and showed that the 'king' bee of a colony is actually a queen, possessing ovaries. His detailed observations of insect physiology helped him to charac-terize the life cycle of insects, from egg to larva, pupa and mature adult. Meanwhile, the Italian physician Marcello Malpighi at the University of Bologna saw that the network of blood vessels explored by William Harvey extends all the way to microscopic capillaries, belying the old idea that the blood circulates by simply soaking into the tissues. Malpighi's *De viscerum structura* (1666), which presented microscopic images of human organs, was a major contribution to seventeenth-century anatomy, and the Royal Society rightly considered Malpighi one of their most learned overseas correspondents. And Nehemiah Grew's microscopic investigations of plants suggested to him analogies with animal bodies – plants, he said, have bowels and lungs, while an animal is essentially 'several Plants bound up in one Volume'. Even if such comparisons were a little fanciful, they began to erode firm distinctions between different domains of the living world.

But successes were uncommon. Hooke's hope that the microscope would reveal the physical structure of memory – a 'Chain of Ideas coyled up in the Repository of the Brain' – came to nothing. Too often the microscopic structures were mute, fuelling the accusation that the experimental philosophers were indulging useless curiosity about irrelevant minutiae. The dream that microscopy would ulti-mately reveal the basic atoms of matter turned instead into a fear that greater magnification would show an endless regress of unintelligible structure. Criticizing the microscopic anatomy of Malpighi, the Bolognese physician Girolamo Sbaraglia wrote in 1689 that:

These studies seem to me to be gardens of Tantalus, where things
which appear to be something are actually nothing, so that they detain
our minds by a kind of sweetness; but after the damage has been done,
it is recognized that for medical practice all these things are inanities.

Some contemporary historians, such as Gaston Bachelard, have argued
that because it showed only images beautiful and grotesque rather
than anything interpretable and quantifiable, the microscope actually
impeded the advancement of learning. Others, such as Catherine
Wilson, say that on the contrary the divorce between observation and
what was immediately useful or meaningful was an essential development in science.

In any event, this barrier of incomprehensibility was surely one
reason why the enthusiasm for the microscope waned. By 1692, Hooke
was lamenting that no one was much interested in it any more, beyond
wealthy gentlemen – and increasingly, ladies – who considered it a
toy for their amusement (they could by then purchase pre-prepared
specimens which might, for a time, entertain their friends):

Much the same has been the Fate of Microscopes, as to their Invention,
Improvements, Use, Neglect, and Sighting, which are now reduced
almost to a single Votary, which is Mr *Leeuwenhoek*; besides whom, I
hear of none that make any other Use of that Instrument, but for
Diversion and Pastime, and that by reason it is become a portable
Instrument, and easy to be carried in one's Pocket.

Hooke was convinced that there was still much more to be discovered
with the microscope, but no one, aside from Leeuwenhoek (Malpighi
died in 1694) seemed any longer interested.

Could it be that the human mind was not designed to comprehend
the world on so small a scale?* John Locke, who was sympathetic to
but bemused by the activities of the Royal Society, stressed that this
understanding had limits: 'The Works of nature are contrived by a
Wisdom, and operate by ways too far surpassing our Faculties to

---

* That remains possible. When objects become so small as to have their behaviour
governed by quantum rather than classical Newtonian mechanics, this behaviour
often eludes intuition, even if it is accessible to precise mathematical description,
because the mind has not evolved under a constant exposure to quantum effects.

discover, or Capacities to conceive, for us ever to be able to reduce them into a Science.' This, he thought, was certainly true of the microscopic world, not least because he could not imagine that God would have placed anything worth knowing in places where it was inaccessible to our everyday senses. Indeed, he felt that if someone were to have vision as acute as that available from the microscope, he would then acquire a way of conceptualizing the world that was incongruent with that of ordinary people – he would 'be in a quite different world from other People', where

> Nothing would appear the same to him, and others: The visible *ideas* of every thing would be different. So that I doubt, whether he, and the rest of men, could discourse concerning the objects of sight: or have any communication about colours, their appearances being so wholly different.

In some degree this is a valid argument against excessive reductionism: we can't tell the time from the atoms of a clock. It also cautions against trying to interpret the microworld with reference to macroscopic concepts and objects: as Joseph Glanvill put it, 'there's but little resemblance between the *Mucous sperm*, and the completed *Animal*'. But it also reiterates the fact that experimental philosophy had revealed a plurality of worlds, while raising doubts about the extent to which they can communicate with one another. Put another way, it seemed to make the microscope an instrument that could serve curiosity but not science.

11 The Light of Nature

There is a *Master* of *Arts* in this *University* that when He shifts
Himself, emits such *sparks* so violently that they have been heard
to *crackle* like the *sparks* of *fire*.

Robert Plot, Natural History of Staffordshire (1686)

In America there are birds so luminous at night that one can use
them to read by.

Bernard de Fontenelle, Conversations on the Plurality of Worlds

Servants in Robert Boyle's household were accustomed to strange
demands at all times of the day and night. But one can only guess
at what passed through the mind of the footboy who was summoned
to attend Boyle's investigations into the way diamonds might be made
to glow. He stood by one night while his master plunged the precious
stone (worth goodness knows how many years' wages) into oil and
acid, spat on it, and (in Boyle's words) took it 'into bed with me, and
holding it a good while upon a warm part of my naked body'.

'The annals of seventeenth-century natural philosophy', say
Lorraine Daston and Katharine Park, 'abound with stories of interrupted meals, forsaken guests, and missed bedtimes as observers
dropped everything to devote themselves to a fleeting, fascinating
phenomenon.' Curiosity demands stamina, for nature will not fit its
timetable of revelations to the conveniences of human affairs.* Boyle

---

* Any scientist who has carried out an experiment involving, say, cell culturing or
cryogenics, knows this too well, recalling alarm clocks set to three in the morning
so that the nutrient will not be consumed or the coolant evaporated by breakfast-time.

might be inclined in the dead of night suddenly to call on his staff to bring a piece of rotten fish and place it inside his air pump.

On one occasion, however, it was the staff themselves who set off Pan's hunt. 'Yesterday, when I was about to go to bed', Boyle wrote to a friend in 1672,

> an amanuensis of mine, accustomed to make observations, informed me, that one of the servants of the house, going upon some occasion to the larder, was frighted by something luminous, that she saw (notwithstanding the darkness of the place) where the meat had been hung up before. Whereupon, suspending for a while my going to rest, I presently sent for the meat into my chamber, and caused it to be placed in a corner of the room capable of being made considerably dark, and then I plainly saw, both with wonder and delight, that the joint of meat did, in divers places, shine like rotten wood or stinking fish; which was so uncommon a sight, that I had presently thought of inviting you to be a sharer in the pleasure of it.

Boyle was at the time suffering from a head cold caught while trying out a new telescope, but nonetheless he stayed up for most of the night investigating this strange phenomenon.

As this passage makes clear, the emission of light (luminescence) from decaying organic materials such as wood and fish was already known. This new manifestation of the phenomenon filled Boyle with 'wonder and delight' – a frank confession of the role that these emotional responses still played in motivating curiosity.

The sheer spectacle afforded by luminescence was indeed one reason why it was awarded such attention by the virtuosi of the Royal Society. The effect raised some fundamental and profound questions, for the production of light without heat (the glow of hot metal being nothing to be remarked upon)* challenged prevailing theories of light itself. But there was no denying the pure wonderment that shining substances induce, and demonstrations of them had the power to captivate and entertain both the society's pleasure-seeking Fellows and the public

---

* That's to say, it was rendered unremarkable by familiarity. It turns out to be very hard to explain, however: a full account demands quantum physics, and it was in fact one of the problems that occasioned the birth of quantum theory at the beginning of the twentieth century.

at large. That sort of marvel had to be handled with great care, however, since it was liable to make the virtuosi seem no different from the cheap conjurors and mountebanks who drew crowds in public spaces.

The play of light has always occupied this ambiguous middle ground between science and miracle. Ever since the *fiat lux* with which all of Creation began, light was a spiritual entity. For early Christian writers it was transcendental, an emanation from God, celebrated in the Neoplatonic mysticism of the fifth-century Syrian writer known as Dionysius. Coloured light held particular significance, flooding from the great windows of the Gothic cathedrals to cast a kaleidoscopic mosaic on to the stones. And what could be more emblematic of the mystery and power of God than the rainbow, ethereal and ephemeral symbol of the Virgin, of the post-diluvian renewal of life, and also of the apocalypse? The shifting hues of the peacock's tail, meanwhile, symbolized the colour changes within the crucible during the Great Work of alchemy, tantalizingly mimicked by the iridescence of hot metal. The chromatic profusion emerging from the white peacock's egg was considered by early Gnostic writers as analogous to the way God created diversity from unity. Such symbolism was in no way distinct from the practical technologies of colour and light – for it was through alchemy, after all, that painters, dyers and glass-makers acquired their colours. When Roger Bacon and Robert Grosseteste investigated optics in the thirteenth century and began to explore how lenses and prisms magnify and elicit colour, their conception of light was still as much theological as it was mathematical. That religious significance had been eroded only a little by the time Robert Boyle came to examine light and colour and Isaac Newton began to unweave the rainbow. While Newton was discovering that the old image of a peacock's many-coloured tail emerging from whiteness applied to sunlight itself, he acquired an alchemical anthology containing a treatise attributed to Albertus Magnus. In this tract he will have read, concerning white, that 'All colours that can be conceived by men in the world appear there, and then they will be fixed and complete the Work in a single colour, that is the white, and in that all colours come together.' Curiosity about light was permeated with mystical and theological associations and assumptions. This was, as it always had been, a divine science.

Darkness visible

Luminescence or 'cold light' was recognized since classical times. In *De anima* (*On the Soul*), Aristotle says that 'Some things, indeed, are not seen in daylight, though they produce sensation in the dark: as for example the things of fiery and glittering appearance for which there is no distinguishing name, like fungus, horn* and the head scales and eyes of fish.' Pliny says that 'the putrified and rotten wood of some old trunk of an oak' glitters in the night, and that there is a type of fungus that 'shine[s] in the night, and by the light that it gives in the dark, men know where and how to gather it'. And of course anyone in the Mediterranean lands couldn't fail to be enraptured by the glimmer of fireflies and glow-worms. In the mid-sixteenth century, the encyclopaedist Conrad Gesner gave an exhaustive account of the 'rare and marvellous plants . . . that shine in darkness', and he also described luminescent animals and stones. The Bolognese naturalist Ulisse Aldrovandi claimed to have drafted a book on 'things that shine at night', although he did not live long enough to get it published. Girolamo Cardano reported seeing luminous fish on a trip to Scotland in his *De rerum varietate*, and the glow of rotten wood was mentioned by Della Porta and Francis Bacon.

Thus luminescence was so well-attested and established a phenomenon that Boyle's willingness to stay up all night in an ailing state to examine glowing meat from his pantry is a little surprising even for one so curious. But previously these things had tended to be considered marvels and wonders: the sort of rare and extraordinary circumstance that would appeal to cataloguers like Pliny and Gesner, but not deemed worthy of much attention from philosophers. Only the professors of secrets had previously attended closely to luminescent phenomena. Della Porta's *Natural Magick* lists the classical examples of glow-worms, fungi, fishes and rotten wood, and describes how to make a glowing liquid (*liquida lucida*) by mashing glow-worms, fermenting them in the heat of manure, and distilling the liquid. (This procedure is quite spurious: the light cannot be 'extracted' into a solution, as both Boyle and Thomas Moffett later pointed out.)

---

* It seems likely that this is a mistranslation: the Greek word for horn (*keras*) is easily confused with that for flesh (*kreas*), and so the reference may be to the luminescence of rotting fishes' flesh.

In *Advancement of Learning*, Francis Bacon explains that these glowing substances show that 'it is not the property of fire alone to give light'. The magical effect of luminescence from rotten wood was appreciated even in the nineteenth century, when Hans Christian Andersen depicts it in his fairy tale 'Ole Luk Oie, the Dream God'. It enchanted Henry David Thoreau one night in Maine Woods. It is caused by light-producing fungi and is a form of bioluminescence in which the light emission is caused by biochemical reactions involving oxygen. Essentially the same process causes the luminescence of fireflies, glow-worms and some microscopic marine creatures (which is what makes seawater itself sometimes seem to shine).

Phosphorescence in the modern sense is a different process. A phosphorescent material absorbs and stores light energy while it is illuminated, and re-emits it subsequently. However, in the seventeenth century any form of light emission without heat was called 'phosphorescence', and any substance that exhibited this effect was called a 'phosphorus', from the Greek for 'bearer of light', a name given to the morning star, Venus. For consistency with the contemporaneous accounts, I will hereafter use 'phosphorescence' in this older meaning.

Some forms of 'phosphorus' then known were, however, truly phosphorescent in the current sense. One was called the Bolonian or Bolognian stone, a mineral found in the vicinity of Bologna which glowed with a blue radiance in the dark after being exposed to light. To acquire this property the mineral must first be heated with charcoal, a process discovered in 1603 by the Bolognese shoemaker and alchemical dabbler Vincenzo Cascariolo.* Galileo knew of this stone, although he disputed the view of the philosopher Fortunio Liceti at Padua that the moon glowed not with reflected light but because it was made of similar phosphorescent material. John Evelyn saw the Bolonian stone during a journey to Italy in 1645; he witnessed a mountebank use it to summon a gawping crowd in a piazza in Rome, which typifies the company such wonders kept.

A sample of the Bolonian stone was sent to the Royal Society in 1673. Three years later the Fellows were pleased to receive a new type

---

* The mineral is barium sulphate, which is converted by chemical reaction with charcoal to the phosphorescent barium sulphide. It's not an easy process to conduct with seventeenth-century methods, and the details of Cascariolo's method are still being studied by chemical historians.

of luminous stone prepared by the Saxon chemist Christoph Adolph Balduin (or Baldewein), which was said to be a better 'phosphorus'. It was apparently made from chalk, and is thought to have been the phosphor calcium nitrate. Once it was tested and confirmed as a genuine phosphorus, Balduin was made an FRS.

The early *Philosophical Transactions* are full of reports of new types of luminescence. In the first issue, a Frenchman who examined '*Shining Worms* in Oysters' says that one had a rather fantastical appearance: 'red, and resembling the common *Glowworms*, found at land, with folds upon their backs, and feet like the former, and with a nose like that of a dog, and one eye in the head'. Robert Boyle was the foremost investigator of these things. He set out to establish by experiment which factors influence the glowing – hence the strange operations attested above on a diamond that was observed to emit light when rubbed vigorously. Boyle could even afford to break diamonds apart, noting how they would emit a flash as they fractured. Some of these gems belonged to the king himself; one, Boyle said, would shine after 'one brisk stroke of a bodkin'. This mechanically induced luminescence (which can be seen at less expense by crushing sugar) is now called triboluminescence.

Enlightened alchemy

In September 1677 a group of virtuosi convened for supper at Boyle's home, Ranelagh House in Pall Mall, after which they were treated to a dramatic demonstration of yet another new type of phosphorus. It was exhibited by a German chemist of Dresden named Daniel Krafft, who had come to London by royal invitation to display the marvel that he had already been touting around the courts of continental Europe for the princely fee of 1,000 thalers. Boyle and his fellow philosophers were delighted at this opportunity to see what they had only previously heard rumoured, and Boyle had invited Krafft to set out his wares before the luminaries of the Royal Society.

It was an entrancing evening. Krafft arrived with various vials, tubes and flasks of liquids and solids.* One of these, a glass sphere sealed

---

* Pure phosphorus (for this is what it was) melts at 44°C, and a little impurity will lower this temperature, so it was common for it to be prepared in both solid and molten form.

with wax, contained a reddish liquid that, according to Boyle, shone 'like a cannon bullet taken red hot out of the fire', albeit paler and fainter. Other vials gave off flashes and smoke when shaken. Then Krafft took a lump of glowing matter and, breaking it into pieces with his fingers, scattered them over Lady Ranelagh's fine Turkish carpet, where they glittered like stars (without, Boyle was relieved to discover, harming the carpet). Finally Krafft dipped his finger into the phosphorus and proceeded to write the glowing word DOMINI, which Boyle considered to possess 'a mixture of strangeness, beauty and frightfulness' – to the pious man, it must have seemed uncanny indeed. The visitor then wiped the residue from his finger on to his host's hand and cuff, leaving them streaked with radiance.

Nobody minded that the final demonstration – the ignition of warm gunpowder with phosphorus – didn't work, perhaps because the powder was old or damp. Krafft returned to Ranelagh House the next Saturday anyway to carry out this experiment with more success, and thereby to anticipate the safety match.

In the 1683 edition of his influential textbook of chemistry *Cours de chymie* (first published in 1675), the French chemist Nicolas Lémery credited Krafft with the discovery of this new phosphorus. It was a common mistake, not least because Krafft had encouraged others to believe it as he took this 'perpetual fire' (in Lémery's words) around Europe. But in fact Krafft hadn't made the stuff himself at all – he had bought it at the knock-down price of 200 thalers (along with a guarantee of a further supply on demand) from a cash-strapped alchemist of Hamburg named Hennig Brandt.

Rather little is known of Brandt before he discovered how to make phosphorus, except that he was a glass-maker and an alchemist of the old school who sought the philosopher's stone. His laboratory was financed by his two wives – by the dowry of the first and then, after she died, by the inherited wealth of the widow he married subsequently. Yet neither of these sources of income was sufficient for Brandt, who remained throughout his life desperate to win gold from alchemy by the more tangible means of selling his discoveries. Convinced that some crucial ingredient of the philosopher's stone might be extracted from urine (his reasoning is not clear: the golden colour is rather too glib an answer, although correspondences of colour were certainly an important part of alchemy), Brandt distilled vast quantities of it in 1669. Heating the solid residue that he had collected, he found that this turned into a glowing liquid that burst spontaneously into flame when it came in contact with air, and which gave off a disagreeable garlic aroma. Brandt was able to collect this glowing, inflammable substance as a soft solid which he kept more or less secret for six years while trying to convert it into the fabled stone.

This substance was the chemical element that today we uniquely call phosphorus. It offers one of the best examples of the counter-intuitive nature of chemistry, for in its pure elemental form it is highly hazardous – both toxic and liable to cause serious burns – and yet it is also an essential nutrient for health. Combined with oxygen as phosphate, it is a key component of bone and makes up part of the backbone of the chain-like nucleic acids DNA and RNA. Excess phosphorus in the body is excreted in urine. Phosphorus glows spontaneously in air because of chemiluminescence, akin to the glow of fireflies: reactions with oxygen in air on the surface of the substance give rise to chemical compounds that emit light.

So, unlike the Bolonian stone, this substance glowed perpetually without exposure to sunlight. It was impossible to keep secret something so extraordinary. When a visitor from Hamburg spoke about it at the University of Wittenberg, the professor of chemistry there, Johann Kunckel, decided to go to that city to see it for himself. He also wrote to his colleague Krafft to tell him of the story, a decision he was later to regret.

When Kunckel tracked down Brandt, he found him to be both untutored and frustratingly secretive about how his phosphorus was made. But Krafft had designs on it too, and he also made his way to Hamburg. His strategy was more to Brandt's liking: he was willing to pay. The rather unlikely story has it that Krafft and Brandt were in the midst of their negotiations when Kunckel turned up again at Brandt's door, imploring him for the secret of the phosphorus. To get rid of the unwelcome visitor, Brandt blurted out that it came from urine, but would say nothing more. Krafft then bought up all the phosphorus that Brandt possessed, and set off to make a fortune with it. Meanwhile, Kunckel began his own experiments on the distillation of urine, and by 1676 he succeeded in extracting some phosphorus. He too sought to capitalize on the wondrous appearance of this glowing substance by parading it before the wealthy and noble, while withholding the secret of its preparation.

Brandt's role in all this might have eventually been forgotten, had Krafft not demonstrated phosphorus before the Duke of Saxony in Hanover in 1677. Among the onlookers was the duke's librarian and historian, the young Gottfried Leibniz. When Leibniz visited Hamburg later that year and heard that there was a man in the city who also knew how to make this stuff, he tracked down Brandt and arranged for him to come to Hanover as court alchemist. This he did in 1678, and apparently he made some more phosphorus there. But he did not stay long and it isn't clear what became of him subsequently.*

Leibniz meanwhile wrote a letter to the Royal Society in 1677 describing the new phosphorus. At much the same time Robert Hooke received a letter from Kunckel which contained similar information. Meanwhile, Henry Oldenburg came across a description by the German chemist Georg Caspar Kirchmeyer of a 'Phosphorus Fulgurans' that shone without being first illuminated. A small piece of this stuff had been said still to be shining after being kept (presumably in the dark) for two years, so that 'if a considerably big piece were prepared of it, it would serve for a perpetual, or, at least, a very long lasting light'. Naturally, then, the London virtuosi were excited to learn that Krafft was coming later in the year with samples of what might be this very substance.

Boyle was obviously desperate to discover from Krafft how to make this marvellous phosphorus, but the German wasn't telling. Even when Boyle offered to trade that information for a chemical secret involving the preparation of a new 'mercury', all Krafft would say was that its source was 'somewhat that belonged to the body of man'. Boyle guessed that this was probably urine, and he set out to make it by distillation. He had no success for over a year, but in 1679 he procured the assistance of a young German chemist named Ambrose Godfrey Hanckwitz (known as Godfrey), who managed to extract phosphorus

---

* Although Brandt is now generally credited with the discovery of phosphorus, this is one of those chemical discoveries that might have been made several times earlier in the course of alchemical investigations. Paracelsus describes a protracted multiple distillation of urine that produces 'icicles, which are the element of fire', although he makes no more of this observation, which one would expect to have been considered remarkable if indeed the product was phosphorus.

the following year. Godfrey's success may have owed something to a trip he made to Hamburg to find Brandt and beg or buy more information – it seems that he discovered there that the final steps of the preparation required high temperatures to vaporize the phosphorus.

Godfrey's duty to supply Boyle with phosphorus compelled him to fill his cramped and meagre lodgings in London's Chandos Street with vats of human urine and faeces, from which the dangerous, flammable element was distilled. At the same time the house overflowed with the unappealing detritus of other experiments conducted on Boyle's behalf: 'the resuscitation of plants, the separation of sulphur in animals, and the mercurification of lead'. Here poor Godfrey, in his own words 'a poor young beginner, and a servant too', lodged as best he could in a single room with his wife. Things did not look good, and they were about to get worse.

Godfrey relates how in 1680 Boyle instructed him to accommodate a fellow countryman, a German alchemist who claimed to know how to make the philosopher's stone. Godfrey does not mention the German's name,* but he was apparently a Rosicrucian. The visitor duly arrived and began his work at Boyle's expense. Before long he complained that he missed his family, so Boyle arranged for him to be joined by his wife and daughter. For a time, says Godfrey, the German family 'were lodged . . . in my own bedchamber at Chandos Street, along with me and my wife all in one room'.

Eventually Boyle agreed to bear the cost of putting up the German alchemist in another house, where he set out his furnace and continued his labours, yet with never any sign of success. When Boyle began to entertain doubts that he would ever see a 'projection' of lead into gold, he cut the alchemist's salary. In a fury, the alchemist vented his spleen at Godfrey, splenetically assisted by his wife. Godfrey tells how

* It has been suggested that the man was the notorious German alchemist Johann Becher. The idea is tempting but seems untenable. It is true that Becher came to England after fleeing Holland in 1679 when he failed to deliver on his lavishly financed promise to turn sand into gold. Christian Philipp of Hamburg wrote to his friend Leibniz at the time saying that Becher, who the Dutch authorities would treat 'most rigorously if he allows himself to be found', had surfaced in England where 'he has found his dupes as always'. Boyle corresponded with him, and Becher later dedicated books to Boyle – for which reason alone it seems unlikely that Godfrey's fractious 'Crosey-Crucian', as he called him, was indeed Becher.

this 'terrible bawling creature' pursued his own wife in the street 'with scolding, spitting at her, and exclaiming at her, though, thank God, all in German, that the people understood her not'. At length Boyle finally accepted that he was being taken for a ride. He dismissed the alchemist from his service, causing the man to let fly another barrage at Godfrey before departing for his homeland. 'This', Godfrey mused, 'is the fruits of meddling with process-mongers.'

Despite these travails, Godfrey was soon providing his master with as much phosphorus as he could desire. Boyle often demonstrated to his guests Krafft's trick of writing with a finger dipped in the glowing substance, and he took to sleeping with a vial of it by his bedside, even musing that it might be used on the dials of clocks (as the yet more hazardous radioactive element radium was in the twentieth century)* so that the time could be read in the dark. He wondered if the glow of phosphorus might explain the will-o'-the-wisp, the lights seen over marshlands at night. (He was not far wrong: it seems this might be caused by the ignition of marsh gas, or methane, by bacterially produced phosphorus-containing gases.) For a few years it preoccupied him night and day – on one occasion he watched a phosphorus preparation crackle and flash for hours, until 'the late time of the night obliged me to go to bed before the experiment was finished'.

In 1680 Boyle deposited a sealed letter with the Royal Society that was to be opened only after his death, in which he described how phosphorus was made. It was a curious decision, given that the matter was already starting to become public knowledge; if nothing else, it indicates how in this chemical business Boyle was still very much habituated to the secretive traditions of alchemy. When he died in 1691, the note turned out to contain nothing at all surprising. The synthesis begins, he wrote, with a *'considerable quantity of Man's Urine'*

---

* The parallels with radium, which glows because of the high-energy radiation it emits, go further. Like phosphorus, radium can burn the fingers of the experimenter, though more insidiously because of radiation damage – radium's discoverers Marie and Pierre Curie were often left with blistered or reddened fingers. Godfrey found that phosphorus burns are extremely painful and slow to heal. And radium was, like phosphorus, initially seen in the wider world as primarily a source of amusement and entertainment, used for example in glowing cocktails. Moreover, it too became the ingredient of quack medical cures, an indication that light-giving substances retained a miraculous aura even in the twentieth century.

– on one estimate, around 5,000 litres are needed to produce one hundred grams of phosphorus. Distil this to the consistency of a 'somewhat thick Syrup', he wrote, and heat it with fine white sand, collecting the vapours. Eventually the mixture releases a substance that emits a 'faint blewish light, almost like that of little burning Matches dipt in Sulfur'. Finally there arrives a solid mass that falls to the bottom of the collecting flask – this is what you want.

Without disclosing these preparatory details, Boyle began to publish papers and booklets on phosphorus, or as he called it with reference to its heatless glow, the 'icy noctiluca'. Its enchanting displays made it ideal for the public demonstrations that the Royal Society staged to gain wider acceptance for the experimental philosophy – except that the evil-smelling fumes could offend the ladies. Boyle solved that problem by immersing phosphorus in oil of cinnamon or cloves, thereby adding 'to the pleasure of a delightful apparition that of a fragrant smell'.

It is nice to know that the beleaguered Godfrey eventually profited most of all. He became so proficient at making phosphorus – and discovered so avid a market for it – that he left Boyle's employ in 1682 and set up a workshop in Covent Garden. Three years later it had grown into a flourishing business. The demand grew as phosphorus became an ingredient in medical remedies (none of them effective, and some positively harmful), and Godfrey was soon exporting it to Europe at the rate of fifty shillings an ounce. By 1707 he was wealthy enough to open new premises on nearby Southampton Street, where he ran a pharmacy until his death in 1741. The German antiquary Zacharias Conrad von Uffenbach, who seems to have gone to every curious location in England during his 1710 visit, gave the following account of the premises:

> We went to the house of the well-known German chemist Gottfried . . . we bought from him a supply of English salt, etc. and saw his incomparably handsome laboratorium, which is both neatly and lavishly appointed, being also provided with all manner of curious stoves. For an ounce of salt we had to pay a shilling, but for the essence of lavender that goes with it . . . five shillings. We also purchased phosphorium at eight shillings, a drachma.

Godfrey studied the substance scientifically, and after publishing several papers on phosphorus chemistry in the *Philosophical*

Transactions he was elected an FRS in 1730. The fortunes of his business fluctuated after his death, but as Godfrey and Cooke (after his great-grandson's assistant Charles Cooke entered into partnership) it continued until 1916.

Wondrous light

The tale of phosphorus illustrates that the conventions of alchemy – secrecy, the quest for gold, the need for patronage, the possibility of exploiting privileged knowledge for financial gain, and the persistent role of wonder – were still ingrained in the fabric of science in the late seventeenth century. There was ambivalence about this. Chemistry was, in comparison to disciplines such as astronomy, mechanics and optics, still one of the least well understood of the sciences and among the most resistant to a mathematical treatment. It also carried the taint of a long history of fraudulent and mystical alchemy. Among the Paracelsians and Hermetic philosophers, said Joseph Glanvill in *Plus ultra: or, the Progress and Advancement of Knowledge Since the Days of Aristotle* (1668),

> Chymistry was very *phantastick*, *unintelligible*, and *delusive*; and the *boasts*, *vanity*, and *canting* of those *Spagyrists* brought a *scandal* upon the *Art*, and exposed it to *suspicion* and *contempt*.

Chemistry offered some of the most spectacular experiments, but that was itself a double-edged sword. These might impress princes and dukes and the less serious Fellows of the Royal Society, but they were also redolent of the conjuring tricks of charlatans (as Evelyn witnessed in Rome) and left the virtuosi open to the accusation that they were no better than these theatrical frauds. Robert Boyle's friend the physician Daniel Coxe wrote in 1674 that 'Chymistry doth daily present those, who are very conversant with her, a great number and variety of objects, highly diverting for their prettiness and curiosity in colour, figure and other accidents', such as dramatic colour-change experiments prepared for the Royal Society by Boyle. But 'diverting prettiness' was precisely the kind of frivolity for which the experimental philosophers were criticized, leading Boyle himself to caution (preaching as he did not always practice) that marvels and wonders should be judged 'by their use, not their strangeness or prettiness'. He claimed it was for mountebanks, not natural philosophers, 'to desire to have their discoveries rather admired than understood'.

What is more, chemical experiments were notoriously unreliable, so that the promise of spectacle could backfire. As Thomas Sprat confessed in his *History*:

> many *Experiments* are obnoxious to failing, either by reason of some *circumstances* which are scarce discernable [*sic*] till the work be over, or from the diversity of *materials*, whereof some may be *genuine*, some *sophisticated*, some *simple*, some *mix'd*, some *fresh*, some may have lost their *virtue*. And this is chiefly remarkable in *Chymical Operations*.

Phosphorus exemplified these problems. It glowed reliably enough, but was apt to burst into flame if it came too close to heat, as Kunckel discovered when he ventured near a fire with a blob of phosphorus in his pocket: the resulting conflagration 'burned all his clothes and his fingers also'. Pieces of the stuff could ignite if exposed even to the mild heat of the body – on at least one occasion Ambrose Godfrey was left with holes in his breeches, which Boyle confessed 'I could not look upon without some wonder as well as smiles'. Its rarity made it an object of desire, and evidently it (or the secret of making it) could become a tradable commodity among chemical adepts. But that made it seem like the kind of 'curiosity' associated both with purely acquisitive antiquarians and with the quasi-magical beliefs attached to *Wunderkammern*. (Its value also made it hard for serious philosophers to get hold of any.) Boyle's friend John Beale warned that the rarity of phosphorescent phenomena in general exposed them to 'suspicion of being Prodigies'. Phosphorus *sensu stricto* was particularly liable to be seen as a supernatural manifestation – Beale worried that experimenting on it would 'raise stories of Ghosts in my house', not unlike the fears among Boyle's servants of glowing meat in the storehouse. 'If we had a mind to act Pageantries', wrote Beale, 'or to spread a story of Goblins, you see how easily it might have been done, by smearing ones hands and face all over with the tincture of light.'

But so astonishing were the demonstrations of phosphorus that the virtuosi could not avoid, if not superstitious awe, then at least a sense of the natural magician's belief in microcosmic-macrocosmic correspondences. There is more than a hint of Paracelsus's alchemical cosmos in the account by John Evelyn of an experiment made by the physician and chemist Frederick Slare at Samuel Pepys's house in 1685:

> Dining at Mr Pepys', Dr Slayer showed us an experiment of a wonderful nature, pouring first a very cold liquor into a glass, and super-fusing in it another, to appearance cold and clear liquor, also; it first produced a white cloud, then boiling, divers coruscations and actual flames of fire mingled with the liquor, which being a little shaken together, fixed divers suns and stars of real fire, perfectly globular, on the sides of the glass, and which there stuck like so many constellations, burning most vehemently, and resembling stars and heavenly bodies, and that for a long space. It seemed to exhibit a theory of the eduction of light out of chaos, and the fixing or gathering of the universal light into luminous bodies.

Here in a glass flask was a new Creation, a miniature universe in the process of forming. The images of various types of 'phosphorus' in *De phosphoris observationis* (1681) by the German Johann Sigismund Elsholtz, physician to the Elector Friedrich Wilhelm of Brandenburg, who saw the demonstrations by Krafft, vividly portray this sense of a bottled alchemical cosmos.

Cold fire

If the Fellows of the Royal Society were to make a persuasive case that their interest in luminescence went beyond the gawping curiosity of the wonder-merchant, they needed to be able to say something about how it was caused. Christoph Balduin could do little than adduce a quasi-alchemical explanation that drew on the occult idea of sympathies:

> There resides in this phosphorus . . . the most inmost soul . . . of the innate and invisible philosophical fire, attracting by magnetical means the visible fire of the sun and emitting and throwing out in the dark its splendour in return.

This kind of speculation was precisely what Boyle wished to avoid, and he characteristically sought recourse in experiment, aiming to

Phosphorus fulgurans   *Phospho=rus*   *Stella=tus.*

Phospho=rus   *Nubi=losus.*   *Phospho=rus*   *Litera)=tus*

In *De phosphoris observationis* (1681), the German physician Johann Sigismund Elsholtz gives an intimation of the cosmic and alchemical associations of phosphorus's luminescence.

establish the facts of the phenomenon before any consideration of the mechanism. He placed some of his shining specimens in the air pump to see whether the withdrawal of air would change them. Glowing meat was dimmed by this procedure, but a luminescent diamond was not. He therefore called the former 'aerial noctiluca', and suggested that the emission depended on the same 'vital spirit' in air that was considered essential for respiration and combustion, which later chemists identified with phlogiston.* Frederick Slare was another of the Fellows who studied luminescence closely, and he proposed that, because phosphorus was extracted from a product of the human body, it might be the flammable matter (*flammula vitae*) that was thought to circulate in the blood – a reason, perhaps, for why it was also thought possible to extract phosphorus from blood.

Robert Hooke also investigated luminescence. As he explained, the puzzling thing about the light from 'rotten Wood, rotten Fish, Sea water, Gloworms, etc.' is that they 'have nothing of tangible heat in them', and yet 'they shine some of them so Vividly, that one may make a shift to read by them'.† The shining of fire and hot metal was generally considered to be explained by something like Descartes' corpuscular theory, in which light was excited by the agitation of the tiny particles of the all-pervasive Cartesian plenum and transmitted to the eye by the mechanical interactions of the intervening particles. But without heat there was no agitation, and so it wasn't easy to see how this idea could work for a phosphorus. Perhaps the luminescent diamond investigated by Boyle was agitated into glowing by being rubbed; but what of fireflies and rotten wood?

Hooke, picking up on the suggestion that phosphorescence might be linked to combustion and respiration, proposed in *Micrographia* that the particles are set in rapid motion by a process of putrefaction or fermentation:

> that all kind of *fiery burning Bodies* have their parts in motion, I think will be very easily granted me. That the *spark* struck from a Flint and Steel is in a rapid agitation, I have elsewhere made probable. And that the Parts of *rotten Wood*, *rotten Fish*, and the like are also in motion, I think, will as easily be conceded by those, who consider, that those parts never begin to Shine till the Bodies be in a State of putrefaction;

---

* As I mentioned earlier, this sort of chemiluminescence involves reactions with atmospheric oxygen, which was eventually substituted for phlogiston as its chemical mirror image. The glow of phosphorus needs air too.

† Hooke's apparently harmless comment, aiming merely to give some sense of the brightness, was seized upon for ridicule by the writer Thomas Shadwell in his satire *The Virtuoso* (see page 362), in which he wilfully misinterprets Hooke as implying that these decaying substances might be put to use for that end.

and that is now generally granted by all, to be caused by the motion of the parts of putrifying bodies.

Moreover, he argued that there is always some agent inducing motion and agitation in all other cases of luminescence:

That the *Bononian stone* shines no longer than it is either warmed by the Sun-beams, or by the flame of a Fire or of a Candle, is the general report of those that write of it, and of others that have seen it. And that heat argues a motion of the internal parts, is (as I said before) generally granted . . . a *Diamond* being *rub'd*, *struck*, or *heated* in the dark, Shines for a pretty while after, so long as that motion, which is imparted by any of those Agents, remains.

Hooke did however begin to draw some elementary distinctions between the various types of luminescence then known. He lists as one class the glowing of dead fish, fireflies, rotten wood and also 'the Rob [concentrate] of Urine found out by Dr Kunkell, and many others'; another class, 'such as shine by Impression of Light made upon them', includes the 'Bononian and Balduinian Phosphorus'; and thirdly, 'Substances which will shine with a degree of Motion or a little rubbing', such as diamonds. That these three groups correspond with what today we would regard as chemiluminescence, phosphorescence and triboluminescence is rather impressive.

Yet behind all these ideas and hypotheses lay a fundamental question that no one seemed able to answer satisfactorily: what is the nature of light?

Rainbows revealed

Isaac Newton's *Opticks* (1704) is considered one of the great scientific works conceived in the seventeenth century, the culmination of decades of thought and experiment on what light is and what it does. Although it did not go through the printing press until 1704, in principle it could have been published long before. The reason it was not is that Newton refused to release the work until his adversary Hooke was dead.

The dispute between these men did not stop at the question of who first deduced the inverse-square law of gravity. At least that was a phenomenon on which they both agreed. But the nature of light divided them, and neither would give ground. For Newton light had to have a material component: it was composed of a stream of corpuscles. But Hooke would contemplate it only as a wave passing through some rarefied medium, the aether.

It's now popular to proclaim that they were both right, for light can only be properly explained by quantum theory, within which the famous wave-particle duality allows light sometimes to behave as particles (called photons) and sometimes as waves (self-propagating electrical and magnetic fields). But Newton cannot really be justified on that account, for there was no experimental reason to invoke photon-like particles in his time. Hooke's view gives by far the better explanation of why light behaves in the ways it was then seen to do, and Newton's preference for particles must be seen more as dogged adherence (in this instance, at least) to the materialism of Descartes.

Yet we can't in honesty paint either Newton or Hooke as the progressive. The views of both men on light are a tangled mixture of new and old, of groundless assertion, acute intuition and careful observation. All the same, we would have to proclaim Newton the victor, for his biographer Richard Westfall is hardly overstating the case to say that *Opticks* 'dominated the science of optics with almost tyrannical authority' until the nineteenth century. Not by any means the least of its achievements was to provide the first complete and mostly satisfactory explanation for that ancient wondrous spectacle, the rainbow.

Newton was able to do so because he understood for the first time that sunlight carries all the rainbow colours within it. It is hardly any surprise that this conclusion was so long in coming, even though the rainbow attracted the attentions of philosophers at least since Aristotle, because it is so counter-intuitive. Who could have imagined that the sun's bright white beams are spun from red and blue, green and purple, especially since artists knew very well that mixing many colours made them only become progressively more muddy and dark? That, indeed, was one reason why Newton's theory of colour and light was long resisted by both artists and philosophers.

The prevailing view in the seventeenth century was Aristotelian: sunlight passing through a transparent substance such as a crystal prism or water emerges as a multicoloured spectrum because of alteration incurred in the passage. Aristotle seems to have perceived no direct connection between this effect and the formation of the rainbow, for in his *Meteorologica* he asserts simply that the rainbow is light reflected from a distant cloud. Moreover, in line with the common notion that sight was something transmitted from the eye to the object rather than vice versa, Aristotle understood the light's trajectory to begin at the observer and bounce off the cloud towards the sun.

Why should this produce a semicircular arc? Aristotle didn't really have an answer. He proposed that the observer stands at the apex of a cone of reflected rays, but seems to imply that the mere assertion of this geometry is sufficient explanation – an example of the classical predilection for arguing from geometric principles rather than physical mechanisms. The colours meanwhile – and for Aristotle there were just three (red, green and purple), the others arising from the 'contrast' at the boundaries – were said to be produced on the assumption that all colour is a mixture of light and dark. The reflections somehow darken the light and therefore tint it.

Centuries of confusion about the mechanism of rainbow formation stemmed from a failure to distinguish two different optical processes: reflection and refraction. In the first, light bounces off a surface. In the second, light passes from one transparent medium to another, for example from air into water or glass.* As it does so, the light rays are bent. Ancient and medieval philosophers had some inkling of these distinctions, but only as murky shapes in their mental landscape. Robert Grosseteste asserted that the rainbow is primarily a refractive phenomenon, but he sometimes treated reflection and refraction as synonyms. His was a complicated model in which sunlight is refracted by its passage through a spherical watery cloud, which acts as a kind of lens, and is there projected on to a second, screen-like cloud, from where it is reflected to the eye. This rescues the empirical fact that we see a rainbow with the sun behind rather than in front of us, but at the cost of a dishevelled jumble of ideas. Grosseteste's protégé Roger Bacon showed his mentor little filial respect when he stated bluntly that 'those are in error who say that the bow is caused by refraction' – his theory depended instead on reflection.

---

* More precisely, refraction happens when light passes between media that have different refractive indices, a measure of how fast light travels in them. One way of understanding the effect is to say that light always finds the fastest possible path, and this implies some deviation of the rays when the speed changes.

For Grosseteste and Bacon the nature of light was not simply a puzzle set by nature for the curious mind. It was as much a theological as a philosophical question, and they surely chose to devote so much energy to optical investigations because the Neoplatonic reverence for light would leave one expecting to find there a manifestation of God's presence. The metaphor of Gnostic divine illumination – of light as a source and vehicle of revelation – is preserved in the Renaissance natural magician's notion of the Light of Nature, from which by degrees it was secularized during the seventeenth century into the notion of an Enlightenment. This juxtaposition of revelatory mysticism and scientific discovery is eloquently expressed in Joseph Wright of Derby's famous depiction of Hennig Brandt's discovery of phosphorus.

Theories of rainbow formation in the early seventeenth century advanced the understanding only a little. William Gilbert's celebrated

The Alchymist in Search of the Philosopher's Stone (1771) by Joseph Wright of Derby offers a dramatic image of Hennig Brandt discovering phosphorus.

insistence on experience is little in evidence in his suggestion that rainbows are caused by light reflected from some dark obstacle like a mountain or cloud on to a misty vapour between it and the observer. Like Aristotle, he attributes the colours to a diminution of light intensity, with red being the most intense. The science historian Carl Boyer offers a harsh but fair judgement in calling this theory 'vague, speculative, and wearisome'. And that of Johannes Kepler is arguably even worse, for he insists on dividing the bow in two at the yellow band, making all the colours on the red side the result of reflection (which again progressively diminishes the brightness) while those towards the purple are caused by the 'tinting' of light through refraction. It doesn't help that he too sometimes makes these two optical processes synonymous, nor that he places the bow between the observer and the sun. All the same, Kepler did eventually formulate the basis of the right approach, for he decided that the optical effect originated not in reflective clouds but in the paths of light rays through individual raindrops. He perceived that a ray entering the spherical prism of a droplet would be redirected on to a new path first by being refracted, then reflected from the inner surface of the drop, and then refracted again as it exits. He figured that, by analogy with a regular prism, refraction has something to do with the production of the rainbow spectrum of colours, although he had no explanation why. He was hampered by, among other things, not knowing the correct angles of refraction; Thomas Harriot later worked those out.

So most of the pieces were in place by the time Descartes decided that the rainbow would be the ideal vehicle for demonstrating the new scientific methodology described in his *Discours de la méthode* (1637). It 'is such a remarkable phenomenon of nature, and its cause has been so meticulously sought after by enquiring minds throughout the ages', he wrote in the short chapter in which he finally lays out the correct underlying optical principles. To judge from a letter he wrote to Mersenne in 1629, he had already solved this problem eight years earlier, although he admitted there that it was 'the matter that has given me the greatest difficulty'.

Descartes states that, as Kepler had hinted, rays of sunlight are diverted by droplets in the rainbow due to 'two refractions and one reflection'. He was able to show why this will produce an arc that always seems to appear face-on to the observer. But the one thing

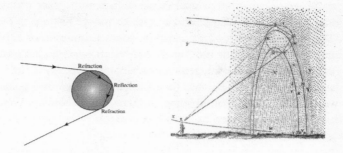

Johannes Kepler understood that the rainbow is created by light reflected from the inside of a raindrop. It is refracted twice along this path: once when it enters the droplet, and once when it exits (left). In 1637 René Descartes used this idea to explain why the rainbow always appears as an arc with the sun behind the observer (right).

Descartes couldn't explain properly is the spectrum of colours. His geometric optics essentially produces a white bow, and Descartes superimposed the colours by appeal to his spurious theory of light in which it is a sort of pressure transmitted from the source to the eye by the particles of his plenum. These particles, he argued, are like little rotating balls, and colour is produced if the balls rotate faster or slower as a consequence of refraction.

Newton finished the job. His famous experiment proving that refraction splits white light into its component colours was performed while he was studying at Trinity College in Cambridge in 1664–5; the exact date is not clear, but if it was before January 1665 then Newton had not yet even graduated. It seems likely, however, that his work on light and optics was refined, if not initiated, not in Cambridge but at his family home in Woolsthorpe, Lincolnshire, to which he retreated in 1665 after the plague forced the university to close. The work is recorded in the notebook *Questiones quaedam philosophicae* (*Certain philosophical questions*) that he began as an undergraduate.

Optical illusions

Newton's *experimentum crucis*, as he called it, had two components. First, he showed that the spectrum produced when a beam of sunlight shines through a slit into a darkened room and passes through a prism can be recombined into a shaft of white light by passing it through a focusing lens. Second, he used a mask to isolate just a single colour from the spectrum, and passed this alone through a second prism, finding that the colour did not change. Both experiments refute the idea that the light is somehow tinted by refraction during its passage through a prism or lens. And they show that the spectral colours are fundamental, irreducible to further subdivisions. Newton concluded that white light is constituted from 'differently refrangible' rays of the spectral colours.

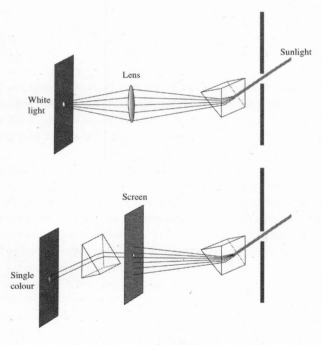

Newton's *experimentum crucis* proved that white light was composed of coloured light. He allowed a shaft of light (on the right) to pass through a hole in the window shutter and fall on a glass prism. This produces a spectrum of light which is then focused by a convex lens to reconstitute the spectrum as a beam of white light. The second component of the experiment involved a screen that allowed just one of the coloured spectral bands to pass. This light was unaltered in colour when it passed through a second prism.

This, as he later showed, accounts for the bands of the rainbow. Refraction in a raindrop bends purple light more strongly than red, and so the purple band appears on the inside and the red on the outside edge. Implicit in this, although not recognized by Newton, is the fact that the edges of the rainbow are the points where the wavelengths of the refracted rays pass beyond the visible spectrum into the infrared (outside) and the ultraviolet (inside), which we cannot see. Newton did not know why light should show this colour-dependent refraction, but he speculated that it was because the particles of red light have more momentum than purple and blue and are therefore less easily deflected from their path. That could be because they are more massive, or because they move faster; he vacillated between these alternatives.

Newton's crucial experiment was not necessarily as definitive as is sometimes implied. For one thing, the lens recombines the spectrum into white only at one spot (the focal point), beyond which they diverge again. So one could conceivably argue that this is an incidental artefact. And modern reconstructions have shown that it is by no means easy to excise cleanly a small, monochromatic band of the spectrum and pass it through the second prism. When the French scientist Edmé Mariotte tried in 1681 to reproduce the experiment for the Académie des Sciences, he found that the second refraction did produce new colours. A successful replication of Newton's result often demands that his recorded procedure be modified or augmented. It therefore seems quite likely that Newton's description involved a certain amount

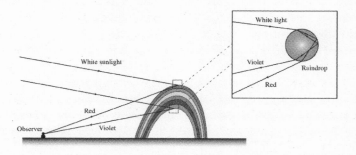

Newton brought light into the rainbow by recognizing that different spectral colours are refracted by the raindrops to different degrees.

of idealization – that, in a quasi-medieval manner, he was describing what ought to happen rather than what did. As historian Rupert Hall says:

> While Newton certainly did make what he called his 'crucial experiment' in one way or another, as his notebooks and lectures reveal, it was never perhaps exactly as he so confidently described it in his letter [published in the *Philosophical Transactions*], and this experiment never carried, in the historical development of Newton's optical ideas, the uniquely decisive role there assigned to it.

All this makes it the more noteworthy that Newton seemed determined to locate his conclusions beyond serious doubt. Eschewing the diffidence about interpretation that was characteristic of Boyle (and encouraged in all the experimental philosophers), he claimed when submitting his work for publication that

> what I shall tell concerning them is not an Hypothesis but most rigid consequence, not conjectured by barely inferring 'tis thus because not otherwise or because it satisfies all phenomena (the Philosophers universal Topick) but evinced by the mediation of experiments concluding directly & without any suspicion of doubt.

In other words, his interpretation was not a proposal consistent with the 'facts' (as Boyle would have described the experimental observations), but was an inevitable and indisputable corollary of them. This confidence in the experimenter's ability to move beyond observation directly to mechanism is something that marks Newton apart from most members of the Royal Society. And it is a style of science that sat uncomfortably with the one adduced by Francis Bacon.*

Soon after he returned to Cambridge when it reopened in 1667, Newton began his rise to fame. Around 1668 the Lucasian Professor of Mathematics at Trinity, Isaac Barrow, recognized Newton's precocity and asked the young man to help him prepare for publication his

---

* It isn't clear that Newton read Bacon with any great attention. His library contained only the anthology *Opuscula varia posthuma* and the *Essays*, apparently unused.

lecture notes on optics. This left Newton well placed to take Barrow's place in late 1669 when he vacated the chair to study divinity. Suddenly the 27-year-old Newton was a somebody at the prestigious university, although that in itself probably meant little to the man who had already decided the place was rife with time-marking, pension-drawing scholastics.

His position, however, gave Newton the confidence to send his 'New Theory of Light and Colours' to the Royal Society in 1672. His self-built miniature reflecting telescope had already brought him to the society's attention the previous year, and as a result Newton had been elected (at the suggestion of Seth Ward) a Fellow in January. Hearing this news, he wrote to say that

> I am purposing . . . to be considered of & examined, an account of a philosophical discovery which induced me to the making of the said telescope, & which I doubt not but will prove much more grateful then the communication of that instrument, being in my judgement the oddest if not the most considerable detection which has hitherto been made in the operations of Nature.

It was Newton's experiments on optics that had led him to consider reflecting telescopes the only way to avoid chromatic aberration (a distortion of images owing to the differential refraction of light of different wavelengths). In any event, his paper on light followed in February.

In London it was scrutinized by Robert Hooke, as unwilling as ever to admit novelty in anyone whose interests intersected with his own (particularly if their opinions did not agree). The experiments may very well be correct, Hooke allowed (he even confessed to being 'not a little pleased with the niceness and curiosity of his observations'), but he insisted that the interpretation was flawed and that only his wave theory of light could fully account for them:

> For all the experiments & observations I have hitherto made, nay and even those very experiments which he alleged, do seem to me to prove that light is nothing but a pulse or motion propagated through an homogeneous, uniform and transparent medium.

Hooke was right that waves are needed to explain refraction properly – but not for the reasons he gave.* Besides, at the root of Hooke's disagreements was not just his wave conception of light but his allegiance to the idea that colour comes from modification of refracted rays. He suggested that the spectrum is produced because the light pulse is distorted as it travels through the prism. The front of the pulse encounters the greatest resistance from the medium, and it prepares the way for the rear of the pulse to pass more easily, like a party of explorers forging in procession through long grass. As a result, the light is somehow 'twisted' in a manner that creates the sensation of colour.

Newton's paper was quickly published all the same, but he was infuriated by Hooke's attack. There is little indication that any of the Fellows discerned its true worth; after all, as Rupert Hall says, 'To challenge the perfectly uniform simplicity of white light . . . was to upset all received wisdom.' Christiaan Huygens was taken by Newton's ideas, however: 'The theory of Mr Newton concerning light and colours appears highly ingenious to me', he said.

Newton's annoyance deepened when Hooke published his critical report in the *Philosophical Transactions*, prompting Newton to send Oldenburg an acerbic letter in June suggesting that his opponent 'knows well it is not for one man to prescribe Rules to the studies of another, especially not without understanding the grounds on which he proceeds'. His letter contained a point-by-point rebuttal of each of Hooke's criticisms, and when it was read to the society, Hooke was deemed to have gone too far and was, doubtless to his own fury, instructed to reconsider his views.

Nonetheless, Newton was so wounded that he vowed not to bother publishing anything again. When it was suggested in October of that year that he prepare his own lectures on optics at Trinity for publication – the material that eventually became the *Opticks* – he retorted that he had found 'already by that little use I have made of the Presse, that I shall not enjoy my former serene liberty till I have done with it'. Better not to publish at all than to do so and

---

* In fact Newton saw closer to the truth of that matter than Hooke did. He admitted that a wave model might explain his results if the waves decrease in wavelength from red to violet, as in truth they do. Hooke did not consider colour to be linked to the wavelength of the rays.

suffer his critics. In March of 1673 he wrote to Oldenburg to say furthermore that

> I desire that you procure that I may be put out from being any longer a Fellow of the R. Society. For, though I honour that body, yet since I see I shall neither profit them, nor (by reason of this distance) can partake of the advantage of their assemblies, I desire to withdraw.

In the end the diplomatic Oldenburg let the storm pass without accepting Newton's resignation, but that did not prevent his 'withdrawal'. His feud with Hooke blew hot and cold in private correspondence for several years, sometimes marked with false civility: 'I do justly value our excellent disquisitions' (Hooke), 'what passes between friends in private usually deserves the name of consultation rather than conquest, & so I hope it will prove between you & me' (Newton). But Newton wrote ever less to the Royal Society, and after Hooke became its secretary following Oldenburg's death in 1677 he resigned himself to work in isolation in Cambridge – a solitude broken only when Edmond Halley goaded him into transforming celestial and terrestrial mechanics by writing the *Principia*.

More interference

The conception of light that Newton revealed in *Opticks* is, perhaps like Newton himself, a blend of the uncannily prescient and the incongruously mystical. His seven spectral colours – the red, orange, yellow, green, blue, indigo and violet even now recited by schoolchildren – are to some degree arbitrary divisions of the rainbow predicated on a spurious analogy with the seven notes of the musical scale: a cosmic correspondence reminiscent of Kepler's Neoplatonism. Newton allowed that light might have a wave aspect too, saying that its particles may excite vibrations in an all-pervasive aether when they strike a surface, and that these vibrations may be involved in vision, carrying sensations from the eye to the brain. The concept of light as aethereal vibrations persisted until Einstein's theory of special relativity dispensed with the concept of a disembodied 'ether' altogether in the early twentieth century. Yet Newton's aether seems rather to belong to the fanciful imaginings of a Robert Fludd or Athanasius Kircher. 'Perhaps', he wrote in 1675,

> the whole frame of nature may be nothing but various contextures of some certain aetherial spirits, or vapours, condensed as it were by precipitation, much after the manner, that vapours are condensed into water . . . Thus perhaps may all things be originated from aether.

You could choose to regard this as an anticipation of quantum theory, in which particles appear as nothing more than energy fields. But it is probably more aptly seen as a relic of Paracelsian alchemical philosophy.

The *Opticks* makes it clear that Newton wanted to make the science of light and optics a branch of mechanics: a question of particles moving in space, interacting via forces. He agreed with Boyle's suggestion in his 1664 treatise on colour that the colours of physical objects were caused by irregularities on their surfaces that 'interrupt' the light in some manner. 'It is not impossible', he wrote, 'but that microscopes may at length be improved to the discovery of the corpuscles of bodies, on which their surface colours depend.'*

Microscopic studies revealed another aspect of colour that Newton struggled to explain. Among the substances that Robert Hooke placed under the lens was the mineral called muscovy-glass or *lapis specularis*, now known as mica, which Hooke professed to have 'as many Curiosities in its Fabrick as any common Mineral I have met with.' It could be easily split into transparent, flexible flakes 'which with care and diligence may be slit into pieces so exceedingly thin as to be hardly perceivable by the eye'. 'I found', wrote Hooke,

> that up and down in several parts of them I could plainly perceive several white specks or flaws, and others diversely coloured with all

---

* Most colours of objects arise from the absorption of some wavelengths of the incident light, stripping away part of the spectrum from the reflected light so that it appears coloured. But some objects are indeed coloured because of their physical textures: butterfly scales, bird feathers and insect cuticle contain arrays of microscopic layers that cause interference between reflected light rays, resulting in the preferential reflection only of particular wavelengths. Understanding the basis of these 'physical colours', however, requires a wave model of light.

the Colours of the *Rainbow*; and with the *Microscope* I could perceive, that these Colours were ranged in rings that incompassed the white speck or flaw, and were round or irregular, according to the shape of the spot which they terminated; and the position of Colours, in respect of one another, was the very same as in the *Rainbow*. The consecution of these Colours from the middle of the spot outward being Blew, Purple, Scarlet, Yellow, Green; Blew, Purple, Scarlet, and so onwards, sometimes half a score times repeated.

Hooke wondered 'whether from these causes . . . may not be deduced the true causes of the production of all kinds of Colours.' He compared the rings to the glorious hues of the peacock's tail and of mother-of-pearl, but felt that they might be more easily understood because they arise in bodies that are 'simpler and more regular'. This led him into his lengthy discussion of light as a consequence of the agitation of vibrating corpuscles. He takes issue, however, with Descartes' idea that colour is imparted to the rainbow by rotation of the 'globules' of aether between the raindrop and the eye. Hooke offers instead his rather vague notion of tinting by a 'confusion' of the rays as they are refracted.

Newton considered this phenomenon at length in the second book of the *Opticks*, thereby seizing possession of it so that poor Hooke was forgotten (again) and the effect became known as Newton's rings. He could reproduce the rings by laying together a convex lens and a flat glass plate, and he made very precise measurements of the distances between the rings, looking at the effects of pressing the glass surfaces together or putting water in the intervening gap. He too tried to offer an explanation based on refraction. But that simply doesn't work, for the rings are caused by reflection, being a consequence of the wave nature of light. Rays reflected off the upper and lower surface of the mica plates, or off the glass surfaces that border the air gap for Newton's lenses, will interfere with one another. When the peaks and troughs of the waves coincide with one another, they reinforce by 'constructive' interference and enhance the brightness of the rays. When they are perfectly out of step, they cancel out and produce darkness. The relationship between the two waves depends on the distance between the two reflecting surfaces, and also on the wavelength, and thus the colour, of the light. So the positions of the regions

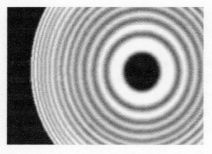
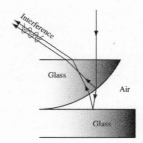

The concentric patterns of light and dark that appear when a lens is placed on a flat glass plate are called Newton's rings (left). They are caused by interference between light waves reflected from the inner surface of the lens and those that pass through to be reflected from the plate (right).

of constructive interference vary with colour, separating the reflected light into rainbow bands. None of this was fully understood until the wave theory of light became accepted in the nineteenth century.

Compared with atmospheric rainbows, Newton's rings seem small beer. That's precisely the point: from here on, everything was relevant. A phenomenon did not need to possess the grandeur of the rainbow before natural philosophers would deem it worth attending to. Two of the foremost virtuosi of the age could thus be found at loggerheads over an effect that almost no one else had ever seen, which seemed unlikely to have even the slightest practical significance, and which in earlier times might, if it could have been noticed at all, have been dismissed as just one of nature's imponderable quirks. In fact this effect held clues to the central puzzle of the whole discipline of optics: the nature of light. If neither Hooke nor Newton was really aware of that, no matter. The fact was that the smallest, most incidental questions now had to be heeded, because one could no longer be sure where they might lead.

12 Chasing Elephants

'Tis below a virtuoso to trouble himself with men and manners,
I study insects.

> Sir Nicholas Gimcrack in Thomas Shadwell's *The Virtuoso*
> (1676)

for his excellence
In height'ning of words, and shad'wing sense,
And magnifying all he writ,
With curious microscopic wit,
Was magnify'd himself no less
In home and foreign colleges.

> Samuel Butler, 'The Elephant in the Moon' (*c*.1670s)

While taking coffee with his drinking friends in late May of 1676, Robert Hooke was told about a new play by Thomas Shadwell, being performed by The Duke's Company in their splendid Dorset Garden Theatre on the riverfront at Whitefriars. The Duke's Company, initially managed by the Poet Laureate Sir William Davenant, was one of the two theatre companies instated by the spectacle-loving Charles II after the dour years of Cromwell's Interregnum, and its patron was Charles's brother the Duke of York, the future king James II. King Charles had been to see the production, Hooke's colleagues told him, and they urged him to go too. Whether or not they revealed much about the play's content is not known, but since one of those present was the gossipy John Aubrey, it's tempting to suppose that they fed Hooke just enough information to spur him on. But perhaps the title alone was sufficient bait: *The Virtuoso*.

Hooke went to the Dorset Garden a week later in the company of his friends, the watchmaker Thomas Tompion, the clerk John Godfrey and the surveyor John Oliver. He did not have a good evening.

It turned out that *The Virtuoso* was a wicked satire on the activities of the Royal Society, personified by the bumptious clown Sir Nicholas Gimcrack. This gentleman is first encountered in his study lying face down on a table and imitating the motions of a frog in a bowl of water. He is, he explains, learning to swim. When his visitors express scepticism that this method is going to be very effective, Gimcrack replies:

> I content myself with the speculative part of swimming; I care not for the practic[al]. I seldom bring anything to use; 'tis not my way. Knowledge is my ultimate end.

That, by implication, was all the Fellows of the Royal Society were doing: wasting their days in absurd experiments and observations that were not of the slightest use to anyone, nor even eludicating the phenomena they supposedly explored. They are constantly claiming to be capable of marvellous things, yet these claims are always hypothetical and never actually realized.

Worst of all, it became increasingly apparent to Hooke that *he* was the model for Gimcrack. Shadwell's ridiculous virtuoso experimented with an air pump and studied and weighed varieties of 'air'. In Gimcrack's claim to be (again in theory only) 'so much advanc'd in the art of flying that I can already outfly that ponderous animal call'd a bustard, nor should any greyhound in England catch me in the calmest day before I get upon wing', there seemed an allusion to Hooke's assertions to have made a flying machine. Most revealing of all was the description of her uncle by Gimcrack's niece: 'A sot that has spent two thousand pounds in microscopes to find out the nature of eels in vinegar, mites in a cheese, and the blue of plums', all of them being Hooke's subjects in *Micrographia*.

Hooke was enraged. 'Damned dogs', he snarled in his diary that night, '*Vindica Me Deus*' (God avenge me). He suspected with anguish that the personal attack was noticed by others. 'People almost pointed', he fumed, and he was sure for weeks afterwards that his acquaintances at the coffee-houses were laughing behind his back.

Shadwell's mockery was but one face of the attacks launched against the new experimental philosophy and its limitless curiosity. He suggests that all this effort, this scribbling and observation and operation of instruments, is ludicrous when directed at the most inconsequential of questions, such as the structure of a gnat's leg or the weight of air. Others responded to such antics not with laughter but with moral condemnation. No longer was the objection so much a supposed impiety at prying into the secret recesses of God's creation (although the perceived materialism of many of the new philosophers led to accusations of atheism); rather, it was seen as a dereliction of one's civil and religious duties to spend one's time so unproductively. And we have heard already of the philosophical misgivings of Thomas Hobbes, who was unconvinced that significant truths about the world could be gleaned from experiments. The critics were thus diverse in motive and method. Some were social, philosophical or political conservatives, some sceptical free-thinkers, others had different ideas about how the world was to be explained. But if they can be said to share one thing in common, it is a distrust of asking questions, or of asking them in the wrong way or for the wrong reasons. If curiosity could no longer be so easily condemned by traditionalists and moralists as the Devil's work, then new objections would have to be found.

These attacks hurt. Between the 1670s and the 1720s the Royal Society struggled to survive. Its membership fell from 187 in 1671 to 116 in 1691, and because its funding came largely from subscription fees, its finances were left in disarray. The decline partly reflects an inevitable cooling of interest from the gentlemen dilettantes who had signed up initially to what seemed like a novel divertissement, only to discover that the virtuosi did not afford their entertainment the highest priority. But the various broadsides at the curious enthusiasms and the enthusiastic curiosity of the new philosophers also took their toll, and by the middle of the eighteenth century the intellectual world was searching elsewhere for its diversions. Experimental science had by then become too well established to wither away, but it developed a very different flavour that partook ever less in the culture of the inquisitive virtuoso.

Disturbing the peace

Given that the new philosophy's declared agenda was to supplant the old, it was inevitable that some of its most vocal critics were traditionalists protecting their revered Aristotle. Their arguments were sometimes rancorous and ad hominem, and they were prepared to enlist any means to discredit their opponents. The Puritan physician and classicist Henry Stubbe has often been placed, perhaps unfairly, in this camp. His polemics against the Royal Society could be interpreted variously as a personal vendetta, a defence of Aristotle, or a subtle form of religious or anti-Royalist dissent (Stubbe nearly got himself hanged for his literary attack on the Duke of York). He was nevertheless a friend of Robert Boyle and often invoked the great scientist in his support; his spleen was reserved mostly for Joseph Glanvill and Thomas Sprat, whom he regarded as arrogant virtuosi who needed cutting down to size. You could at least see why. In *Plus ultra*, Glanvill pulled no punches, accusing the universities of being filled with 'subtile, Eagle-ey'd Schoolmen, who see Conclusions so far off through the more unerring telescopes of their own piercing Understandings', and with complacent physicians who 'never brought the World so much practical, beneficial knowledge, as would help towards the Cure of a Cut Finger'. Stubbe responded in *The Plus ultra reduced to a Non Plus* (1670), accusing Glanvill of endangering the whole tradition of medicine. (It slightly weakened his case that he was paid to make it by an irate member of the Royal College of Physicians.) Meanwhile, Stubbe called Sprat's *History of the Royal Society* a threat to the Church, the state and the universities, and for good measure he added the common charge that the mechanical philosophy was inevitably atheistic. The assessment of the English antiquary Anthony Wood that Stubbe had a 'hot and restless head' seems to have been fairly accurate, whether or not Wood was right to ascribe it to his 'carrot-coloured' hair.

Stubbe can't be dismissed as a mere reactionary, however. Historian James Jacob claims that he was an advocate of civic reform, support of the poor and religious toleration, and challenged the Church's self-serving assertions of divine intervention in human affairs. He differed from the Royal Society not so much in terms of a commitment to change but over the form it should take.

The Anglo-Swiss scholar Meric Casaubon was another influential critic of the new philosophy. Sharing Stubbe's concern that it could lead to atheism and undermine the social order, Casaubon railed against a seventeenth-century version of what we now call scientism, which sets up science as the sole arbiter and vehicle of truth. He was scandalized by the suggestion of men like Sprat and Glanvill that the interpretation of holy Scripture should be based on reason alone. As he saw it, the obsession for novel experiences and phenomena evident in the experimental philosophy was actually antithetical to true learning, and in particular to the way that tradition and scholarship provided the religious foundation of moderation and humility. Thus he questioned how well they 'that would reduce all learning to natural experiment or at least would have all learning . . . regulated by them . . . provide for Religion, the peace and tranquillity of publick Estates'. In other words, society needs more for its stability than clever experiments. In *Reflections upon the Conduct of Human Life* (1690), the Cambridge Platonist John Norris demanded in stentorian cadences to know if there is 'anything more Absurd and Impertinent [than] a Man who has so great a concern upon his Hands as the Preparing for Eternity, all busy and taken up with *Quadrants*, and *Telescopes*, *Furnaces*, *Syphons* and *Air-Pumps?*'

Casaubon did not accept the defence of experiment in terms of natural theology – he considered that merely an excuse for 'hunting after Novelties'. The new philosophy, he complained, induced a fixation on 'what is unusual, far fetch'd, and seldom seen', rather than leading men to recognize the work of God in such mundane processes as the tides and the passages of the sun and moon. These complaints came close to the old accusation of hubris, lately rehearsed in Marlowe's *Doctor Faustus*. In his *An Essay on Man* (1732–4) Alexander Pope advises us to respect our place and our limitations. We must understand both God and the universe with what resources we have been given, and not go snooping about with artificial aids. He is clearly thinking of the new astronomy's plurality of worlds when he says that

> Through worlds unnumber'd though the God be known,
> 'Tis ours to trace him only in our own.

Only God, says Pope, can and should contemplate the universe in all its immensity:

> He, who through vast immensity can pierce,
> See worlds on worlds compose one universe,
> Observe how system into system runs,
> What other planets circle other suns,
> What varied being peoples ev'ry star,
> May tell why Heav'n has made us as we are.

This applies equally to the microcosm as to the macrocosm: there is an implicit rebuke to Robert Hooke in Pope's insistence that

> The bliss of man (could pride that blessing find)
> Is not to act or think beyond mankind;
> No pow'rs of body or of soul to share,
> But what his nature and his state can bear.
> Why has not man a microscopic eye?
> For this plain reason, man is not a fly.
> Say what the use, were finer optics giv'n,
> T' inspect a mite, not comprehend the heav'n? . . .
> Then say not man's imperfect, Heav'n in fault;
> Say rather, man's as perfect as he ought:
> His knowledge measur'd to his state and place,
> His time a moment, and a point his space.

Besides, what practical good had yet come of it all? You didn't have to be a staunch Aristotelian to wonder whether the experimental philosophers were putting their prodigious brains to good use. The writer and politician Joseph Addison, who co-published the newspaper *The Tatler*, suggested that

> There are some men whose heads are so oddly turned this way, that though they are utter strangers to the common occurrences of life, they are able to discover the sex of a cockle, or describe the generation of a mite, in all its circumstances. They are so little versed in the world, that they scarce know a horse from an ox; but at the same time, will tell you with a great deal of gravity, that a flea is a rhinoceros, and a snail an hermaphrodite.

It was not just perverse but also morally lax to indulge in such trivia. For Addison this is no satirical jibe but a serious matter: 'the mind of man . . . is capable of much higher contemplations' and 'should not be altogether fixed upon such mean and disproportionate objects'. The Duchess of Newcastle, Margaret Cavendish, elaborated on her opposition to the experimental philosophy with this notion that it led to nothing of practical value and nothing that extended the capabilities of human art beyond what the ancients had achieved:

> But could experimental philosophers find out more beneficial arts than our forefathers have done, either for the better increase of vegetables and brute animals to nourish our bodies, or better and commodious contrivances in the art of architecture to build us houses . . . it would not only be worth their labour, but of as much praise as could be given to them: But, as boys that make play with watery bubbles or fling dust into each other's eyes, or make a hobbyhorse of snow, are worthy of reproof rather than praise, for wasting their time with useless sports; so those that addict themselves to unprofitable arts, spend more time than they reap benefit thereby . . . they will never be able to spin silk, thread, or wool, etc. from loose atoms; neither will weavers weave a web of light from the sun's rays; nor an architect build an house of the bubbles of water and air . . . ; and if a painter should draw a louse as big as a crab, and of that shape as the microscope presents, can anybody imagine that a beggar would believe it to be true? but if he did, what advantage would it be to the beggar? for it does neither instruct him how to avoid breeding them, or how to catch them, or to hinder them from biting.

In her essay 'The Character of a Virtuoso' (1696), Mary Astell, a seventeenth-century advocate of women's rights and another of the Royal Society's critics, defines a virtuoso as one who 'has Abandoned the Society of Men for that of *Insects, Worms, Grubbs, Maggots, Flies, Moths, Locusts, Beetles, Spiders, Grasshoppers, Snails, Lizards,* and *Tortoises.*' Astell expands on the stock criticism that these men did nothing of any real value to anyone, but merely acquired items and tidbits of knowledge valued only for their rarity and strangeness:

> To what purpose is it, that these Gentlemen ransack all Parts both of *Earth* and *Sea* to procure these *Triffles*? . . . I know that the desire of knowledge, and the discovery of things yet unknown is the Pretence; but what Knowledge is it? What Discoveries do we owe to their Labours? It is only the Discovery of some few unheeded Varieties of Plants, Shells, or Insects, unheeded only because useless; and the Knowledge, they boast so much of, is no more than a Register of Names, and Marks of Distinction only.

To some degree these attacks are transparently rhetorical. Even if the experimental philosophy was not, at its inception, an abundant source of technological innovation, it was obvious that the Aristotelian or Galenic traditions were no more generous in their yield of useful 'works' either. It was largely from the artisan crafts, rather than from philosophies of any sort, that practical advances tended to arrive in areas such as chemistry, medicine and metallurgy. Besides, hindsight now reveals how myopic it is to spurn the accumulation of information about the world because it lacks immediate applications. Yet the very fact that their absence could be used as an offensive weapon suggests that the Baconian spirit had infused the philosophy of the age: the aim and justification of knowledge, it now seemed, was to relieve mankind's estate.

The complaint wasn't simply that the experimental philosophy was of little practical use. Rather, Cavendish and others imply that unbridled curiosity was being permitted to *divert* men's minds from more valuable work. Those who invented microscopes and other optical instruments, she says, have done

> the world more injury than benefit; for this art has intoxicated so many men's brains, and wholly employed their thoughts and bodily actions about phenomena, or the exterior figures of objects, as all better arts and studies are laid aside.

If there's little here to distinguish the scientist who wastes his time investigating ephemera and irrelevant oddities from the collector who wastes his money in amassing them, that is a reminder of how the two traditions were still entwined. But they were already beginning to diverge, whereas it helped the critics' case to keep them yoked together. Good Baconians were themselves losing patience with the hoarding antiquaries whose aim was ownership rather than understanding. Thomas Browne's *Musæum clausum* (1684) is a mock-catalogue of lost rare books and objects that parodies the obscure items collectors tended to acquire. Among them are 'The Skin of a Snake bred out of the Spinal Marrow of a Man', 'A neat Crucifix made out of the cross Bone of a Frogs Head' and 'An exact account of the Life and Death of Avicenna confirming the account of his Death by taking nine Clysters [enemas] together in a fit of the Colick'. The references are largely a series of recondite in-jokes, but the point is plain enough. Browne's tract is just one example of what became an entire literary sub-genre of the 'curiosity-spoof' in the seventeenth century, and it would have been hard, given their attachment to collections and curiosities, for experimental philosophers to distance themselves from any association with the object of these satires.

These accusations were all contesting the issue of what curiosity is for. Is knowledge worth having for its own sake, or because of what can be done with it? Bacon saw a value in both – in 'experiments of light' as well as 'experiments of fruit' – but he clearly leaned towards a preference for knowledge that yielded a technological harvest. The historian Walter Houghton has suggested that already in the seventeenth century a distinction was starting to appear between what we would now call the pure and the applied scientist, or what could at that time be regarded as the virtuoso and the Baconian.* Whether individual scientists could be classified this way is open to question – Hooke was clearly more the technologist, Boyle the virtuoso, but neither is a pure type. But the point is that the conceptual dichotomy was emerging, and that critics of the experimental philosophy found the figure of the unworldly, unproductive virtuoso – who would have

---

* In fairness, Mary Astell does recognize that there is a distinction between the idle dabbler and those who have a genuine desire to learn about and understand the world. She excepted Boyle in particular from her comment about the unproductive virtuosi, and allowed that 'the *Royal Society*, by their great and celebrated Performances . . . highly merited the *Esteem*, *Respect* and *Honour* paid 'em by the Lovers of Learning all *Europe* over.' She explains that 'though I have a very great Veneration for the *Society* in general, I can't but put a vast difference between the particular Members that compose it.' And in truth it was still overstocked with dilettante gentlemen.

been ridiculed by Bacon – to be the most valuable for their purposes. Not least, it could be guaranteed to raise a laugh.

Theatre of the absurd

The Virtuoso is in many respects a typical Restoration farce, full of the bawdy sexual comedy and cross-dressing that delighted the monarch. That it derives its humour at the expense of the Royal Society makes it no different from any topical satire poking fun at the issues of the day. The very fact that it chooses this target, and had an enthusiastic reception, illustrates how familiar the well-to-do audiences in London were with the activities of the experimental philosophers.

Shadwell was a celebrated poet and prolific playwright, who succeeded John Dryden as Poet Laureate in 1689. His play spoke to the many people in polite London society who were bemused to the point of scorn or hilarity by the way in which the Royal Society seemed to have made the most trifling or absurd questions the subjects of intense scrutiny: How heavy is air? Why are plums blue? What is the geography of the moon? Gimcrack's niece Miranda exclaims that he 'has broken his brains about the nature of maggots [and] has studied these twenty years to find out the several sorts of spiders'.

It is the study of tiny animals that draws most of Shadwell's ridicule, as though this is a self-evidently silly venture – the attempts of Moffett and his fellow entomologists to justify their passion had, it seems, little impact. Gimcrack says that he has even tamed a spider named Nick:

he knew his name so well he would follow me all over the house. I fed him indeed with fair flesh flies. He was the best natur'd, best condition'd spider that I ever met with.

'Not a Creature so little', says Gimcrack's friend, the ridiculously pompous Sir Formal Trifle, 'but affords him great Curiosities.'

To the urbane Bruce and Longvil, the 'gentlemen of wit and sense' who represent the voices of sanity, the virtuosi study these creatures simply because they are there to be studied – an unworthy reason indeed!

Bruce: What does it concern a man to know the nature of an ant?

Longvil: O it concerns a virtuoso mightily; so it be knowledge, 'tis no matter of what.

Shadwell also has a great deal of fun with the experimental philosophers' studies of air. The virtuoso might reasonably complain that he couldn't win, being ridiculed as much for his interest in the obscure and rare as in the prosaic. To which the 'common sense' response might have been that there is nothing in the rare and marvellous worth knowing, and nothing in the prosaic that is not already known. And indeed, what could possibly be so diverting in something so ubiquitous and familiar as air, that the experimentalist should wish to bottle it, squeeze it, evacuate, weigh and measure it? But it is clear that Shadwell has no real understanding of why Boyle and his colleagues were studying air, no notion that there is a science of pneumatics, nor deep and perplexing questions about the chemistry and life-giving agency of air, the solution to which in the late eighteenth century gave birth to chemistry as a modern science. His mockery – typical of the satirists – stems from how an experiment *looks*, without regard to its purpose.

Thus, Gimcrack is bottling air not to understand it but to capture a kind of essence of place, so that the ambience of one location can be recreated elsewhere: 'Choose your air, you will have it in my chamber: Newmarket, Banstead Down, Wiltshire, Bury air, Norwich air, what you will.' If he weighs it, that is merely to secure a quantitative signature of its identity:

O yes, I have sent one [assistant] to weigh air at the peak of Teneriffe; that's the lightest air. I shall have a considerable cargo of that air. Sheerness and the Isle of Dogs' air is the heaviest.*

---

* In reality, the last laugh does not belong to the public at this absurd spectacle of bottled air, but is at their expense. 'Most certainly', jokes Longvil, 'the world is very foolish not to snuff up bottled air as they drink bottled drink.' But now they will pay a premium for another ubiquitous fluid in bottles, advertised for its geographical origin in picturesque mountain springs. One suspects that the lack of a market for bottled air is merely for want of good marketing.

Why weigh it at all, he is asked? 'To know what it weighs', he explains. 'O knowledge is a fine thing. Why I can tell to a grain what a gallon of any air in England weighs.'

Pomposity is always fair game for satire, and laughter a valuable defence against too much dour self-importance. What takes the edge off *The Virtuoso* now, however, is its reliance on nothing more than a superficial knowledge of its subject: it laughs much as one culture laughs at what it takes to be the arcane, pointless antics of another. Moreover, this mockery comes at the expense of a coherent plot. The action culminates when Gimcrack's house is besieged by 'ribbon weavers' incensed that the virtuoso has invented a mechanized 'engine loom'. Naturally, Gimcrack protests his innocence by citing his inability to do anything so useful: 'We virtuosos never find out anything of use, 'tis not our way.' Given how the situation is set up, it seems likely that Gimcrack is indeed innocent, but it is never made clear why the ribbon weavers imagine otherwise in the face of abundant evidence of the virtuoso's practical incompetence. We don't discover the truth of this matter in any case. The irony is, of course, that the experimental philosophy was later to merge with the well-established craft of engineering – in the persons of such men as Robert Stephenson, Thomas Telford and Henry Bessemer – to transform industry and society in just the ways that the ribbon weavers deplore. The production of commodities, the transportation of people and goods, the power of communications and warfare, all were to be altered irrevocably by the applications of scientific understanding. Within scarcely more than a hundred years, the harmless, comical curiosity of Gimcrack was to be recast as the Faustian lust for knowledge of Victor Frankenstein.

Such a crowd-pleasing attack on their enterprise was bound to unsettle the experimental philosophers. William Petty grumbled that 'The endeavours to improve Natural knowledge the best and most pleasant of all others have been turned into ridicule by the late Comedye called the Virtuoso, only by comparing the ends they seek and the means they use to the vile and ridiculous matters and operations.' It was clear enough who was the target of this ridicule: Shadwell took care to mention Gresham College, who had turned down Gimcrack's application for membership because, his wife explains bitterly, 'they envied him'.*

The narrative and thrust of *The Virtuoso* were aped in another play that premiered at the Duke's Theatre in 1687: *The Emperor of the Moon* by the female playwright Aphra Behn. Virginia Woolf's assessment of Behn – that her career was more significant than her literary works – is surely apposite, for her Restoration farces are generally rather humdrum fare, whereas her life outshines any of them. To be a successful woman writer in the seventeenth century was impressive enough, but her frankness about female sexuality and her exploits as an English spy in the Second Anglo-Dutch War of 1665, when she became the lover of a Dutch royal, mark her as a personality of extraordinary force. Yet she was precisely the sort of person who would scoff at the virtuosi: a sexual libertine but a social conservative, devotedly Royalist and passionate in human affairs, she would have seen little point, and perhaps even something emasculating, in the experimental philosophers' obsessions with microscopes and moons. *The Emperor of the Moon* was the last of her plays to be performed before her death in 1689, and it tells of an Italian philosopher of Naples, Baliardo, who, like Gimcrack, neglects his worldly obligations to pursue scientific studies. Baliardo's observations of the moon have left him

---

* That Gimcrack is not actually a member of the Royal Society is one of the reasons why literary scholar Joseph Glide questions whether the society was in fact Shadwell's real target. Glide points out that Gimcrack's determination to ignore, in the pursuit of his obsessions, any practical benefits that might accrue make him the kind of idle virtuoso that the Royal Society claimed to disapprove of. Moreover, the florid 'philosophical' language of his friend Sir Formal Trifle contrasts with the society's expressed intention to speak and write as plainly as possible. In such respects, says Glide, Shadwell's play can be considered wholly in accord with the Royal Society's principles. But this argument does not seem very robust. For one thing, *The Virtuoso* is very much of a part with the explicit attacks on the Royal Society by other literary wits such as Butler. And why would Shadwell draw so heavily on the actual experiments of Hooke and his colleagues, if he did not intend to ridicule them? Exaggeration to the point of caricature is surely the standard technique of satire. Besides, there seems no question that *The Virtuoso* was interpreted in its time as a broadside against the Royal Society. If Shadwell held back from actually naming them as the objects of his derision, perhaps he was simply being careful not to offend their royal patron (although it seems unlikely that Charles II would have cared in the slightest).

convinced that it hosts a civilization like ours. Like Don Quixote with his chivalrous romances, Baliardo has become besotted with the stories of lunar journeys and inhabitants: with Lucian's *True History*, John Wilkins' *The Discovery of a World in the Moone*, Francis Godwin's *The Man in the Moon*, and, in the words of his despairing daughter Elaria, 'a thousand other ridiculous Volumes, too hard to name'. Scaramouche, the *commedia dell' arte* stalwart who is one of Elaria's confidantes,* says of her father, 'lunatic we may call him, without breaking the Decorum of good Manners; for he is always travelling to the Moon'.

The nub of Elaria's frustration is that her father will not drag himself from his studies to consent to her marriage to her lover Don Cinthio, nor that of his niece Bellemante to the young gallant Don Charmante: a neglect of paternal duties (and of the exigencies of romantic love and sexuality) that again echoes Gimcrack's. Behn's contempt for the new experimental science is clear when Elaria's governess Mopsophil commands Scaramouche to attend to Baliardo: 'my Master's conjuring for you like mad below, he calls up all his little Devils with horrid Names, his Microscope, his Horoscope, his Telescope, and all his Scopes'. Not only is there no distinction between a microscope and an astrological forecast, but both have the suspicious flavour of diabolical magic.

There follows a routine farce of sexual intrigue and deception, culminating in the decision of the suitors and their friends to disguise themselves as visiting dignitaries from the moon so that they might gain Baliardo's attention and approval in asking for the hands of the young women. The doctor is mightily flattered by the honour and agrees to the unions. But as the ceremonies are underway, he discovers the deception – only then to realize the error of his ways:

Doctor: Are not you then the Emperor of the Moon? And you the
            Prince of Thunderland?
Cinthio: There's no such Person, Sir.
            These Stories are the Fantoms of mad Brains,
            To puzzle Fools withal—the Wise laugh at 'em—
            Come, Sir, you shall no longer be impos'd upon.

---

* Behn's play takes much of its shape from a French Harlequinade called *Arlequin empereur dans la lune*, which was performed in Paris by the great Italian Harlequin Giuseppe-Domenico Biancolelli, and published in 1684.

Doctor: No Emperor of the Moon, and no Moon World!
Charmante: Ridiculous Inventions.
              If we 'ad not lov'd you you 'ad been still impos'd on;
              You had brought a Scandal on your learned Name,
              And all succeeding Ages had despis'd it.
[Doctor leaps up.]
Doctor: Burn all my Books and let my study blaze, Burn all to Ashes,
              and be sure the Wind Scatter the vile contagious monstrous
              Lyes . . . Come all and see my happy Recantation of all the
              Follies, Fables have inspir'd till now . . . I see there's nothing
              in Philosophy.

For all its knockabout joviality, the play's dismissal of the new science
could hardly have been more harsh. 'There's nothing in philosophy',
its books fit only for burning and its pretensions deserving the laughter
of the wise.* Nothing better illustrates the gulf that had opened up
between the new philosophy and the fun-loving aristocratic Restoration
society than this cavalier mockery launched from the stage of the
king's favourite theatre.

Of mice and elephants

Shadwell was not the first to use wit, rather than morality or tradition, as a rapier to prick the virtuosi. But while Ben Jonson's *The
Alchemist* supplied the model, his target was the older stereotype of
the alchemical charlatan. John Donne – who, we saw earlier, was by
no means antagonistic to the new science† – also lampooned the
courtly virtuoso in *The Courtier's Library* (c.1603–11), a fictitious catalogue of books of the kind that might make up the collection of this

---

* It is possible that the old maxim 'curiosity killed the cat' has its origins in Behn's
The Lucky Chance (1686), which proclaims rather more grandly that 'too much curiosity lost paradise'. But the phrase 'care killed the cat' – apparently referring to the
dangers of smothering over-attentiveness – had currency in the Renaissance, wherein
'curiosity' could be substituted in the now archaic sense. That, of course, would
have totally changed the meaning – but if this is what happened, its cultural resonance seems to have preserved it.
† There's reason to suppose that Donne may, however, have been unsympathetic to
Baconian experimental philosophy, since he idolized the Earl of Essex, who was
rather coldly betrayed by his protégé Bacon during the earl's fateful trial for treason.

older style of virtuoso (Thomas Browne's *Musæum clausum* was indebted to it). Among these invented works was one by John Dee that explained the mystic significance of the pattern of hairs on the tail of the dog owned by the biblical Tobit, a treatise by Cardano called *On the Nullibiety of Breaking Wind*, and an absurdly pompous volume allegedly inspired by the collector of curiosities Ulisse Aldrovandi, called (in abbreviated form) *Quis non? Or, a Refutation of all the errors, past, present, and future, not only in theology but in all other branches of knowledge, and the technical arts of all men dead, living, and as yet unborn.*

Mocking the Royal Society quickly became something of a fad with Restoration wags, an anonymous one of whom got in an early blow with the 'Ballad of Gresham Colledge', which circulated from around 1662. It lampooned the occasion in 1661 when the Danish ambassador had been shown an experiment with Boyle's air pump:

> To the Danish Agent late was showne
> That where noe Ayre is, there's noe breath.
> A glasse this secret did make knowne
> Where a Catt was put to death.
> Out of the glasse the Ayre being screwed,
> Pusse dyed and ne'er so much as mewed.

One of the first poets of note to join the fray was Samuel Butler, whose anti-Puritan satire *Hudibras*, the first two parts of which were published in 1663 and 1664, ingratiated him to Charles II. The poem had a broad target in which the Royal Society featured strongly, in particular the microscopic observations of Robert Hooke, which Butler mocked by having the poem's virtuoso astrologer Sidrophel study 'Maggots breed [bred] in Rotten Cheese'.

Butler subsequently discharged several broadsides at the Royal Society, among them the explicitly titled 'Satire Upon the Royal Society', where the virtuosi ask such profound questions as these:

> What is the nat'ral cause why fish
> That always drink do never piss;
> Or whether in their home the deep
> By night or day they never sleep?

Butler's most barbed assault came in his poem 'The Elephant in the Moon', written in the 1670s, which anticipates Behn's play in ridiculing the readiness of the virtuosi to spin elaborate and unlikely theories on the back of mistaken observations. In one sense, then, the poem challenges not so much the value but the reliability of the experimental approach, implying that it erodes the critical faculties so that men find marvellous explanations where mundane ones will do.

Butler presents the observation of the moon not as an idle endeavour (like Baliardo's) but as a rapacious one: it is a prelude to imperialistic conquest, which was after all how real astronomers were apt to portray it:

> A learn'd society of late,
> The glory of a foreign state,
> Agreed, upon a summer's night,
> To search the Moon by her own light;
> To make an invent'ry of all
> Her real estate, and personal;
> And make an accurate survey
> Of all her lands, and how they lay,
> As true as that of Ireland, where
> The sly surveyors stole a shire:
> T' observe her country, how 'twas planted,
> With what sh' abounded most, or wanted;
> And make the proper'st observations
> For settling of new plantations,
> If the society should incline
> T' attempt so glorious a design . . .
> And all stood ready to fall on,
> Impatient who should have the honour
> To plant an ensign first upon her.*

---

* It feels obligatory to point out that impatience for the honour of planting the first flag (and thus at least notionally staking a territorial claim) was the reason why, three centuries after Cyrano de Bergerac's fictional voyage, it was made in reality. It remains alarming how both curiosity and the colonialist Columbus narrative are so readily enlisted as a cover and justification for the space race. I need only add that the 'inventory of all her real estate' continues apace with a view to mining the lunar mineral resources, such as they are.

This last, needless to say, suggests also the sexual metaphor of violation so problematically expounded by Bacon.

In the course of this investigation, the philosophers are astonished to discover that the moon's surface is populated by armies waging war. These they ascribe to two populations, named the Privolans and Subvolans after the races described in Kepler's *Somnium* (see page 229), who live above and below the ground respectively. Then an even more amazing sight comes into view:

> A wonder, more unparallel'd
> Than ever mortal tube beheld;
> An elephant from one of those
> Two mighty armies is broke loose . . .
> It is a large one, far more great
> Than e'er was bred in Afric yet.

The virtuosi are overjoyed. No longer will they have to endure the scorn and scepticism of the populace, nor be the subject of coffee-house gossip, for this discovery will silence all critics. And so they begin, in the manner approved by the Royal Society, to draw up an announcement for publication 'in the next *Transaction*'. But as they do so, the telescope is left for the society's footboys to toy with. And one of these lads sees something that none of the virtuosi has noticed: the 'elephant' seems rather to be a field mouse that has become trapped in the tube. No sooner has this suggestion been whispered than it plunges the learned gentlemen into turmoil. Butler implies that they have abandoned faith in the transparent evidence of the senses and instead have adopted arbitrary theories for interpreting what they see – theories to which they cling with stubborn and disputatious tenacity: some swore 'that they never would recant/One syllable of th' Elephant', while 'Others began to doubt and waver'.

This suggestion that astronomers make absurd interpretations of what they see in the telescope appeared also in *Hudibras*, where Sidrophel sees a boy flying a kite and exclaims, with allusion to the 'new stars' of Tycho Brahe and Kepler:

> A comet, and without a beard!
> Or star, that ne'er before appeared!

The silly implausibility of this rather blunts the satire, all the more so now we know that the new stars, the lunar landscape, the Medicean moons of Jupiter and the rings of Saturn were anything but blemishes on the lens.

Eventually Butler's astronomers agree to open up the 'optic tube' and look inside, whereupon they find that not only does it indeed harbour a mouse but also 'prodigious swarms of flies and gnats', which they had mistaken for the armies of the moon. Their shame at being so mistaken puts them into 'a worse and desperater brawl', from which they can take away but one stark lesson:

> That those who greedily pursue
> Things wonderful instead of true,
> That in their speculations choose
> To make discoveries strange news,
> And natural history a Gazette
> Of tales stupendous and far-fet;
> Hold no truth worthy to be known,
> That is not huge and over-grown;
> And explicate appearances,
> Not as they are, but as they please;
> In vain strive Nature to suborn,
> And, for their pains, are paid with scorn.

It is a crushing indictment: the arrogance and exclusivity of the philosophers is made all the worse in being accompanied by credulity. These men who pride themselves on their learning are ready to believe the most ridiculous things that common folk could see through in an instant.

The accusation is mostly unfair. Like the professors of secrets, the new philosophers aimed to sift through recipes, experiments and observations, by practical testing where possible, to distinguish the true from the fanciful. They turned scepticism, previously a mark of the ignorant bumpkin, into the tool of the scientist. But that is apparently not how it seemed from the outside, and it is not difficult to see why. The *Philosophical Transactions* still swelled with oddities, marvels, strange sightings and events – signs that the worst habits of the collectors of curiosities were far from expunged. John Wilkins reported on 'a maid in Holland, who voided seed by urine, which being sown grew'. Walter Charleton described the monstrous birth of a 'baby' weighing twenty-three pounds, without a head or bones. Another FRS, Timothy Clarke, reported a woman who had been pregnant for eighteen years, while Oldenburg passed on news of a woman in France with four breasts. Boyle was one of the worst offenders, describing glass that could be hammered like silver, a liquid that expanded when the moon was full, and swallows found encased in ice that flew off when it melted. Even the supposedly hard-headed Christopher Wren described a creature born of the union of a male cat with a female rabbit, and enthusiastically investigated the 'phantom drummer of Tidworth'.

The Royal Society exposed itself to charges of credulity simply by attempting to gather and sift reliable information. 'Some of the questionnaires they sent overseas', says historian Rosemary Syfret, 'read more like . . . extracts from Lewis Carroll than serious scientific inquiries' – a point that she illustrates with a letter to Sir Philiberto Vernatti in Batavia in which he is kindly requested to verify

> Whether in the Island of Sambrero, which lyeth northwards of Sumatra, there be found such a vegetable as Mr James Lancaster relates to have seen, which grows up to a Tree, shrinks down, when one offers to pluck it up, into the Ground, and would quite shrink, unless held very hard? And whether the same, being forcibly pluck'd up, hath a Worm for its Root, diminishing more and more, according as the Tree groweth in Greatness; and as soon as the Worm is wholly turned into a Tree, rooting in the Ground, and so growing great?

In the event, Vernatti's reply assured the society gravely that 'I cannot meet with any that ever have heard of such a vegetable.'

The curious investigator was in a double bind. For nature was evidently become so strange under close scrutiny, thronging with invisible animals, wondrous new medicines, and odd lights in the sky, that one could not rule out any possibility a priori, least of all because it was not verified by the ancient authorities. Yet to investigate such reports was to risk seeming a superstitious gull. Would not 'common sense' have laughed at the rings of Saturn and the intricacy of the louse? So the *Philosophical Transactions* became a treasure trove for the satirist. What, after all, was one expected to make of an experiment to test the belief that spiders became immobile if placed inside a ring of powdered unicorn horn? When a spider was so situated, the report soberly explained, it 'immediately ran out'. Such examples seemed to vindicate the notion advanced by Shadwell's Longvil that, as far as the virtuosi were concerned, 'As there is no lie too great for their telling, so there's none too great for their believing.' (One has to admit that no one in the Royal Society seems to have wondered where the powdered unicorn horn came from.)

The world ablaze

As Butler illustrates, another accusation against the virtuosi was that they did not always agree with one another. Today it is usually seen as a virtue of science that anyone is free to disagree with their colleagues and to argue a different case. But disagreements were apt to be taken by the opponents of the new philosophy as a sign that there was no consensus on anything.*

The same complaint of intellectual discord was made by Margaret Cavendish in her fictional polemic against the experimental philosophy, *The Description of a New World, Called The Blazing-World.* This took the familiar form of an imaginary voyage to a new land. The heroine of the tale is a noblewoman who is abducted from her father's house near the sea by a merchant who has fallen in love with her but is of insufficient social standing to ask openly for her hand. The ship is blown off course to the North Pole, where all those aboard die of cold except the young woman, preserved 'by the light of her Beauty, the heat of her youth, and Protection of the Gods'. The boat is forced by the strong winds to cross into another world, the Blazing-World, which touches our own at the pole but has its own sun. Here the heroine finds a world populated by people with the features of animals – Bear-men, Fox-men, Bird-men, Ant-men and so on, as well as human-like beings of lurid colours who make up the governing class. The noblewoman is taken to the emperor of the Blazing-World in his

---

* Of course, this is still the case today when it suits the critics' agenda. Scientific consensus, meanwhile, is regarded by them as prima facie evidence of institutionalized dogmatism and censorship.

palace of gold and precious stones on the island of Paradise, where he marries her at once.

Here she instigates a programme of science and technology, making the Bear-men her experimental philosophers, the Bird-men her astronomers, the Fly- and Fish-men her natural philosophers, and so forth. Everything proceeds well until the empress asks the Bear-men to observe the heavens: she finds that their telescopes 'caused more differences and divisions amongst them, then ever they had before'. Some said the sun stands still and the Earth moves, others the opposite. They couldn't agree on how many stars there are, or on whether the moon's dark patches are land or ocean.

Seeing this confusion, the empress grows angry. 'Now I do plainly perceive', she says,

> that your Glasses are false Informers, and instead of discovering the Truth, delude your Senses; Wherefore I Command you to break them, and let the Bird-men trust onely to their natural eyes, and examine Cœlestial Objects by the motions of their own Sense and Reason. The Bear-men replied, That it was not the fault of their Glasses, which caused such differences in their Opinions, but the sensitive motions in their Optick organs did not move alike, nor were their rational judgments always regular: To which the Empress answered, That if their Glasses were true Informers, they would rectifie their irregular Sense and Reason; But, said she, Nature has made your Sense and Reason more regular then Art has your Glasses; for they are meer deluders, and will never lead you to the knowledg of Truth; Wherefore I command you again to break them; for you may observe the progressive motions of Cœlestial Bodies with your natural eyes better then through Artificial Glasses.

The Bear-men implore her to spare their telescopes, saying that it is their delight and sport to argue and dispute and thereby lay claim to one being wiser than another, which would not be possible 'if there were nothing but truth, and no falshood'. They insist that their spyglasses 'are our onely delight, and as dear to us as our lives'. The empress relents, so long as the experimental philosophers keep their petty disputes to themselves and don't let them cause unrest in the state.

It is easy to see how unfair all of this is to what astronomers were actually doing. But underlying it is an aristocratic disdain for argument which would seemingly prefer ignorance to a fractious pursuit of knowledge. It is when the Bear-men, overjoyed at their reprieve, pull out their microscopes, sure that these will better please Her Majesty, that we see the truth of what the astute reader must already have begun to suspect. For what the philosophers show the empress is precisely what Robert Hooke observed – fly's eyes consisting of 14,000 'pearls', the microscopic pores of charcoal, the fine needles of a nettle. And what the empress says in response is precisely what Cavendish herself said to refute Hooke's claims about these structures. In other words, this divinely beautiful, all-powerful and near-immortal empress is none other than the Duchess of Newcastle, effortlessly dismissing the nonsensical theories and futile instruments of her social inferiors.

One might object that this can't be literally so, since the 'Duchess of Newcastle's soul' makes an appearance later in the tale, when she is summoned by the empress to be her 'spiritual Scribe'. But that only adds to the evidence: the 'Duchess' – described with false modesty as 'not one of the most learned, eloquent, witty and ingenious, yet . . . a plain and rational Writer' (at least, the narrator attests, in comparison with Galileo, Gassendi and Descartes) – is thus the empress's alter ego. The fictional Duchess even confesses that 'I would fain be as you are, that is, an Empress of a World, and I shall never be at quiet until I be one . . . though I cannot be Henry the Fifth or Charles the Second, yet I endeavour to be Margaret the First.' The literary scholar Anna Battigelli offers the rather convoluted defence of Cavendish that 'by parading her own allegedly boundless will and ego, [she] questions the degree to which the subjective interference of the self can be eliminated, thus problematizing the premise on which Hooke's objective certainty was based'. To have confidence in that sort of double bluff requires quite an act of faith, particularly when it is seen against the preening, solipsistic backdrop of *The Blazing-World*.

The empress's other scientists, being natural historians, chymists, anatomists and other naked-eye recorders of nature rather than the Bear-men with their deceitful contrivances, fare better in her favours. But all are reminded of their station by her incisive questioning, being much surprised that she 'had such great and able judgment in Natural

Philosophy'. Only the Lice-men displease her, because they too are pedantic experimentalists who

> endeavoured to measure all things to a hairs-breadth, and weigh them to an Atom; but their weights would seldom agree, especially in the weighing of Air, which they found a task impossible to be done; at which the Empress began to be displeased, and told them, that there was neither Truth nor Justice in their Profession; and so dissolved their society.

Oh, if only a duchess might do the same to a royal society.

Artists of the floating world

Arriving at *Gulliver's Travels* via Francis Godwin, Cyrano de Bergerac, Samuel Butler and Margaret Cavendish makes it seem a decidedly derivative work on a well-worn theme. And in many ways it is. But of course that is never an impediment to literary success, especially if the author has a pen as sharp as Swift's. Besides, like all fictions that become modern myths, *Gulliver* has been transformed into a different story from the one its author told. How else, indeed, could this satire on early Georgian English culture have survived – no, flourished – within the stark presentism of Hollywood, and even to have become regarded as a 'children's classic'? Of course, what recommends it to those purposes is precisely what is hackneyed about it, and not what is original – which is to say, that it describes a voyage to strange lands with odd and sometimes comical inhabitants whose peculiar customs hold up a mirror to our own foibles. And if we suspect that we are now missing their jokes, we can always give them new ones. But read in context, *Gulliver's Travels* is the culmination of the Utopia-manqué, the jaded antidote to the optimistic fantasies of More, Bacon and Campanella. Here *New Atlantis* is turned upside down and Bensalem has degenerated into farce.

The experimental philosophy was just one of Swift's targets, at which he aims in Gulliver's exploits in the land of Laputa. Here the most powerful inhabitants are philosophers who are 'always so wrapped up in cognition, that [they are] in manifest danger of falling down every precipice, and bouncing [their] head against every post, and in the streets, of jostling others or being jostled [themselves] into the kennel'. To ward against these hazards, the Laputans employ servants who, like medieval jesters, must strike their mouths and ears with blown-up bladders to bring them back to reality. The Laputans live in abstraction, capable of conceiving the world only in geometric terms. In the king's palace Gulliver is served bread cut into 'cones, cylinders, parallelograms, and several other mathematical figures'.

These savants are, like all caricatures of the virtuoso, utterly impractical. 'Although they are dexterous enough upon a piece of paper in the management of the rule, the pencil and the divider', says Gulliver, 'yet in the common actions and behaviour of life, I have not seen a more clumsy, awkward, and unhandy people, nor so slow and perplexed in their conceptions upon all other subjects, except those of mathematics and music.'*

And so Laputa is a shambles, because the virtuosi are always concocting ingenious new schemes for making life better while proving incapable of realizing any of them. At the Academy of Projectors in Lagado, the kingdom's capital on the neighbouring island of Balnibarbi, the professors

> contrive new rules and methods of agriculture and building, and new instruments and tools for all trades and manufactures, whereby, as they undertake, one man shall do the work of ten; a palace may be built in a week, of materials so durable as to last for ever without repairing. All the fruits of the earth shall come to maturity at whatever season we think fit to choose, and increase an hundred fold more than they do at present, with innumerable other happy proposals. The only inconvenience is, that none of these projects are yet brought to perfection and in the meantime the whole country lies miserably waste, the houses in ruins, and the people without food or clothes.

The parallels with the claims of the scientists of Solomon's House in *New Atlantis* are clear; what we have here is a travesty of Bacon's Utopia, as the technocrats bungle every project. Several of their efforts

---

* Don't imagine that this imputes any artistic sensitivity to the Laputans; Swift classes music here, in the medieval style, as a branch of natural philosophy, governed by harmony and proportion.

are either aimless to the point of surrealism – an attempt to make a 'naked sheep' from gum, minerals and vegetables, and of building houses from the roof downwards – or so contrived that there is no profit in carrying them out, such as the technique of ploughing with pigs by burying acorns, dates, chestnuts and vegetables in the ground and letting the pigs root for them.

A satire works only if its victim is accurately portrayed. Swift, like Shadwell, took pains to tie his caricature to the real projects of the Royal Society. Thus, for example, the Laputan professors are said to have developed a 'universal language', like that sought by John Wilkins, by replacing words with objects with which they are obliged to encumber themselves: 'such *things* as were necessary to express the particular business they are to discourse on'.* They have discovered two moons of Mars with orbits that fit Kepler's laws: the squares of their orbital periods are proportional to the cubes of their radii, 'which evidently shows them to be governed by the same law of gravitation, that influences the other heavenly bodies'.

But there isn't much of the ridiculous in that, and Swift's determination to make it so leaves him at times wringing laughs from an outdated stereotype: the unkempt, obsessive alchemist satirized in word and image by Chaucer and Brueghel. At the academy at Lagado, says Gulliver,

> The first man I saw was of meagre aspect, with sooty hands and face, his hair and beard long, ragged and singed in several places. His clothes, shirt and skin were all of the same colour. He had been eight years upon a project for extracting sunbeams out of cucumbers, which were to be put into vials hermetically sealed, and let out to warm the air in raw inclement summers.

Another academician has 'hands and clothes daubed over with filth', and his goal is a grotesque hybrid of Hennig Brandt's extraction of phosphorus and Paracelsus's theories of digestion and nutrition:

---

* There may also be an allusion here to the way Comenius believed that children should be educated by showing them real and tangible things, rather than by a verbal description of words and concepts.

His employment since his first coming into the Academy was an operation to reduce human excrement to its original food, by separating the several parts, removing the tincture which it receives from the gall, making the odour exhale, and scumming off the saliva. He had a weekly allowance from the Society of a vessel filled with human ordure, about the bigness of a Bristol barrel.

These parodies of the alchemists had little to do with the experimenters of the Royal Society (one could not accuse Hooke of being proud of his appearance, but he was unusual). All the same, they betray a sense that one objection to the experimental philosophy is that it engages in an unseemly manner with grime and toil. It was Bacon, after all, who professed that the advancement of knowledge might involve the study of 'mean and filthy things'.

Of all the satirical assaults on the Royal Society, Swift's may have wrought the greatest damage, for it was the most popular. The book's initial run of 10,000 copies sold out in three weeks, and according to Swift's friend Alexander Pope and the poet John Gay, 'from the highest to the lowest it is universally read, from the cabinet council to the nursery'. Swift was a tenacious thorn in the side of the experimental philosophers. His famous lampoon of English policy in Ireland in which he recommends that the poor Irish might be encouraged to eat their surplus children was inspired by the cold social calculus proposed by William Petty. And his first book *The Tale of a Tub* (1704), while a satire with primarily political and theological targets, ridicules a hunger for all new knowledge that is indulged without discernment.

The sting

It was the satirists who gave the experimental philosophers the most trouble, for there is of course no keener weapon than laughter. In his *History of the Royal Society*, Thomas Sprat admitted that

We ought to have a great dread of the power of these terrible men: I confess I believe that the New Philosophy need not (like Caesar) fear the pale or the melancholy as much as the humorous and the merry: For they perhaps by making it ridiculous because it is new . . . may do it more injury than all the arguments of our severe and frowning and dogmatical adversaries.

While John Evelyn could complain in 1664 merely of 'a few Ignorant and Comical Buffoons [who] with an Insolence suitable to their Understanding [are] still crying out, and asking, "What have the Society done?"', by 1694 William Wotton was forced to say that

> the sly insinuations of the *Men of Wit*, That no great Things have ever, or are ever like to be performed by the *Men of Gresham*, and, That every Man whom they call a *Virtuoso*, must needs be a *Sir Nicholas Gimcrack*: together with the *public Ridiculing* of all those who spend their Time and Fortunes in seeking after what some call Useless Natural Rarities; who dissect all Animals, little as well as great; who think no part of God's Workmanship below their strictest Examination and nicest Search: have so far taken off the Edge of those who have opulent Fortunes and Love to Learning, that Physiological Studies begin to be contracted among Physicians and Mechanics. Nothing wounds more effectually than a Jest; and when Men once become ridiculous, their Labours will be slighted, and they will find few Imitators.

'Howe far this [ridicule] may deaden the industry of the philosophers of the next age', he fretted, 'is not easie to tell.'

Sprat attempted to call a truce by telling the '*Wits* and *Railleurs*' that the work of the Royal Society was in their interests too, since it could provide a new stock of metaphors and images to supplant the tired clichés of old (see page 220). By jeopardizing the new philosophy, the wits risked depriving themselves 'of the most fertil subject of *Fancy*'. This was a somewhat desperate gambit.

To the grave discomfort of the Royal Society, one of the mockers was their patron. Charles II had few intellectual aspirations, but while he is often caricatured as a libertine fop, he was nobody's fool. His interest in natural philosophy seems to have been genuine, and he kept his own laboratory, used telescopes and apparently made machines, although John Evelyn did not have a high opinion of his abilities. Yet the Royal Society was for him more of an entertaining diversion than a serious commitment. All they got from him in material terms was a side of venison for their annual dinner, and after triggering frantic debate among the Fellows about what to show him on a proposed visit in 1662, he decided in the end not to come. His delight in the wonders of the experimental philosophy would have made him interested in expensive 'toys' like the air pump, but this enthusiasm could quickly to turn to ennui and disdain. According to Samuel Pepys, William Petty felt the edge of this amused scorn when he visited the king in February 1664 to explain a new boat design:

> To Whitehall, where in the Duke's chamber the King came and stayed an hour or two, laughing at Sir W. Petty, who was there about his boat, and at Gresham College in general. At which poor Petty was I perceive at some loss, but did argue discreetly and bear the unreasonable follies of the King's objections and other bystanders with great discretion – and offered to take oddes against the King's best boats; but the King would not lay, but cried him down with words only. Gresham College he mightily laughed at for spending time only in weighing of ayre, and doing nothing else since they sat.

Clearly, Shadwell's play would have only fed his prejudices.

The early nineteenth-century writer Isaac Disraeli, father of the future prime minister Benjamin, records an anecdote about Charles II's mockery that puts the king in a more wry light. While dining with the Royal Society one evening, the king asked the Fellows why it was that, if two pails of water of equal weight are set on a balance and then a fish is placed in one bucket, it does not weigh more than the other. All those present proceeded to explain the phenomenon, each with a different version, until the king demanded to hear an explanation murmured in private to his neighbour that had elicited a laugh. The speaker took a deep breath and boldly suggested to the king that his claim was in fact untrue. Whereupon the king himself laughed and said 'Odds fish, brother, you are in the right!' According to Charles II's recent biographer Jenny Uglow, there are many such tales, all lacking proper documentation and almost certainly apocryphal, but this one certainly illustrates how the popular caricature of the virtuosi makes them disputatious at the expense of common sense.

Fair criticism?

It is clear that the likes of Stubbe, Casaubon, Cavendish and even Hobbes were in various ways seeking to defend an older mode of philosophy. This was a matter of propriety, custom and social order as much as it was a question of how to philosophize. Medieval intellectuals tended to insist on a strict hierarchy of signification, according to which it would be laughable if not blasphemous to imagine that time spent weighing air was as justified as time spent reading the Bible. Small and distant things were small and distant precisely because they were not meant to concern us. That parochial view of what is worth studying was readily translated into secular terms, whereby the virtuoso peering at crystals in urine is little better than the bumpkin searching for fairies, neither of them engaged in anything that remotely respects the responsibilities and dignity of humankind. A culturally structured way of ranking significance in nature is evident in the poet William King's satirical take on the habit of visiting virtuosi in foreign countries to seek out the local curios. King's French traveller in *A Journey to London* (1699) says 'I found myself better disposed and more apt to learn the physiognomy of a hundred weeds, than of five or six princes.' From this distance we are better able to judge which alternative is of more lasting value.

But what provoked poets and writers like Butler, Shadwell and Swift so aggressively to satirize the curiosity of the experimental philosophers? It is somewhat futile to complain that their accusations and portraits are unfair; of course they are, for most humour and satire depends on exaggeration. These writers must have known that their mockery relied on a biased picture. The real question is why, beyond the mere desire to amuse, they wished to present this caricature, which goes well beyond wicked teasing of a ripe target by being laced with genuine ire and disapproval.

The capacity of comedy and satire to defend, as much as to attack, the comfortable status quo is often under-acknowledged. To be captivated by a fly's eye was a notion incompatible not only with medieval scholasticism but also with the newly emerging bourgeois capitalism. Behind their clever and often entertaining words, the literati were expressing profound social conservatism.

That's not all. There is also a sense of affront in these attacks at an emerging culture of self-appointed specialists and experts, a disdain for which continues (largely unrecognized) to fuel much anti-scientific comment today. Who are these arrogant elite who alone assert the right to interpret what they see? Look, says Butler: they are outraged that a mere footboy has questioned the existence of the lunar elephant:

> By giving footboys leave to interpose,
> And disappoint whatever we propose . . .
> For which they have deserved to run the risks
> Of elder-sticks, and penitential frisks.

To the timeless accusation that false authority surrounds itself with impressive titles, Butler adds a somewhat more pertinent complaint: the right to interpret nature and experiment is bought by subscription, and one enjoys that right in direct proportion to one's financial investment:

> those who 'ave purchased of the college
> A share; or half share, of knowledge.
> And brought in none, but spent repute,
> Should not be admitted to dispute;
> Nor any man pretend to know
> More than his dividend comes to.

Although it was undoubtedly far too simplistic to portray the Royal Society as a clique of philosophers who bought their right to pronounce on natural philosophy, the criticism was not wholly gratuitous. In deciding the reliability of testimony and evidence, there was a legacy of privilege: a gentleman's word carried more weight than that of a tradesman, to whom the Royal Society's substantial membership fees made it inaccessible in any case.

But the society's motto *Nullius in verba* – take no one's word for it – spoke at least of the right intentions; and the importance of engaging artisans, merchants and the like in its affairs was paid more than lip service. Reports from unschooled observers such as Leeuwenhoek undoubtedly benefited from an introduction by a known quantity such as Constantijn Huygens, but at least they might be taken seriously.

Increasingly, the criterion for belief was shifting towards reproducibility of phenomena in the company of reliable witnesses and the transparency with which an experiment was described, rather than just a gentleman's testimony.

Of course, the question of who was 'reliable' remained in debate, and to outsiders it could all too easily appear that one's dependability was simply a function of one's acceptance in the right circles. 'The Elephant in the Moon' reveals that there was still an appearance of exclusivity, not just in terms of who was permitted to evaluate and hypothesize about experience but also how knowledge was guarded and distributed. The technical language of the proto-scientists, ridiculed by Shadwell, could easily look like the intentionally obscure jargon of the alchemist, which Jonson vigorously lampooned. Butler accuses his virtuosi of explicitly seeking to hide their knowledge from the rabble in the manner of natural magicians who feel that it is too dangerous for public consumption:

> For Truth is too reserved and nice,
> To appear in mix'd societies.

These criticisms are more plainly stated in one of Butler's celebrated 'Character' portraits, called simply 'A Virtuoso' – whom he identifies as 'A Curious Man'. Here he laments the virtuoso's self-referential and self-determining authority, his impenetrable, specialist language, his craving for praise, and the way his curiosity turns into hoarding of rarities. The last of these shows that the scientist was still not clearly distinguished from the antiquary or collector, stocking his cabinet with ostentatious wonders; Alexander Pope makes the same complaint, with reference to Hans Sloane, in his 'Epistle to Burlington' (1731):

> 'Tis strange, the miser should his cares employ
> To gain those riches he can ne'er enjoy:
> Is it less strange, the prodigal should waste
> His wealth to purchase what he ne'er can taste?

The idea that only science can pronounce on truth, and that what cannot be measured and made scientific is not worth measuring – scientism in its nascent form – was attacked by Butler for its circularity.

The issue of specialist language is more tricky. Baconians like Boyle sought explicitly to write as plainly as possible, discarding the obscure codes of the alchemists. But although their critics weren't convinced, the complaint was not so much about secrecy and exclusivity as vacuity. In his essay *The Transactioneer* (1700), William King rather pithily skewered the style of writing that he perceived to characterize the *Philosophical Transactions*:

> in the late Transactions most of the subjects are indeed most prodigiously sublime, and penned too in a wondrous manner: so that it is a hard matter for the generality of Virtuosi who imitate Bacon, Boyle, or men of that character, to write in the language observable in most of those papers: for there the expressions are suitable to the sublimity of the subjects, and consequently mighty mysterious and above the reach of those Gentlemen.

In other words, the intellectual content is as tenuous as the language. Shadwell riffs on the same theme when Gimcrack explains how plums acquire their blueness:

> it comes first to fluidity, then to orbiculation, then fixation, so to angulization, then crystallization, from thence to germination or ebullition, then vegetation, then plantanimation, perfect animation, sensation, local motion, and the like.

Here he mimics the hypnotic verbiage displayed by Jonson's sham alchemist Subtle, who in turn seems to have learnt his verbal gymnastics from Rabelais. But whereas Subtle's convoluted terminology has the sole purpose of dazzling and impressing his clients – he knows full well that it is a litany of nonsense – Gimcrack seems in contrast to think he *is* saying something meaningful. Indeed, he is repeating almost verbatim Robert Hooke's account, in his microscopic observations of blue mould, of how organic matter might become progressively more organized and sentient. Shadwell invites us to laugh at the emperor's new clothes, at the presumed emptiness behind this verbal smokescreen.

Tortuous terminology is all the more risible when accompanied by pomposity, as implied by Butler's imitation of the typical style of the virtuoso in a skit called *The Royal Society: An Occasional Reflection upon Dr Charleton's feeling a Dog's Pulse at Gresham College*. Attributed to one 'R. B. Esq.', this was very evidently a parody of Robert Boyle's *Occasional Reflections upon Several Subjects* (1665), even to the extent of being similarly addressed to an imaginary interlocutor named Lyndamore:

> From this, Lyndamore, we may learn, that as in general Nature there is neither higher nor lower, but Zenith and Nadir are equally on a Plane, as well as the Poles; so we may receive Matter of Instruction from Objects of the meanest and most contemptible Quality as well as from Things of higher and more sublime Condition; even as the most industrious and elegant Mr Hook, in his Microscopical Observations, has most ingeniously and wittily made it appear, that there is no difference, in point of Design and Project, between the most ambitious and aspiring Politician of the World, and of our Times especially, and that most importune and vexatious Insect, commonly called a Louse.

This is a dangerous game for the satirist. For one thing, it requires real knowledge to be done well (as Ben Jonson did). And imitation of a soporific or convoluted style runs the risk of simply reproducing the effects it seeks to mock.

Jargon remains of course a real problem for science – but not only for science, nor indeed are scientists any longer the worst offenders. Certain words, terms, modes of expression insinuate themselves into academic writing as codes of membership: it becomes simultaneously a learnt habit and a defensive reflex to assert, for example, that one is going to 'problematize' or 'foreground' a topic. These expressions have a meaning, but it is not a meaning in desperate need of a neologism in the same way that 'superconductivity' is. Nonetheless, scientists too sometimes use jargon (often unconsciously) simply to demonstrate that they know it – and that by implication they are familiar with the associated ideas – in the same way that they display their equations and calculations in academic lectures not because anyone can follow them but because these things function like a Masonic handshake. That is all fair game for satire, but not of a sort that strikes very deeply at the integrity of the basic enterprise. Yet perhaps there is still a suspicion akin to the one Butler sought to arouse: that any knowledge requiring so many obscure and newly minted words cannot really be worth knowing. In his bloated and opaque lexicon, is the virtuoso, and no less the contemporary 'boffin', compensating for an absence of real knowledge and trying to glorify pointless curiosity?

The last laugh

The activities of the early scientists mocked by the satirists as self-evidently absurd now seem uncontroversially valuable, even fruitful, and their sneering dismissal the result of a failure of imagination. When we read now of Gimcrack's attempts to make a 'hearing trumpet' by which men may be heard over great distances, even 'from one nation to another', we do not snigger but feel the frisson of prescience (even though long-distance communication was in fact a much older dream). What's more, he appears to appreciate what telecommunications held in prospect: 'When I have perfected it, there needs but one parson to preach to a whole country.' His reply to Bruce's straight-faced mockery ('By this princes may converse, treat, congratulate, and condole, without the great charge and trouble of ambassadors') today sounds dignified rather than self-deluding: 'I hope to effect it.' But in the Restoration they all laughed, rather more than they did at Christopher Columbus.

And who today scorns the mapping of the geography of the heavenly bodies – as Shadwell parodied Galileo's cartography of the moon – especially now that the results are made freely available on the Internet as an educational and recreational tool? Who, faced with a pollination crisis, now questions the urgent need to understand the finer points of the bee's biology? Who doubts the value of elucidating the inner workings of the aquatic 'animalcules' that poison millions of people annually? It's true that the seventeenth-century virtuosi cannot claim to have foreseen such significance in their curiosity – but that, as I argue in the next chapter, is the whole point.

Yet by the time Jonathan Swift was firing barbs at the virtuosic culture, his target had already been deflated. The mockers, satirists and wits seemed to have won. Well-bred gentlemen feared that taking up natural philosophy would expose them to ridicule as a pedant or fool like Gimcrack, and so many of them paraded their ignorance instead as fops and gallants, initiating a tradition of aristocratic disdain for learning that does not show much sign of abating. More serious men turned instead to matters of increasing importance to the burgeoning bourgeois society: to trade and business. It is for this audience that John Locke writes when he accuses the universities of filling people's heads with trash and discounts the value of the arts while expressing no more than tepid enthusiasm for mathematics and natural philosophy. By the time Isaac Newton died in 1727 (by which time he was not so much the author of the *Principia* as a servant of the state, having become Master of the Royal Mint), the age of the virtuoso was over. For a time, at least, the Enlightenment curtailed its curiosity about the natural world and cast its light instead on the affairs of business, politics and empire.

The literary scholar Claire Preston reads also in the antipathy of literary culture towards experimental philosophy an early premonition of the notorious Two Cultures: 'the reluctance of late-humanist culture to move away from textual authority made the antiquary's and the experimentalist's interest in things, rather than books or manuscripts, difficult to assimilate into the prevailing model of learning'. There was, perhaps, a sense of something vulgar and undignified in allowing the world to speak to us rather than fitting it into our preconceptions.

But we can't deny that the scientists brought some of this ridicule upon themselves. For all their aspiration towards gentlemanly discourse free from scholastic acrimony and dogma, the new philosophers became locked in bitter quarrels about the veracity and priority of one another's ideas, often accompanied by a degree of self-importance ripe for parody. They could not honestly evade charges of elitism, even if that was partly a consequence of financial necessity (at least in England). And although the most celebrated among them had outgrown a rather childish delight in wonders, exotic rarities and ingenious contrivances, that breed of leisured, marvel-craving curiosity was still not hard to find. What *Gulliver's Travels* really speaks to is not what the most able scientists of the age were attempting and achieving, but the emerging public image of the scientist and his new, somewhat uneasy accommodation with curiosity. It is to this legacy of the 'Scientific Revolution' that I now turn.

13 Professional Virtuosi, or Curiosity Served Cold

The Universe is so full of Wonders, that perhaps Eternity alone can be sufficient to survey and admire them all.

Henry Baker, *The Microscope Made Easy* (1744)

For Art and Science cannot exist
Except in minutely organized particulars

William Blake, *Jerusalem* (1804)

It's a sign of how confident the modern virtuosi are in the social sanctioning of curiosity that they may be ahead of the Shadwells and Swifts in the satirical game, colluding with ironic self-awareness in the notion that their inquisitiveness leads them to comic excess. They celebrate this comedic aspect with annual awards called the Ig Nobel Prizes, sponsored by Harvard-affiliated associations and given for research deemed particularly noteworthy in its extravagant whimsicality.

The 2005 laureate, chemical engineer Ed Cussler of the University of Minnesota, was rewarded for his study, *pace* Gimcrack, into the 'speculative part' of swimming. Because an increase in the viscosity of a liquid negotiated by a swimmer adds equally to the push she can achieve with the swimming stroke and the resistance to her motion, she should in theory be able to swim precisely as fast in syrup as in water. It sounds a sober enough, if recondite, deduction – except that Cussler and his student Brian Gettelfinger then decided to do the experiment. Theirs was not the sort of simplified lab model that scientists commonly employ, but the real thing. They filled a swimming pool with syrup – a solution of edible thickener that, according to Cussler, 'looked like snot' – and enlisted volunteers to swim in it. Other Ig Nobel recipients have magnetically levitated a frog, studied 'the forces required to drag sheep over various surfaces', and, in an experiment that seems to have strayed right out of the Restoration, studied the effect of ale and garlic on the appetite of leeches.

This is a delicate game. Cussler's work is a gloriously absurd piece of science as theatre, but it might have been greeted less indulgently had it not been so cheap. There can, after all, be few less useful and more arcane questions than those confronted, at vast expense, by the Large Hadron Collider. Where does mass come from? Is there a hidden relation between families of subatomic particles? Are there other dimensions that we cannot perceive directly? The fact that these questions were accepted, with only a little bewilderment, as valid and even profound by a media habitually cynical about academic enquiry suggests that Shadwell's satire has not prevailed. It has become legitimate for science to be interested in anything and everything.

But how, and at what price, was that apparent victory won?

Transforming wonder

Curiosity, the wild-haired, dangerous Amazon depicted by Cesare Ripa in 1593, had a makeover during the course of the next century. Thomas Jefferys copied Ripa's image for the design of a masquerade dress in 1772, but she was tamed in the translation, her hair tidied, her furious features remoulded into a serene smile. This, says literary scholar Barbara Benedict, is curiosity as a 'licentious indulgence, as well as . . . the emblem of legitimate consumption'. Here curiosity titillates rather than threatens.

That transformation was in fact already begun in Ripa's day. A 1592 book of nature engravings called *Archetypa* prepared by Jacob Hoefnagel from the drawings of his father Joris, at that time an illustrator for Rudolf II, gives a mixed moral message about curiosity. 'Let us not too curiously examine divine works with human reasoning, but having been led along let us admire their artificer', the caption to one plate advises us with the familiar pious formula. And yet how alluringly does Hoefnagel lay out those works here, showing us in luxuriant

Thomas Jefferys' depiction of Curiosity in *A Collection of the Dresses of Different Nations, Ancient and Modern* (1772). Compare with the late sixteenth-century version (page 22).

detail flowers, shells, eggs, and the baroque carapaces of insects. Who, seeing these things, could fail to become curious about them? Who could resist a bite of that apple?

In the late sixteenth century the biblical admonition against pride (and some said, curiosity), *noli altum sapere* (see page 12), was associated with Prometheus and Icarus: a warning against the danger of over-reaching ambition. But that sort of ambition came to be seen as a virtue a hundred years later, when Columbus's voyages were compared to the flight of Icarus. The title page of the collected letters of Antony van Leeuwenhoek to the Royal Society, published in 1719, announced the new spirit of the age in its slogan *Dum audes, ardua vinces* – if you dare, you will conquer difficulties. It was a call for the exploration of new lands: not just those over the seas, but the worlds opened up by scientific investigation.

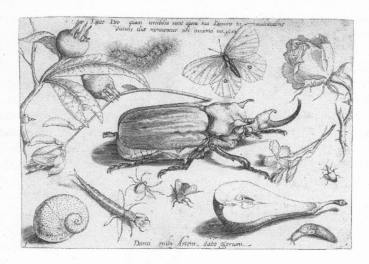

Joris Hoefnagel catalogues nature as a pageant of wonders in *Archetypa* (1592).

While supporters of the new experimental philosophy worked hard to effect this transition of curiosity from a vice to a neutral or even virtuous tendency, the shift in moral status was by no means accepted in other disciplines. In moral philosophy and theology, curiosity tended to remain something to be abjured. As Neil Kenny says, in the early modern period 'the point of invoking curiosity was almost always to regulate knowledge or behaviour, to establish who should try to want to know what, and under what circumstances'. The chameleonic nature of the word meant it could be enlisted to attack or defend almost any mode of thought – as we saw, even Francis Bacon was ready to use it as a label for 'bad', shallow forms of inquisitiveness. Church discourse generally maintained that curiosity was something to be avoided, but began to hint that there might also be a good form of curiosity. And just as in modern times sexual liberalism became as much a marketing tool as an expression of personal liberty, so curiosity was used to sell books and pamphlets, building on the titillating promise of the literature of secrets while also (paradoxically, given the culture of collecting) implying that the item was accessibly priced. If we say that curiosity became sanctioned in the seventeenth century, we are talking about specific spheres of activity and a specific usage of the word that is close to the modern one.

A complex and in some respects complementary shift of status and meaning was experienced by curiosity's cousin, wonder. Commended by medieval theologians, it was for the experimental philosophers a dubious ally at best. They accepted its value as a snare for curiosity and a spur to diligent enquiry: a kind of recruiting agent for the experimental philosophy. But it was apt to degenerate into the kind of shallow spectacle that the virtuosi sought to pin on charlatans and mountebanks. At its worst, wonder was thought to induce stupefaction. Descartes was careful to distinguish useful wonder (admiration) from useless (astonishment, literally a 'turning to stone' that 'makes the whole body remain immobile like a statue'). For Descartes, useful wonder was 'a sudden surprise of the soul which makes it tend to consider alternatively those objects which seem to it rare and extraordinary'. And if that sounds a little like the stimulus to enquiry that Newton attributed to curiosity, this was no doubt because the two were in the process of exchanging places.

Sometimes scientists would place wonder on the far side of curiosity: something that emerges only after the study is complete and understanding has been gained, itself a dour and arduous business that few could be expected to sustain. In this way, wonder could be safely and dutifully channelled away from the phenomenon itself and directed towards God. 'Wonder was the reward rather than the bait for curiosity,' say Lorraine Daston and Katharine Park, 'the fruit rather than the seed.' It is only after he has carefully studied the behaviour of ants to understand how elegantly they coordinate their affairs that Jan Swammerdam admits to his wonder at how God could have arranged things thus. 'Nature is never so wondrous, nor so wondered at, as when she is known', wrote Bernard de Fontenelle. This is a position that most modern scientists, even those of a robustly secular persuasion, are comfortable with: 'The science only adds to the excitement and mystery and awe of a flower', said physicist Richard Feynman.

Moreover, this kind of wonder is reserved for the economy and design and exquisite craftsmanship evident in nature's normal behaviour. In this view, the strange and bizarre, the marvels and oddities are not appropriate stimuli – it is the snowflake, not the monster, that properly elicits wonder. Comets are wondrous not because they are extraordinary but precisely because they are recognized to follow the same predictable gravitational laws as the planets.

This kind of wonder is not an essential part of scientific philosophy, but may constitute a form of post hoc genuflection. It is *informed* wonder that science generally aims to cultivate today. The medieval alternative, seen as ignorant, gaping wonder, was and is denounced and ridiculed. *That* wonder, says Mary Baine Campbell, 'is a form of perception now mostly associated with innocence: with children, the uneducated (that is, the poor), women, lunatics, and non-Western cultures . . . and of course artists'. Since the Enlightenment, Daston and Park concur, wonder has become 'a disreputable passion in workaday science, redolent of the popular, the amateurish, and the childish'.

The Romantic movement sought to harness the informed, sober variety of wonder, whether in the *Naturphilosophie* of Schilling and Goethe or the passion of English Romantics such as Samuel Coleridge, Percy Shelley and Lord Byron, who had a more than passing interest in science. Now it was not God but nature herself who was the object of veneration. While natural theologians such as William Paley and the authors of the Bridgewater Treatises* discerned God's handiwork in the minutiae of natural history, the grander marvels of the Sublime – wonder's 'elite relative' as Mary Baine Campbell nicely calls it – exposed the puny status of humanity before natural forces of altogether vaguer origin. The author of the Sublime was no intricate craftsman who wrought exquisite marvels, but worked only on a monolithic scale, with massive and inviolable laws. He (if he existed at all) was an architect not of profusion but of a single, awesome order. Just as wonder was here enlisted to instil awe not in God but in the secular mystery of nature, so Keats's objection to scientific curiosity in 'Lamia' – never was a celebration of ignorance more ravishingly voiced – was predicated on the damage it allegedly did to abstract poetic wonder rather than to religious authority:

---

* These works were commissioned in 1829 by the eighth Earl of Bridgewater, Reverend Francis Henry Egerton, to display 'the Power, Wisdom and Goodness of God as manifested in the Creation . . . as, for instance, the variety and formation of God's creatures, in the animal, vegetable and mineral kingdoms'. The authors included leading scientists such as William Prout, William Whewell and William Buckland. Despite their theological agenda, they were significant works of science popularization.

Do not all charms fly
At the mere touch of cold philosophy?
There was an awful rainbow once in heaven:
We know her woof, her texture; she is given
In the dull catalogue of common things.
Philosophy will clip an Angel's wings,
Conquer all mysteries by rule and line,
Empty the haunted air, and gnomed mine –
Unweave a rainbow

Most famously, the ingenious mechanism of natural selection that Darwin discerned behind life's 'grandeur' did not need to be attributed to any Divine Mechanic, but showed that life could take care of its own elaboration and exuberance. Yet Darwin himself professed that 'I worked on true Baconian principles, and without any theory collected fact on a wholesale scale.' He seems to have held back from publishing his evolutionary theory not so much because of misgivings about the theological implications as because he was distrustful of whether his facts fully supported it. Surely one of the reasons why Darwin still looms so large in the minds of many scientists, particularly those who find no place for God in the universe, is that he personifies the modern struggle with curiosity. We marvel and puzzle at how Darwin could delay the 'greatest idea in history' for the sake of an exhaustive tract on barnacles, and at how the formation of soil and the curling of climbing plants seemed to command as much of his attention as the great chain of being that was unfolding in the fossil record. Was Darwin one of Isaiah Berlin's foxes who found himself unwillingly thrust into the position of a hedgehog? I suspect Darwin's eulogizers experience more than a little anxiety at the thought that he would have been as contented if he had published only his books on earthworms and barnacles and left *On the Origin of Species* and *The Descent of Man* as manuscripts on his shelf. Is a genuine love of all the detours and diversions of one's subject really compatible with the urge to create synoptic theories of the world? And which makes one happier?

There is no need to answer that, and probably no point in doing so. It's the question that matters.

Taming curiosity

Whether you call it wonder or curiosity, the enjoyment of 'the false, the unreal, the surreal, the perverse and the impossible' is, for Campbell, a necessary prerequisite for the creation of art. If that is so (and I suspect it is), then the determination of Francis Bacon and the virtuosi to eliminate such pleasures, and to make the aesthetics of curiosity at best something of a refined, gentlemanly matter of good taste, sounds like a betrayal of the human spirit. Of course, it would be wrong to lay that accusation solely at the door of the experimental philosophy, for the scholastics and medieval theologians were equally disapproving of sensuous 'irrational' pleasures. Many societies have always been suspicious of these all-too-human passions. We have no better an accommodation today, although the suppressive impetus tends to come far more strongly now from social, political and religious conformism than from science. But what did the emergent modern science try to do to bring such instinctive exuberance under control?

There was never any doubt that curiosity alone is not a sound basis for an experimental programme. It must be disciplined, for example by coupling it with a method for establishing reliable 'facts' empirically and using them to formulate hypotheses and interpretations. Otherwise one is just a collector of curios. We've seen that this harnessing and managing of curiosity involved the explicit social construction of a prototypical scientific procedure and an associated literary system for conveying the results. As science expanded, it became specialized, and its practitioners were no longer faced with the huge, quasi-random 'to do' lists of Leonardo, Boyle and Hooke. If science could be curious about everything, individuals were obliged to focus and narrow their attention, with the consequence that fewer of them saw the big picture. Even nature herself was subjected to a more rigorous discipline, as natural laws replaced tendencies. 'In order to be made useful', say Lorraine Daston and Katharine Park, 'nature had to be made uniform.' Parsimony, not abundance, was what became praiseworthy in her ways.

Ultimately, this process threatened to end in disenchantment with the world. As Catherine Wilson says, the result is a vision in which

> the world is not there to delight us; the scientist is one who unmasks the delusions of self-indulgent human consciousness and replaces them

with the hard facts: nature in itself is morally and aesthetically neutral, neither benevolent nor cruel, neither beautiful nor ugly. In place of a sympathetic cosmos, whose members are bound together by analogies, harmonies, and sympathies and kept distinct by metaphysical individuality and antipathy, we have only one kind of matter forming one pattern, and, in place of love and strife, little structures and machines producing all the illusions of subjectivity.

'The coldness and aloofness of the scientist', she adds, 'are the price that has to be paid for [their] exceptional vision.' Here is Blake's ambivalent view of Newton: awed by the scope of his sight, chilled and outraged by the coldness of his gaze.

If this seems a depressing spectacle, we should nonetheless resist the temptation to contrast it with a romanticized past in which the medieval man stands bathed in the awesome, nurturing splendour of nature. That way lies only the utopian mythology, the self-deluding lyricism, of the William Morris print. Making use of both curiosity and wonder in the Baconian quest to relieve man's estate, without compromising their broader virtues, is a remarkably difficult balancing act. This is surely implicit in the extremes and contrasts with which curiosity was exhibited in the seventeenth century. The most hard-headed men betray themselves as credulous fantasists. The virtuosi draw up comically surreal or grotesque experimental plans. They abuse their own bodies with hair-raising abandon: Newton nearly blinds himself by staring into the sun or inserts a bodkin 'betwixt my eye and the bone' to deform the eyeball's lens. They profess high principles but squabble like schoolboys. They are pious, they are atheists, they denounce childlike wonder and capitulate to it.

What nonetheless emerges is a code of practice and a philosophical stance along the lines that Wilson sketches, and which we can still recognize today, in which the scientist bleaches all passion from his or her professional persona. At its best, this instils (or at least creates an impression of) calm and objective authority; at its worst it is a form of psychosis, a suppression of sexuality and emotion, an apparent (mostly just that) determination to rule the world without a shred of compassion.

As ever, the caricature is revealing precisely because it is a caricature. Real scientists are almost never like this. They would regard Hooke as odd and Newton as odder still. On the whole they are like the rest of us, except perhaps with different interests: snails, maybe, or gravity. But the taming of curiosity that was the price of its licensing is still apparent in the bloodlessness of the scientific literature, mostly stripped of adjectives and pronouns (especially the first person singular) and browbeaten into the passive voice. These are hard habits to shake off; indeed, they offer something of a protective carapace. But scientists who believe that curiosity counts for anything might now reasonably start thinking of shedding them, if only to throw a few Shadwells off the trail.

Why be curious?

The idea that curiosity as we understand it today is a natural and irrepressible human instinct is just an extension of the reflexive tendency to make of ourselves a model for all humanity, everywhere and at all times. The evolutionary benefit of an impulse to find things out surely has its limits, and those limits do not extend to wanting to know the crystal structure of calcite, the mechanism of supernovae or the gastric anatomy of the housefly. Yes, 'you never know when it may be useful' – but that alone hardly seems likely to root a boundless curiosity securely in evolutionary psychology. And this is before we even start to consider how curiosity has had so many different meanings, such differing social valence and functions, over the centuries.

So we should be wary of general statements about the value, appeal and significance of curiosity. Like many human attributes, it seems less a 'fundamental' than a feature with a particular and contingent history. The 'meanings' of curiosity today must thus lie in the question of how it arrived at this place, and in this book I can pretend to offer no more than a part of that story. All the same, I hope this incomplete tale forces us to think again about what we mean when we use the word, and about the conditions, consequences and corollaries of exercising our curious *need to know*.

I suspect most scientists would agree with Robert Boyle that one of the virtues of the experimental philosophy, both then and now, is the sheer 'pleasure of making physical discoveries'. But he warned that one who gives himself wholly to this approach should also expect to be provoked and even tormented by an endless curiosity. For each time we glimpse a little more of nature's system of order, we are compelled to wonder how it operates in yet other circumstances: each discovery raises more questions, and so discovery is always accompanied by 'both anxious doubts, and a disquieting curiosity'. And this is precisely because nature is not an assembly of disconnected facts but a system of broad general principles: her behaviour in one place has implications for that in others. In that respect, said Boyle, science differs from art and literature, which in his view present self-contained narratives that do not interact. If you are reading Aesop's fables, he says,

> or some other *collection* of apologues of differing sorts, and independent one upon another; you may leave off when you please, and go away with the pleasure of understanding those you have perused, without being solicited by any troublesome itch of *curiosity* to look after the rest, as those, which are needful to the better understanding of those you have already gone over, or that will be explicated by them, and scarce without them. But in the book of nature, as in a well-contrived romance, the parts have such a connection and relation to one another, and the things we could discover are so darkly or incompleatly knowable by those, that precede them, that the mind is never satisfied until it seems to come to the end of the book . . . And yet the full discovery of life's mysteries is so unlikely to fall to any man's share in this life, that the case of the pursuers of them is at least like theirs, that light upon some excellent romance, of which they shall never see the latter parts.

There is no rest for the curious experimenter, then – but neither is there any prospect that he or she will ever hear the whole story.

For Boyle, the ultimate motivation for curiosity was religious: a sacred duty to examine every niche and corner of God's creation. The scientist thus becomes a 'priest of nature', embarked on a holy quest. It was a position shared by many of his contemporaries. In his *Religio medici* (1643), Thomas Browne wrote that

> The world was made to be inhabited by beasts, but studied and contemplated by man: 'tis the debt of our reason we owe to God, and the homage we pay for not being beasts . . . Those highly magnify him, whose judicious inquiry into his acts, and deliberate research into his creatures, return the duty of a devout and learned imagination.

Browne attached a powerful and ambivalent image to this natural theology in saying that it was the aim and obligation of the scientist to 'suck divinity from the flowers of nature'.

It was for this reason that Boyle, while reticent about conjuring up mechanical 'causes' for his experimental observations, had no hesitation in seeking metaphysical, divine Final Causes. That was not to the taste of Descartes, who felt that we have proceeded far enough when we can account for material phenomena in terms of mechanical principles such as his theory of vortices. Thomas Hobbes held that view so strongly as to attract accusations of atheism, especially from the Cambridge Platonist Ralph Cudworth,* who could not accept causes 'without any intention for ends and good, that is, without the direction of any mind'. Of Descartes' argument that Final Causes are too mysterious for the human mind to grasp, Cudworth could only say that this was 'an hypocritical veil of humility'.

The Cartesian view is now the one preferred by most scientists, who consider it unnecessary and untestable to leap from physical to theological explanations for physical phenomena. Faith may demand them, but logic finds them superfluous. But perhaps Francis Bacon offered the most astute position. Ever concerned to avoid being branded an atheist, he insisted that trying to shackle science to theology did neither of them any favours, for it made faith a prisoner of reason and therefore contingent, under perpetual threat of revision to fit with new discoveries. 'Not only fantastical philosophy, but heretical religion spring from the absurd mixture of matters divine and human', he wrote. One need only witness today's sad attempts to traduce or manipulate scientific understanding in order to preserve literalist fundamentalism – that is, to read holy books as natural history – to see that here Bacon had a more salutary view than the pious Boyle.

However, even Cudworth was troubled by the idea that God should have to administer to every trivial detail of nature. This, he feared, would make him 'operose, solicitous and distractious' – a sort of fiddling pedant. Instead, Cudworth postulated that nature had a degree

---

* Cudworth was briefly a member of the 'Royal Society' before it received its royal charter. His daughter Damaris was another of those women who, in a more enlightened age, would surely have been an important scholar herself. She married the nobleman Francis Masham, and was a close friend of John Locke, a correspondent of Gottfried Leibniz, and published works on moral philosophy and theology.

of autonomy, allowing her to act as God's assistant without his having
to intervene. Strikingly, Cudworth elected to describe this artisan-
nature in terms borrowed from the chemical philosophy of Paracelsus,
who had suggested that the human body contains an alchemical
agency, the Archeus, which separates the good from the bad in the
food and drink we ingest. Cudworth spoke of an 'artificial nature,
which, as an Archeus of the whole world, governs the fluctuating
mechanism thereof, and does all things faithfully, for ends and purposes
intended by its director'.

While sharing Cudworth's determination to preserve the hand of
the divine in the variety and wonder of the world, Boyle found the
Archeus too personified a representation. In *A Free Enquiry into the
Vulgarly Received Notion of Nature* (1686) he attacked the view that
nature is a kind of wise and benevolent being; he asserted instead
that 'creative nature' was more like a machine, or, in his extraordinary
expression, a 'pregnant automaton'. Even so, he felt that nature could
reach beyond mechanical drudgery to invest the world with the splendour of true artistry, for example in birdsong and ornate plumage,
'particularly those that make up the peacock's train'. It became a point
of much debate, and some discomfort, where the line should be drawn
between the agencies of God and of nature. Leaving too much to
God seemed to make him a tinkerer, while ascribing too much to
nature risked depriving him of the credit. This question of how clearly
(if at all) God's hand can be discerned in the ordering of nature's
particulars was to become (and remains) the central theme in arguments about the necessity for divine intervention in the natural world,
whether this concerns the living world or the beginnings and fundamental laws of the universe.* We can now see that this matter first
became problematic not as nature's grander principles were elucidated

---

* There remains today a secular analogue of this question: do nature's laws themselves permit, or even demand, a kind of creativity? To what extent does natural
selection determine the fine details of, say, animal ornamentation or modes of
communication, and to what extent might they be adaptively neutral or even inescapable consequences of the contingent exuberance of physical laws acted out in
complex systems? Are these just examples of what Cudworth's colleague Henry
More, using the popular theatrical imagery of the age, called 'the admirable windings
of Providence in her Dramatick Plot which has been acting on this Stage of the
Earth from the beginning of the World'?

– the mathematical order of the Copernican cosmos could readily enough be interpreted as a manifestation of divine wisdom and planning – but as curiosity led philosophers to confront the dizzying scope, diversity and *intricacy* of creation.

No ideas

It probably goes without saying that Darwin was never as Baconian as he imagined. On the *Beagle* he collected specimens, samples and notes tirelessly, but his mind was constantly on the general principles: how and why speciation happens, why species die out, what are the forces that govern geological and biological evolution. 'Isolated facts soon become uninteresting', he averred.

Has any scientist ever truly kept hypothesis at arm's length while 'heaping up facts'? True, the naturalists of the late eighteenth century seem sometimes content to classify and catalogue with never a care for any grand synthesis. But even classification presupposes a certain order and scheme of nature. It is hard to imagine a curiosity worthy of the name (at least in today's sense) that will be satisfied with unregimented observation.

In fact it is worse than that. As long as the Baconian method remains frozen at the stage of data collection, it is not really science at all. Not only is nothing explained, but one is not even in the foothills of explanation, taking the first small steps towards the summit – or rather, one is perhaps there but wandering aimlessly, with next to no chance of finding the route that leads upwards. Data cannot be meaningfully collected without a prior hypothesis, simply because there is too much of it. You can't be sure ever of assembling a coherent body of facts from which a hypothesis might be framed, unless you have some notion of where to look at the outset. The early virtuosi seem to have intuited this. Robert Boyle's experiments on luminescence in the air pump, for example, focused quite quickly on the discovery that some light emission required air and some did not: a fundamental distinction that, as Hooke suggested, identifies at least two classes of this phenomenon and offers hints about the relationship of 'aerial phosphorescence' to other processes that depend on air. Things often can't help but fall into place, largely because we are the kind of creatures that look for places to put them.

This is not to deny that sometimes science is forced to wander with blind optimism in the foothills of data, for lack of any better option. When it does so, however, one detects a marked lack of curiosity, as well as of imagination and vigour. In such circumstances, scientists risk making pointless measurements because they do not know what else to do. They might persuade themselves that the data are surely going to be useful one day, and that the mere filling up of databanks must correspond to progress of a sort. That is evident, for example, in some aspects of the modern extension of Darwin's exploration of inheritance which seeks to find and chart out the intersections not of new species but of new genes. While there are some very soundly motivated questions to be investigated from the decoding of genomic sequences in human individuals and populations, the limitations of simplistic notions about how genes work, and the absence of sound theoretical principles to put in their place (as well as the money and human-power thrown at the enterprise), means that sometimes this collecting of genetic information has been thoughtless. It is as if the researchers, their wits slowed by Bacon's recommended 'leaden weights', share his rather desperate hope that one can feed all the data into a processing machine that will churn them into theories. 'During the twentieth century', says biologist Robert Weinberg of the Massachusetts Institute of Technology, commenting on the genomics fad,

> biology – traditionally a descriptive science – became one of hypothesis-driven experimentation . . . Implicit here was the notion that observations should only be made to support or attack hypothesized mechanisms of action, and that simple observation – phenomenology for its own sake – is of relatively little use . . . Now the dominant position of hypothesis-driven research is under threat.

Small-scale hypothesis-driven projects in genetics, says Weinberg, are being neglected for the sake of massive data-collecting exercises. One of their main conclusions so far has been simply to show that what we thought we knew about what the data mean and how they fit together is in fact only a small part of the puzzle. That's worth knowing, but the data themselves have not yet furnished any better ideas.

The endless frontier

The 'discovery machine' that is the Large Hadron Collider is arguably a Big Science project at the other extreme: hypotheses and theories constrain its data collection to an almost unprecedented degree. The physicists working at CERN know exactly what they are looking for – or at least, each of them has very strong ideas about what they hope to find and where to look for it, even if these presumptions don't all agree. But then one must ask: to what extent is this, too, research driven by curiosity? Is it not, at best, curiosity by consensus and committee, whereby all have decided en masse and years if not decades before the experiment is done (the planning began in the 1980s) what questions they should be asking? If the genome projects can look rather like the automated, random matching of pieces in a hundred vast jigsaws, for none of which is the final picture known, the LHC appears uncomfortably like assembling a huge team to put the sole remaining piece in place (with the knowledge that, if it doesn't turn out to fit, the whole assembly needs rethinking). Meanwhile, even naming NASA's new Mars rover vehicle (launched in November 2011)

NASA's Mars rover *Curiosity* embodies meanings of the word that would have been resonant in the seventeenth century. It is the sort of ingeniously 'curious' mechanical device that would have delighted Robert Hooke, and like the telescope and microscope it extends by artificial means the reach of human inquiry. To what extent, however, can its distant agenda be guided by spontaneous and responsive impulses?

Curiosity does not alter the fact that its programme of inquisitive exploration is preordained in fine detail: this is by necessity the curiosity of careful, exquisite contrivance.

It seems to me that, in contrast, that brand of curiosity which quickens the spirit with an 'eagerness to know' – the curiosity that Einstein called 'holy' – is a spontaneous gesture. That's perhaps one reason why the publicity machine that supports the LHC has enlisted the services of curiosity's old partner wonder: a motive force to fill the sails and sustain this alleged voyage of curiosity. Mindful that the real questions being probed by the LHC are too refined for most laypeople even to know what they mean, publicists at CERN and in the physics community cultivated a more generalized sense of wonder at this opportunity to look far back into the beginning of everything. We might not understand *what* was being probed, but we could appreciate the majesty of the issue.

Nonetheless, the LHC experiments and the *Curiosity* rover still exemplify what scientists today generally mean when they talk about curiosity-driven research: it involves questions that are investigated purely out of intellectual interest. This is generally contrasted with research that is driven by some economic or applied incentive – to develop a useful technology or device or medicine, say. There is a growing feeling that market forces are producing a science dominated by the demand for applications – by the economic bottom line, with its ever-contracting time horizon. The argument is that curiosity-led or 'blue skies' research has always proved to generate economically valuable discoveries and inventions, such as the laser and the transistor, that would never have been anticipated in a purely market-led approach: you often just don't know what you *want* to invent. '[The] curiosity-driven approach seems increasingly old-fashioned and underappreciated in our modern age of science', says chemistry Nobel laureate Ahmed Zewail. 'Some believe that more can be achieved through tightly managed research – as if we can predict the future. I believe this is an unfortunate misconception that affects and infects research funding.'

There is a lot to be said for Zewail's position, not least in an age that has become so highly goal-driven and has seen the decline of several major industrial laboratories that conducted basic research. But we can see now that this presumed role for and implication of 'curiosity' in science are recent ones. When scientists of former times studied problems that seemed spectacularly useless or recondite, this was no open-ended enquiry about what might emerge. There was generally a deeper agenda, or rather, several agendas. For the late-Renaissance virtuosi, curiosity was an expression of courtly conventions and the rituals of patronage; in the *Wunderkammern* it functioned as an emblem of status and power, enacted through the principles of natural magic. For Francis Bacon, gathering as much information as possible about the way the world worked was a political enterprise ultimately serving the purposes of state power. For Robert Boyle, asking questions about all aspects of creation was a religious duty. For Robert Hooke, curiosity was (however genuine) always linked to priority, reputation and fame. Some sought to find the divine code, the cosmic harmony imposed by God. Some sought to control the occult forces of nature; others were mere collectors of strange and rare items. Perhaps it's worth thinking carefully about what are the real agendas of curiosity today.

I began this book by contrasting Michel Foucault's fervent advocacy of curiosity as a way of making what surrounds us appear strange with the scientist's standard defence of curiosity as a mechanism for practical, Baconian discovery and invention – and also, by advancing knowledge, as a way to banish credulity and 'make wonders cease'. But I believe that the justification for scientific curiosity sought from unforeseen applications is largely a code, an accepted formula adopted by science's early pioneers from Bacon's bureaucratic model of state-sponsored research. It makes curiosity amenable, in today's terms, to a kind of cost-benefit appraisal, just as some seek to justify the preservation of the biosphere by appealing to its hidden economic value. The fact is that the link between curiosity and wonder cannot be severed, for true curiosity (as opposed, say, to obsessive pedantry, acquisitiveness or problem-solving) grinds to a halt when deprived of wonder's fuel. What changes is our readiness to talk about these things. We first emancipated curiosity at the expense of wonder, and then re-admitted wonder to take care of public relations. In the fear of the subjective that marks scientific research, wonder is one of the casualties; excitement and Foucault's fervour remain banished from the official records. But this does not mean they do not exist. Indeed, the passions involved in wonder and curiosity, as an aspect of the motivations for research, are a part of the broader moral economy of science that, as Lorraine Daston says, 'cannot dictate the products of science in their details [but is] the frame-work that gives them coherence and value'.

Whatever science might prefer, it is probably true that in popular culture wonder is 'hot' and curiosity is 'cool'. But cool curiosity looks merely odd, even comical, and in that guise it supplies candidates for the Ig Nobel Prizes. Because such self-mockery helps to defuse the tension and suspicion, it is welcomed by most of the scientific commu-nity, as indeed it should be. That, however, is not enough to alleviate the stereotypes of the unworldly boffin bean-counting in an ivory tower, or, more damagingly, of the heartless Fausts and Strangeloves plotting a soulless future.

It would be naïve to think that science's image could be improved overnight by making wonder a more openly permissible attribute of the whole enterprise (and not just a marketing tool). But pretending that science is performed by people who have undergone Bacon's purification of the emotions only adds to the danger that it will seem alien and odd to outsiders, something carried out by people who do not think as they do. Daston believes that we have inherited a 'view of intelligence as neatly detached from emotional, moral, and aesthetic impulses, and to a related and coeval view of scientific objectivity that branded such impulses as contaminants'. It was the caution of Robert Boyle, fearful of being read as an enthusiast, that spoke down the ages and told James Watson not publicly to call the X-ray structure of DNA 'beautiful' in his classic paper with Francis Crick in 1953. This was perhaps sound advice to a young scientist who wanted his work to be taken seriously in those austere times. But it was nevertheless misguided even from a rationalistic point of view. It amounts to suppressing information (the author's emotional and aesthetic response to an idea) that would inform the reader of the context in which it is being reported. It means that the report involves a certain degree of pretence. 'The instant I saw the picture [Rosalind Franklin's photograph of the X-ray diffraction pattern]', Watson later wrote in *The Double Helix*, 'my mouth fell open and my pulse began to race.' That comment speaks volumes about what science is and how it is done. You could even call it part of the data.

On the other hand, true wonder cannot be manufactured. In the Renaissance courts, natural philosophers would use wonder and spec-tacle to sell their work to their patrons. Francis Bacon felt that it should be enough to advertise instead the practical benefits of science; as the Royal Society discovered, this was not necessarily an effective strategy for eliciting state support. Today, we use a bit of both. CERN operates a website on the 'science' of Dan Brown's *Angels and Demons* (in which the laboratory featured); pictures from the Hubble Space Telescope are coloured and cropped to recall Romantic imagery of the Sublime. It would be foolish to deny, or to refuse from some misplaced purism, the value of wonder and entertainment in selling science. But it's important that we are clear what we are doing. When we dress up the

The wonder of the sublime: images from the Hubble Space Telescope echo the tropes of the Romantic era, such as the North American landscapes of Frederic Church.

spectacle and exuberance of human space flight by pretending that it is serving important scientific ends, we have lost our way. Similarly, there is a danger that scientists may start to believe their own advertising rhetoric and slogans – to think, for example, that in sequencing genomes they really *are* 'reading the book of life'. And when selling becomes outright prostitution – pizza deliveries and golf antics on the International

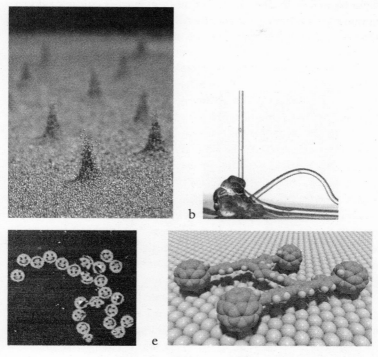

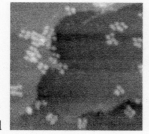

'Small science' can be impulsive, whimsical, led more by instinct than agenda. It can uncover solitary waves in shaken grains (*a*) or bouncing streams of shampoo and cooking oil (*b*); it exults in exploits of curious, intricate craftsmanship, such as DNA origami (*c*) and molecular cars, as seen under a microscope (*d*), and (*e*), as depicted in a computer-graphic impression.

Space Station, silly equations commissioned by commercial companies for the perfect penalty kick or the best beach – we are in trouble.

We already have the ingredients for a better recipe than that. It is not simply the case that modern science, when liberated from the dictates of the marketplace (as it ever more rarely is), conducts itself with a mixture of institutionalized curiosity and blind Baconian data-harvesting, all hidden behind a carefully regulated discourse of dispassionate enquiry. For one thing, scientists are too human for that, as Watson showed. They can disguise their excitement to conform to their professional conventions, but they can't fool us. The love, the awe, the passion they feel for their work leaks out; even the destructive 'mad scientist' caricature acknowledges this. Curiosity is undoubtedly seen as having a Faustian frisson, but no one would (or should) mistake the emotionally guarded Bacons, Hookes and Newtons for a Faust.

The flip side of dangerous Faustian obsession is the endearing quixoticism of an experiment like Ed Cussler's. Although this is surely an eccentric example, it is in the inventive and even playful bench-top experiment, in the delightful curiosity of small science, that its true spirit lies. Here is a curiosity that draws vigour – as it did in the seventeenth century – from an intimacy with the 'curious' object itself: curiosity not as rhetoric but as tactile, sensuous experience. Largely unheralded among the juggernauts of particle physics and genomics, countless little investigations using cheap, often home-made apparatus provide an unceasing flow of surprises about the world. If you shake a tray of little metal balls, they will arrange themselves into patterns of waves with beautiful geometries. DNA can be designed and built that will fold up into world maps and smiley faces, each no bigger than a protein molecule, generating what one scientist has called 'the most concentrated solution of happiness ever made'. Molecules can be fashioned into cars, complete with molecular wheels that rotate. Streams of shampoo can be made to leap and arch. Some of these experiments tell us about the world as it is; some tell us about the world we can make. Some have deeply serious objectives; others provide modest answers to modest questions, or rely on nothing more than to ask 'what if . . . ?' They cost negligible amounts; they may turn out to be fantastically useful, or spectacularly useless. They are each little acts of homage to curiosity, and in consequence, to our humanity.

Cast of Characters

Adelard of Bath (c.1080–c.1152): English scholar and translator of scientific works, and one of the most original thinkers of the twelfth century.

Heinrich Cornelius Agrippa (1486–1535): German physician and alchemist, and a major writer on the occult philosophy during the Renaissance.

Johann Valentin Andreae (1586–1654): German theologian and philosopher, widely considered to have written the tracts that launched the Rosicrucian movement.

Elias Ashmole (1617–1692): English antiquarian and politician, whose collection provided the foundation of the Ashmolean Museum in Oxford.

Mary Astell (1666–1731): English writer and early feminist who sought to expand women's participation in civic society.

John Aubrey (1626–1697): English writer; a biographer of many of the leading figures of his age.

Francis Bacon (1561–1626): English philosopher and statesman, who called for a new philosophy based on experiment, empiricism and practical applications of knowledge.

Roger Bacon (c.1214–1294): English Franciscan friar and philosopher, who championed the use of experiment in studying nature.

John Beale (c.1608–1683): English clergyman and writer, friend of Samuel Hartlib and an early member of the Royal Society.

Aphra Behn (1640–1689): English playwright, popular during the Restoration, and one of the first female professional writers.

Cyrano de Bergerac (1619–1655): French soldier and dramatist who authored two works of early 'science fiction' about travel to other worlds.

Robert Boyle (1627–1691): Anglo-Irish scientist; one of the leading figures in the new experimental philosophy of the seventeenth century.

Tycho Brahe (1546–1601): Danish astronomer who was unrivalled in his time for the accuracy and detail of his observations of celestial motions.

Hennig Brandt (c.1630–c.1710): German alchemist, credited with the discovery of phosphorus around 1669.

William Brouncker (1620–1684): Anglo-Irish aristocrat who held several governmental appointments under Charles I and was favoured by Charles II after the Restoration. He became the first president of the Royal Society.

Thomas Browne (1605–1682): English writer and collector, sympathetic to the experimental philosophy and author of the debunker's manual *Pseudodoxia epidemica*.

Giordano Bruno (1548–1600): Italian philosopher and astronomer, a Copernican whose religious unorthodoxy led to his condemnation by the Inquisition and execution by burning.

Tommaso Campanella (1568–1639): Italian philosopher and theologian, who opposed Aristotelianism, advocated a quasi-mystical empirical philosophy, and was imprisoned for his alleged political activism against Spanish rule.

Girolamo Cardano (1501–1576): Italian mathematician and physician with heterodox inclinations, who pioneered the study of probability.

Meric Casaubon (1599–1671): Swiss classicist educated in England. A Royalist, he defended the classical humanist tradition against the challenges of the new experimental philosophy.

Margaret Cavendish (1623–1673): The Duchess of Newcastle-upon-Tyne. A Royalist and prolific writer with philosophical and scientific aspirations, she rejected the mechanical philosophy.

Federico Cesi (1585–1630): Italian nobleman who founded the scientific society Accademia dei Lincei, members of which included Giambattista Della Porta and Galileo.

John Amos Comenius (Jan Amos Komensky) (1592–1670): Moravian writer and educator. A Protestant exile from Habsburg Bohemia, he was involved in educational reform in England, Sweden and Hungary.

Ralph Cudworth (1617–1688): English philosopher, one of the Cambridge Platonists, largely sympathetic to the new experimental philosophy.

John Dee (1527–1609): English mathematician, astronomer and occult philosopher, and sometime adviser to Queen Elizabeth I.

Daniel Defoe (*c*.1659–1731): English writer, novelist, merchant and political pamphleteer, best known for his novel *Robinson Crusoe* (1719).

Giambattista Della Porta (*c*.1535–1615): Italian natural philosopher, scholar and playwright, and one of the foremost authorities of the occult philosophy in the late Renaissance.

René Descartes (1596–1650): French philosopher who made seminal contributions to the mechanical philosophy, mathematics, and philosophical methodology.

John Donne (1572–1631): English poet and writer with a strong interest in natural philosophy.

John Dury (1596–1680): Scottish Protestant preacher and writer, an advocate of religious toleration and a member of the intellectual circle of Samuel Hartlib.

John Evelyn (1620–1706): English diarist, writer, and a founding member of the Royal Society.

Robert Fludd (1574–1637): English physician, mystic and supporter of the Paracelsian chemical philosophy.

Bernard le Bovier de Fontenelle (1657–1757): French writer, a publicist for the new science, Cartesianism and Copernicanism, and long-standing secretary of the French Academy of Sciences.

Galileo Galilei (1564–1642): Italian astronomer and mathematician who established key principles of mechanics and supported the heliocentric cosmology of Copernicus – for which he was forced to recant and placed under house arrest by the Catholic Church.

Pierre Gassendi (1592–1655): French philosopher who, although an empiricist and proponent of the mechanical philosophy, clashed with René Descartes.

William Gilbert (1544–1603): English natural philosopher best known for his work on electricity and magnetism and his theory that Earth is itself a giant magnet.

Joseph Glanvill (1636–1680): English writer and clergyman, and although not a scientist himself, one of the leading propagandists for the Royal Society and experimental philosophy.

Ambrose Godfrey Hanckwitz (known as Godfrey) (1660–1741): German chemist who came to England as an assistant to Robert Boyle and later established a successful apothecary business.

Francis Godwin (1562–1633): Bishop of Hereford whose book *The Man in the Moone* (published posthumously in 1638) was one of the first fantasies about space travel.

Nehemiah Grew (1641–1712): English botanist, pioneering microscopist, and early member (and sometime secretary) of the Royal Society.

Theodore Haak (1605–1690): German scholar and member of the Hartlib circle and of the London group that was one of the precursors of the Royal Society – of which Haak became a Fellow in 1661.

Edmond Halley (1656–1742): English astronomer and mathematician, who became the second Astronomer Royal, and is credited with persuading Isaac Newton to publish his *Principia*.

Thomas Harriot (c.1560–1621): English astronomer, mathematician and linguist who made the first telescope observations of the lunar surface, and recorded the natural history of the New World when he joined Walter Raleigh's expeditions to Virginia.

Samuel Hartlib (c.1600–1662): Prussian scholar and supporter of Francis Bacon's call for a reformed natural philosophy. His wide-ranging interests, from natural science to education, placed him at the centre of an international network of intellectuals, including several founders of the Royal Society.

William Harvey (1578–1657): English physician who, while somewhat conservative by inclination, made pioneering discoveries in anatomy, particularly the circulation of the blood.

Thomas Hobbes (1588–1679): English philosopher best known for his political theory of the state expressed in *Leviathan* (1651). Hobbes's interests were much broader, however, including science – but he was sceptical of the new experimental philosophy.

Robert Hooke (1635–1703): English scientist and inventor, who became the first curator of experiments at the Royal Society. Hooke's prodigious mechanical skills were accompanied by a perceptive intellect that led him to contribute to microscopy, watchmaking, astronomy and wave mechanics, among other things.

Christiaan Huygens (1629–1695): Dutch scientist, mathematician and inventor who made important discoveries in astronomy, optics, mechanics and clock-making.

Ben Jonson (1572–1637): English playwright and poet, considered a rival to Shakespeare.

Edward Kelley (1555–1597): Assistant to John Dee who claimed to be able to converse with angels via Dee's crystal ball. Kelley is now generally considered a charlatan.

Johannes Kepler (1571–1630): German astronomer who served as court mathematician to Emperor Rudolf II in Prague, where he discovered the laws of planetary motion.

Athanasius Kircher (c.1601–1680): German Jesuit with wide-ranging interests that included geology, medicine, mechanical devices and linguistics. Influenced by Neoplatonic mysticism, he lived and worked for most of his life at the Collegio Romano in Rome.

Johannes Daniel Krafft (early/mid-17th century): German alchemist who toured Europe displaying phosphorus, with the discovery of which he was sometimes wrongly credited.

Johann Kunckel (c.1630–1703): German chemist who served as apothecary and chemical technologist to various German nobles and to the King of Sweden.

Leonardo da Vinci (1452–1519): Italian polymath, whose skills at painting, sculpture, architecture, mechanics and hydraulic engineering earned him the patronage of many Italian princes and rulers, including the Duke of Milan Ludovico Sforza and Cesare Borgia.

Marin Mersenne (1588–1648): French Jesuit priest and theologian and supporter of the philosophy of René Descartes, who made contributions to mathematics and acoustics.

John Milton (1608–1674): English poet and writer, best known as the author of *Paradise Lost* but also an important political thinker and sympathetic to the 'new philosophy'.

Thomas Moffett (or Muffet) (1553–1604): English physician and botanist, best known for his study of insects.

Robert Moray (1608–1673): Scottish soldier and statesman, whose intimacy with the British monarchy was instrumental in establishing the royal charter of the Royal Society.

Henry More (1614–1687): English philosopher and a leading figure of the Cambridge Platonist school, who rejected Cartesian materialism.

Isaac Newton (1642–1727): Often considered the greatest scientist who ever lived, Newton transformed the mechanical philosophy by describing the fundamental laws of mechanics and developing a theory of gravity. Newton also made crucial contributions in optics and pure mathematics (inventing calculus), and had a profound interest in alchemy and prognostication.

Henry Oldenburg (1619–1677): German émigré to England who became the first secretary of the Royal Society and exploited his continental contacts to establish an international network of correspondents.

Paracelsus (Philip Theophrastus Bombastus von Hohenheim) (1493–1541): Swiss physician and alchemist who challenged the prevailing reliance on the works of classical writers such as Aristotle and Galen in favour of an empirical form of natural philosophy.

Samuel Pepys (1633–1703): English diarist, politician, naval administrator, and FRS.

William Petty (1623–1687): English physician and polymath, and a founding member of the Royal Society.

Robert Plot (1640–1696): English naturalist and chemist, first 'keeper' of the Ashmolean Museum, and FRS.

Alexander Pope (1688–1744): English poet, essayist and satirist, and a friend of Jonathan Swift.

Henry Power (1623–1668): English physician and advocate of the experimental philosophy.

Walter Raleigh (*c*.1554–1618): English writer, courtier, explorer and voyager to the New World, sometime prominent in the court of Elizabeth I.

John Ray (1627–1705): English naturalist and FRS, with a strong interest in classification of flora and fauna.

Laurence Rooke (1622–1662): English astronomer, professor of astronomy at Gresham College, and a founding member of the Royal Society.

Emperor Rudolf II (1552–1612): Habsburg King of Hungary and Bohemia, Archduke of Austria, and Holy Roman Emperor. Prone to depression, Rudolf made his capital Prague a centre for the study of the occult sciences.

Girolamo Ruscelli (1500–1566): Italian writer and humanist who claimed to have established the Accademia Segreta to study the 'operations of Nature' through experiments.

Thomas Shadwell (*c.*1642–1692): English poet and playwright, made Poet Laureate under William III in 1689.

William Shakespeare (1564–1618): generally regarded as England's greatest writer, who flourished during the reign of Elizabeth I.

Hans Sloane (1660–1753): Irish-Scottish physician and collector, and president of the Royal Society after Isaac Newton. Sloane's prodigious collection provided the basis for the establishment of the British Museum.

Edmund Spenser (*c.*1552–1599): English poet whose work *The Faerie Queene* was an allegory on the reign of Elizabeth I.

Thomas Sprat (1635–1713): English ecclesiastic who was commissioned to write the *History of the Royal Society of London* (1667).

Francesco Stelluti (1577–1652): Italian mathematician, astronomer and microscopist and one of the most active members of the Accademia dei Lincei.

Henry Stubbe (1632–1676): English physician, religious reformer, scholar and polemicist, and an opponent of the Royal Society.

Jonathan Swift (1667–1745): Anglo-Irish writer, satirist and pamphleteer, best known now for his fantasy *Gulliver's Travels* (1726).

Bernardino Telesio (1509–1588): Italian philosopher who advocated an experimental and empirical approach to understanding nature, in defiance of the authority of Aristotle.

Antony van Leeuwenhoek (1632–1723): Dutch cloth merchant who made important advances in microscopy, in particular discovering microorganisms in natural waters using instruments that he built himself.

John Wallis (1616–1703): English mathematician, and a central figure of the London group of philosophers who initiated the Royal Society.

Seth Ward (1617–1689): English mathematician and astronomer, founding member of the Royal Society, and Bishop of Exeter and Salisbury.

John Webster (1610–1682): English clergyman and physician, whose interest in the occult sciences and Paracelsianism did not prevent him from cautiously approving of the Royal Society.

John Wilkins (1614–1672): English natural philosopher and clergyman and one of the key figures behind the Royal Society – his Experimental Club convened at Oxford has the strongest claim to being its immediate predecessor.

Francis Willoughby (or Willughby) (1635–1672): English ornithologist, a student of John Ray, whose *History of Birds* was published posthumously by the Royal Society.

Christopher Wren (1632–1723): English polymath – an astronomer, architect, surveyor, inventor and mathematician, and founder member of the Royal Society, best known now for his rebuilding of churches (including St Paul's Cathedral) after the Great Fire of London.

Notes

Chapter 1

'To whatever object the eye first turns': G. Dondi, *De fontibus calidis agri Patavini consideratio*, 2, in T. Giunta (ed.), *De balneis omnia quae extant apud Graecos, Latinos, et Arabas* . . . fol. 95v. Giunta, Venice, 1553. • 'The important thing': A. Einstein, quoted in *Life*, 2 May 1955, p. 64. • 'The Large Hadron Collider is a discovery machine': R. Aymar, CERN press release, 10 September 2008. Available at http://public.web.cern.ch/press/pressreleases/Releases2008/PR08.08E.html. • 'We are constantly being told that we live': R. Aymar, *Symmetry* 3(6), August 2006. Fermilab/SLAC publication, available at http://www.symmetrymagazine.org/cms/?pid=1000353. • 'modern society is based on advances': S. Hawking, in *Khaleej Times*, 27 November 2009. • 'Curiosity is a vice that has been stigmatized in turn': M. Foucault, in *Foucault Live (Interviews 1966–84)*, transl. J. Johnston, ed. S. Lotringer, p. 198. Semiotext(e), New York. • 'Remember the use of a Coach': The notebooks of Robert Boyle, BP 9, fol. 22r, BP 10, fol. 72r & 102r, BP 26, fol. 225r. All available at http://www.bbk.ac.uk/boyle/boyle_papers/boylepapers_index.htm. • 'Desire, to know why, and how, CURIOSITY': T. Hobbes (1651/1985), p. 124. • 'the continuall and indefatigable generation': ibid. • 'a more than ordinary curiosity': I. Newton (1672), 'New theory about light and colours', *Philosophical Transactions of the Royal Society* 7, 3075–87. • 'of knowledge there is no satiety': F. Bacon (1605/1944), p. 37. • 'Curiosity was understood in so many ways': N. Kenny (2004), p. 5. • 'not of the same emotional species': L. Daston & K. Park (2001), pp. 304–5. • 'the hinder part of its body is cover'd': R. Hooke (1665/2007), p. 322. • 'fraught with much fabulous matter': in P. Turner (ed.) (1962), p. 7. • 'as if Nature had provided to save their honesty': ibid., p. 84. • 'On the top of Mount Palombra there is a wonderful fountain': in P. Rossi (1968), p. 4. • '[I] take it upon me to speak of everything': in P. Turner (ed.) (1962), p. 20. • '20,000 things, all worthy of regard': ibid., p. 21. • 'There was nothing left untried or unattempted': Pliny, *Natural History* 25.1, in W. Eamon (1994), p. 26. • 'It is owing to their wonder that men both now begin': Aristotle, *Metaphysics*, Book I, Part 2, transl. W D. Ross, in J. Barnes (ed.) (1984), *The Complete Works of Aristotle: Revised Oxford Translation*, vol. 2, p. 1554. Princeton University Press, Princeton. • 'It was not her cunning or wiliness': W. J. Verdenius, *A Commentary on Hesiod, Works and Days, vv1–382*, p. 65. E. J. Brill, Leiden, 1985. • 'these truly are the secrets of nature': Plutarch, *Moralia: Twenty Essays*, transl. P. Holland, intro. E. H. Blakeney, p. 138. J. M. Dent, London. • 'We want no curious disputation after possessing Christ Jesus': Tertullian, *The Prescription Against Heretics*, transl. P. Holmes. Available at http://www.tertullian.org/anf/anf03/anf03–24. htm#P3208_1148660 • 'The secret things belong to the Lord our God': Deuteronomy 29:29. • 'with much wisdom comes much sorrow': Ecclesiastes (Sirach: Apocrypha) 1:18. • 'Do not pry into things too hard for you': Ecclesiastes (Sirach: Apocrypha) 3:21–3. • 'Seek not to know high things': Romans 11:20. • 'Do not take pride in the arts or sciences': Thomas à Kempis, *De imitatione Christi*, in C. Ginzburg, 'The High and the Low' (1990), p. 64. • 'the words do not condemn learning': Erasmus, in ibid., pp. 60–1. • 'It is in divine language called the lust of the eyes': Augustine (AD 397–8/1907), *Confessions*, Book X, 35.54, transl. E. B. Pusey. Everyman, London. See http://www9.georgetown.edu/faculty/jod/Englishconfessions.html. • 'mangled corpses, magical effects': L. Daston & K. Park (1998), p. 123. See Augustine, op. cit., Book X, Chapter 35. • 'Nor dost [God] draw near': Augustine, op. cit., Book V, 3.3. • 'the beginning of all sin': Bernard of Clairvaux (*c.*1127/1929), *The Twelve Degrees of Humility and Pride*, transl. B. R. V. Mills, Chapter 10, p. 70. Macmillan, London. • 'Seek not what is too high for you': Bernard of Clairvaux (*c.*1127/1974), *The Steps of Humility and Pride*, transl. A. A. Conway, in *The Works of Bernard of Clairvaux*, Vol. 2, p. 60. Cistercian Publications, Washington DC. • 'fell from truth by curiosity': ibid. • 'The Seraphim set a limit to impudent': Bernard of Clairvaux, (*c.*1127/1985), *The Twelve Steps of Humility and Pride and on Loving God*, ed. H. C. Backhouse, p. 47. Hodder & Stoughton, London. • 'passion for knowing unnecessary things': William of Auvergne (13th century/1963), *De legibus* 1.24, in *Opera omnia*, Vol. 1, p. 69. Minerva, Frankfurt am Main. • 'O curiosity! O vanity!': Alexander Neckam (*c.*1180/1863), *De naturis rerum libri duo . . .*, ed. T. Wright, 2.172, pp. 281–2. Longman, Green, Longman, Roberts & Green, London. • 'No wrongful curiosity can attend': Thomas Aquinas, *Summa theologiae*, Part 2.2, Question 167, Article 1, in *Summa theologiae*, ed. Institute of Medieval Studies, Ottawa, Vol. 3, col. 2245b. Ottawa University Press, Ottawa, 1940–4. • 'Curiosity is the investigation of matters': in H. Blumenberg (1983), *The Legitimacy of the Modern Age*, transl. R. M. Wallace, p. 290. MIT Press, Cambridge, MA. • 'pleasureable for the student to know': Albertus Magnus, *De vegetabilibus libri VII*, 7.1.1, in *Opera omnia*, ed. A. Borgnet, Vol. 10, p. 590.

Ludovicus, Paris, 1890–9. • 'satisfying the curiosity of students': ibid., 6.1.1, pp. 159–60. • 'Ignorant themselves of the forces of nature': in M. D. Chenu (1968), *Nature, Man, and Society in the Twelfth Century: Essays on New Theological Perspectives in the Latin West*, ed. J. Taylor & L. K. Little, p. 11. University of Chicago Press, Chicago. • 'One will say that it conflicts with divine power': in E. Jeanneau (2000), *L'age d'or des ecoles de Chartres*, transl. by the author, p. 50. Editions Houvet, Chartres. • 'Certainly God can do everything': in J. Le Goff (1993), *Intellectuals in the Middle Ages*, transl. T. L. Fagan, p. 51. Blackwell, Oxford. • 'It is worth while to visit learned men': in L. Thorndike (1923), Vol. II, p. 20. • 'Thus when I have a new idea': in J. Le Goff (1993), *Intellectuals in the Middle Ages*, transl. T. L. Fagan, p. 55. Blackwell, Oxford. • 'The senses are reliable neither in respect': in L.Thorndike (1923), Vol. II, p. 29. • 'It is difficult for me to discuss animals with you': ibid. • 'I know that the darkness that holds you': in M. Müller (ed.) (1934), *Die Quaestiones naturales des Adelardus von Bath*, Beiträge zur Geschichte und Philosophie des Mittelalters 31, Heft 2, pp. 58–9. Aschendorff, Münster. • 'the greatest natural secrets to which man': R. Bacon, *Secretum secretorum cum glossis et notulis*, in R. Steele (ed.) (1937), *Opera hactenus inedita*, Vol. 5, p. 1. Oxford University Press, Oxford. • 'by the paths of knowledge Aristotle': in R. B. Burke (transl.) (1962), *The Opus majus of Roger Bacon*, Vol. 2, p. 634. Russell & Russell, New York. • 'Because the young greatly enjoy *experimenta*': in L. DeMaitre (1980), *Doctor Bernard de Gordon: Professor and Practitioner*, pp. 130–1. Pontifical Institute of Medieval Studies, Toronto. • 'Curiosity is the unbridled desire': in B. Benedict (2002), p. 25. • 'to get to know that which they should not know': in R. J. W. Evans & A. Marr (eds) (2006), p. 152.

Chapter 2

'It must certainly be allowed': in D. Hume (1748/2008), p. 24. • 'Where secrecy or mystery begins': in G. B. Hill (ed.) (1897), *Johnsonian Miscellanies*, Vol. II, p. 1. Clarendon Press, Oxford. • 'an intricate synthesis of observations': M. Kemp (2004), p. 78. • 'a subtle *inventione* which with philosophy': in M. Kemp (1981), *Leonardo da Vinci: The Marvellous Works of Nature and Man*, p. 162. J. M. Dent, London. • 'a quick glance into the most elementary': E. Gombrich, in C. Farago (ed.) (1999), *Leonardo's Science and Technology*, p. 320. Garland, New York. • 'Each painting is, in a sense, a proof': Kemp (2004), p. 3. • 'At this point . . . the opponent says that he does not want': E. Gombrich, in C. Farago (ed.) (1999), op. cit., p. 314. • 'The answer to this is that': ibid. • 'those who fall in love with practice': in I. A. Richter (ed.) (2008), p. 10. • 'although nature begins with the cause': ibid. • 'Before you base a law on this case': ibid. • 'Nature does not break her law': ibid., p. 8.

• 'O marvellous necessity': ibid. • 'There is no certainty where one can neither apply': ibid., p. 9. • 'First I shall test by experiment': ibid., p. 8. • 'Nature is full of infinite causes': ibid. • 'whose reasons are not so clear': Tommaso Garzoni (1588), *Piazza universale di tutte le professioni del mondo*, fol. 80v. Venice. In W. Eamon (1994), p. 135. • 'attend to this profession of secrets': ibid. • 'for his own profit and honour': in W. Eamon (1994), p. 142. • 'to make the most diligent inquiries': Ruscelli, *Secreti nuovi di maravigliosa virtù*, fol. 3v. Venice. In W. Eamon (1994), p. 149. • 'In addition to our own pleasure': ibid. • 'it would be manifested and publicized': ibid., fol. 5r in W. Eamon (1994), p. 149. • 'the first of the moderns': in N. C. van Deusen (1923), 'Telesio: The First of the Moderns'. PhD dissertation, Columbia University, New York. See W. Eamon (1994), p. 156. • 'are very repetitious [and] ramble on and on': L. Thorndike (1941), p. 563. • 'they contain much that would seem': ibid. • 'It is called The Magic Of Reality': *Guardian* supplement, 11 September 2010. • 'the science that attempted to give rational': W. Eamon (1994), p. 194. • 'Magicians are like careful explorers of nature': Cornelius Agrippa (1530), *De vanitate et incertitudine et omnium scientiarum et artium*, Chapter 42. In P. Rossi (1968), p. 19. • 'nature teemed with hidden forces': W. Eamon (1994), p. 194. • 'the most diligent scrutinizer of the secrets': J. Faber (1627), quoted in G. Gabrieli (1927), 'Giovan Battista Della Porta Linceo', *Giornale critico della filosofia Italiana* 8, 360–96 & 423–31, here p. 425. • 'Magic is nothing else but the survey': Giambattista Della Porta (1658), *Natural Magick . . . in twenty books*, Book I, Chapter II. Thomas Young & Samuel Speed, London. See http://homepages.tscnet.com/omard1/jportat2. html. Quoted in J. Henry (2008), p. 8. • 'Just as the Church was intensifying': W. Eamon (1994), pp. 209–10. • 'took the whole seventh book of my *Magiae naturalis*': in J. J. O'Connor & E. F. Robertson (2010), 'Giambattista Della Porta', http://www-history.mcs.st-and.ac.uk/Biographies/Porta.html. • 'Concave Lenticulars will make one see': G. Della Porta (1658), *Magiae naturalis*, Book 17, Chapter X. • 'Unhappy the land where heroes are needed': B. Brecht (1980), *Life of Galileo*, transl. J. Willett, ed. J. Willett & R. Manheim, p. 98. Eyre Methuen, London. • 'I never wanted for at my House': G. Della Porta (1658) (op. cit.), Preface. • 'vied with one another to add new discoveries': G. Della Porta (1677), *Della magia naturale*, quoted in M. Gliozzi (1950), 'Sulla natura dell' "Accademia de' Secreti" di Giovan Battista Porta', *Archives internationales d'histoire des sciences* 12, 536–41, here pp. 539–40. • 'The ignorant philosophers': in G. Paparelli (1956), 'Giambattista Della Porta: Della Taumatologia', *Rivista di storia delle scienze mediche e naturali* 47, 1–47, here p. 19. • 'ever changing, ever joking, subtle, ingenious': W. Eamon (1994), p. 217. • 'True Things be they never so small': G. Della Porta (1658), op. cit., Preface.

Chapter 3

'They say miracles are past': *All's Well That Ends Well*, Act 2, Scene iii. • 'The Heavens, the Seas, the Whole Globe': John Evelyn, letter to Cowley, 12 March 1667, in Letter book No.185 (Evelyn's letter book, 2 vols, Evelyn MS 39, Evelyn Collection, British Library). Quoted in D. Chambers, '"Elysium Britannicum not printed neere ready &c": The "Elysium Britannicum" in the correspondence of John Evelyn', in T. O'Mallet & J. Wolschke-Buhlman (eds) (1993), *John Evelyn's 'Elysium Britannicum' and European Gardening*, p. 115. Dumbarton Oaks Colloquium on the History of Landscape Architecture XVII. Trustees for Harvard University, Washington DC. • 'not only refused, but would not listen to music': B. Castiglione (1528/1967), *The Book of the Courtier*, transl. G. Bull, Book I, p. 58. Penguin, Harmondsworth. • 'If you would have your works appear': G. Della Porta (1658), op. cit., Book I, Chapter 3. • 'to give a greater grace to the practice': H. van Etten (1653), *Mathematicall Recreations*, sig. A5. William Leake, London. • 'A man of *virtù*, acting rationally': A. C. Crombie (1990), p. 161. • 'was a rational artist in all things': ibid., p. 162. • 'The possession of such rarities': H. Peacham (1634), *Compleat Gentleman*, pp. 104–5. quoted in W. E. Houghton (1942) (I), p. 52. • 'Among the qualities of the idealized "learned prince"': W. Eamon (1994), p. 223. • 'the seventeenth-century rehabilitation': ibid., p. 301. • 'Because you know everything': in A. Brown (1986), 'Platonism in fifteenth-century Florence and its contribution to Early Modern political thought', *Journal of Modern History* 58, 383–413, here p. 395. • 'goodly huge cabinet': F. Bacon (1594), *Gesta Greyorum*, in J. Spedding (ed.) (1861), *The Letters and Life of Francis Bacon*, Vol. I, p. 335. Longman, Green, Longman & Roberts, London. • 'Thus when your Excellency shall have': ibid. • 'All the world's a stage': *As You Like It*, Act 2, Scene vii. • 'The purpose of playing . . . is to hold': *Hamlet*, Act 3, Scene ii. • 'His Museum is like a compendium': in E. Battisti (1962), *L'Antrinascimento*, transl. C. Swan, p. 302. Milan. • 'There are to be found in the sea': A. Paré (1573/1971), *Des monstres et prodiges*, ed. J. Céard, p. 117. Librairie Droz, Geneva. • 'a regal gallery adorned with a hundred': in T. DeCosta Kaufmann (1993), p. 174. • 'a panel inlaid with silver ore': J. Kepler (1611/1966), *The Six-Cornered Snowflake: Strena, seu de nive sexangula*, transl. & ed. C. Hardie, p. 43. Clarendon Press, Oxford. • 'Disturbed in his mind by some ailment': in R. J. W. Evans (1973), p. 45. • 'he delights in hearing secrets about things': ibid., p. 196. • 'It is generally agreed amongst Catholics': in H. Holzer (1974), *The Alchemist: The Secret Magical Life of Rudolf von Habsburg*, p. 85. Stein & Day, New York. • 'make it rain, lighten, and thunder': in W. E. Houghton (1942) (I), p. 71. • 'penetrate into the inside of things': in G. Gabrieli (1935), 'Spigolatura Dellaportiane', *Rendiconti della R. Accademia Nazionale dei Lincei. Classe di scienze morali, storiche e filologiche*, ser. 6, 11, pp.

491–517, here p. 507. • 'most sagacious investigators': in D. Freedberg (2002), p. 67. • 'I hate the court': letter from Cesi to Stelluti, 17 July 1604, in ibid., p. 65. • 'can penetrate through trees': quoted in S. J. Gould (1998). • 'tyrant of minds': T. Campanella (1998), *Le Poesie*, ed. F. Giancotti, p. 278. Einaudi, Turin. • 'without effort, merely while playing': in C. Wilson (1995), p. 27. • 'I shall make the heavens a temple': ibid., p. 327. • 'learned men, magicians, cabalists': *Fama fraternitatis*, transl. T. Vaughan (1615). Reprinted in *Rosicrucian Digest* 82(1), p. 14, 2004. • 'a perfect method of all arts': ibid., p. 13. • 'concern themselves with quantitative shadows': in W. Pauli (1955), 'The influence of archetypal ideas on the scientific theories of Kepler', in C. G. Jung & W. Pauli (1955), *The Interpretation of Nature and the Psyche*, transl. P. Silz, p. 196. Pantheon Books, New York. • 'No less than Fludd did [Kepler] believe': A. Debus (2006), p. 107. • 'An Art of Arts, a Science of Sciences': J. A. Comenius (1668/1938), *The Way of Light*, transl. E. T. Campagnac, pp. 32–3. Hodder & Stoughton, London. • 'Torch Bearers of this Enlightened Age': in R. H. Syfret (1947), p. 116. • 'that immodest thirst to know': in N. Kenny (2004), p. 111. • 'knew not whether it were the stone': L. Jardine & A. Stewart (1998), p. 503. • 'The End of our Foundation': F. Bacon (1620/1626/1980), p. 70. • 'for the making of the earth fruitful': ibid. • 'brew-houses, bake-houses, and kitchens': ibid., p. 73. • 'sweeter than any you have': ibid., p. 77. • 'strange and artificial echoes': ibid. • 'glasses and means to see': ibid., p. 76. • 'all colourations of light': ibid., p. 76. • 'several colours: not in rain-bows': ibid. • 'a mathematical house, where are represented': ibid., p. 78. • 'sway[ing] the imagination more than the reason': F. Bacon (1605/1620/1944), p. 19. • 'ancient and honourable': F. Bacon, *De augmentis scientiarum*, in *The Works of Francis Bacon* (1815), Vol. VI, p. 116. M.Jones, London. • 'a sublimer wisdom': ibid. • 'that science, which leads to the knowledge': ibid., p. 100. • 'gives us only some childish': ibid. • 'the nature of gravity, colour': ibid., p. 101. • 'an Agent for the advancement': in R. H. Syfret (1947), p. 94. • 'a collecting together of everything': ibid., p. 99. • 'if all men understand each other': in ibid., p. 112. • 'rarities and things of nature': J. Evelyn, letter of 1 September 1659 to Robert Boyle, in E. S. De Beer (ed.) (1955), *The Diary of John Evelyn*, Vol. 3, p. 232. Clarendon Press, Oxford.

Chapter 4

'Men have entered into a desire of learning': in W. E. Houghton (1942), p. 56. • 'God has given enough for use': in ibid., p. 51. • 'the dominant philosophy of the Elizabethan age': F. Yates (2001), p. 88. • 'To read Shakespeare's fairy scenes': ibid., p. 175. • 'By number, a way is had': in ibid., p. 94. • 'may find an easier passage': in P. Rossi (1968), p. 87. • 'This fable is, perhaps, the noblest': F. Bacon (1605/1620/1944), p. 65. • 'it relates to the state of the world': ibid. • 'exquisite knowledge of all natural things': in J. Henry (2002), p. 73. • 'That part of the fable which attributes': F. Bacon (1605/1620/1944), p. 65. • 'as an allegorical account': Campbell (1999), p. 74. • '*Nature* works by an *Invisible Hand* in all things': J. Glanvill (1665/1978), *Scepsis scientifica*, p. 133. Garland Press, New York. • 'secret and subtile Actors': R. Hooke, *A General Scheme or Idea of the Present State of Natural Philosophy*, in R. Waller (ed.) (1705/1969), *The Posthumous Works of Robert Hooke*, introd. R. S. Westfall, pp. 8, 6. Johnson Reprint Corp., New York. • 'to make many *doublings* and *turnings*': R. Hooke (1665/2007), p. 17. • 'For like as a man's disposition is never well known': in F. Bacon (1824), p. 80. Compare Bacon (1605/1620/1944), p. 49: 'For as a man's temper is never well known until he is crossed, in like manner the turns and changes of nature cannot appear so fully, when she is left at her liberty, as in the trials and tortures of art.' • 'pry[ing] into her secrets by breaking open': R. Hooke (1665/2007), p. 325. • 'there cannot be two passions': D. Hume (1739–40), *A Treatise of Human Nature*, Book II, Part III, in D. Hume (1854), *The Philosophical Works of David Hume*, Vol. II, p. 208. Little, Brown & Co., Boston, MA. • 'The utility or importance of itself': ibid., p. 209. • 'The universe to the eye of the human understanding': F. Bacon (1620/1626/1980), p. 13. • 'That wisdom which we have derived': ibid., p. 8. • 'The discoveries which have been hitherto': ibid., pp. 13–14. • 'who has dwelt upon experience': ibid., p. 12. • 'the magnetic virtue of iron was not first': Bacon (1605/1620/1944), p. 49. • 'who, while he gazed upwards to the stars': ibid. • 'to suit his opinions . . . captive and bound': F. Bacon, 'The refutation of philosophies', in B. Farrington (1964), *The Philosophy of Bacon: An Essay on Its Development from 1603 to 1609*, p. 130. Chicago University Press, Chicago. • 'From the wonders of nature is the nearest': F. Bacon (1857), p. 331. • 'All the bizarre objects and rarities': W. Eamon (1994), p. 299. • 'For we are not to give up the investigation': F. Bacon (1858), p. 168. • 'Our steps must be guided by a clue': ibid., p. 18. • 'In other hunts the prey is only caught, but in this it is killed': in W. Eamon (1994), p. 287. • 'This form of experimenting': F. Bacon (1858), p. 420. • 'an unbroken route through the woods': ibid., p. 81. • 'The mark of genuine science': Bacon, *Refutation of the Philosophers*, in B. Farrington (1964), op. cit., p. 123. • 'What the sciences stand in need of': F. Bacon (1980), p. 23. • 'Heat is an expansive motion restrained': F. Bacon (1620), *Novum Organum*, Book II, Aphorism 20, in F. Bacon (1605/1620/1944), p. 394. • 'the more details of his method he provides': J. Henry (2002), p. 132. • 'passes all understanding': ibid. • 'Our method of discovering the sciences': F. Bacon (1605), *Novum Organum*, Book I, Aphorism 122, in F. Bacon (1605/1620/1944), p. 362. • 'I am emphatically of the opinion': in P. Rossi (1968), p. 33. Compare F. Bacon

(1605), *Novum Organum*, Book I, Aphorism 104, in Bacon (1605/1620/1944), p. 353: 'We must not then add wings, but rather lead and ballast to the understanding, to prevent its jumping or flying . . . whenever this takes place, we may entertain greater hopes of the sciences.' • 'plain, diligent and laborious observers': in M. Hunter (1982), *The Royal Society and its Fellows, 1660– 1700: the Morphology of an Early Scientific Institution*, p. 18. British Society of the History of Science, Chalfont St Giles. • 'axioms once rightly discovered': F. Bacon (1620/1626/1980), p. 27. • 'manifests its worth in enabling us': S. Gaukroger (2006), p. 166. • 'whatever God himself has been pleased': R. Boyle, *Considerations Touching the Usefulness of Experimental Natural Philosophy*, in T. Birch (ed.) (1772), *The Works of the Honourable Robert Boyle* Vol. 2, p. 13. J. & F. Rivington, London; reprinted by Georg Olms Verlag, Hildesheim, 1965–6. • 'For it was not that pure and uncorrupted': F. Bacon (1858), p. 20. • 'we do not presume by the contemplation': F. Bacon (1605/1620/1944), p. 4. • 'if any man shall think, by his inquiries': ibid., p. 5. • 'As at first, mankind *fell* by *tasting*': R. Hooke (1665/2007), p. 20. • 'God has framed the mind like a glass': F. Bacon (1605/1620/1944), p. 4. • 'absurd and impertinent': quoted in M. Hunter (1981), p. 175. • 'long and close intercourse with experiments': F. Bacon (1605), *Novum Organum*, Book I, Aphorism 83, in F. Bacon (1605/1620/1944), p. 340. • 'too curious and irresolute': F. Bacon (1605/1620/1944), p. 5. • 'defends the mind against idleness': ibid., p. 9. • 'consists either in matters or words': ibid., p. 14. • 'cobwebs of learning, admirable indeed': ibid., p. 17. • 'out of a natural curiosity': ibid., p. 23. • 'artificial, chaste, severe': quoted in C. Wilson (1995), p. 48. • 'false theories, or philosophies': F. Bacon (1605/1620/1944), p. 156. • 'men of young years and superficial understanding': F. Bacon, 'An advertisement Touching the controversies of the Church of England', in J. Spedding (ed.) (1861), op. cit., p. 82. • 'leap from ignorance to a prejudicate opinion': ibid., pp. 82–3. • 'who aspire not to guess and divine': F. Bacon (1620/1626/1980), p. 26. • 'But for his *Method*: I tooke him for one': in S. Gaukroger (2006), p. 221. • 'There are not two waies in the whole World': S. Ward (1654), *Vindiciae academiarum*, p. 240, in H. R. Trevor-Roper (1985), *Renaissance Essays*, p. 191. Secker & Warburg, London. • 'has been said to be of greater importance': F. Yates, 'A great magus', *New York Review of Books*, 25 January 1973.

Chapter 5

'Is it not evident, in these last hundred Years': J. Dryden (1668), 'An Essay of Dramatic Poesy', in J. Mitford (1836), *The Works of John Dryden: In Verse and Prose, with a Life*, Vol. II, p. 225. George Dearborn, New York. • 'These two Subjects, *God*, and the *Soul*': T. Sprat (1667/1734), p. 83. • 'burned Mathematical bookes': J. Aubrey (1982), *Brief Lives*, ed. R. W. Barber, p. 322. Boydell & Brewer, Woodbridge. • 'a performance of the Scarabeus': J. Dee, 'Compendious Rehearsal', in J. Crossley (ed.) (1851), *Autobiographical Tracts of Dr John Dee, Warden of the College of Manchester*, Chetham Society Publications (Manchester) 24, pp. 5–6. • 'after reading Dee's exciting preface': J. P. Zetterberg (1980), pp. 94–5. • 'more than a dull study required': quoted in W. E. Houghton (1942) (II), p. 192. • 'how to steale fire from Heaven': in J. P. Zetterberg (1980), p. 89. • 'a petty God, more than a man': ibid. • 'an hollow Statue which gave a Voice': in E. S. De Beer (ed.) (1955), *The Diary of John Evelyn* Vol. 3, pp. 110–11. Clarendon Press, Oxford. • 'related to be seen in the sepulchre': in F. Yates (1972), p. 225. • 'fabulous experiments, idle secrets': W. E. Houghton (1942) (I), p. 195. • 'a large pendant candlestick': in ibid., p. 193. • 'seemed then and was a rarity': ibid. • 'was a very ingeniose man': J. Aubrey (2000), p. 340. • 'the principall Reviver of Experimental': ibid., p. 342. • 'Persons inquisitive into Natural Philosophy': in R. H. Syfret (1947), p. 76. • 'the circulation of the blood': in B. J. Shapiro (1969), *John Wilkins, 1614– 1672*, p. 25. University of California Press, Berkeley. • 'so extremely greedy, that he steales flowers': J. Aubrey (2000), p. 328. • 'as much a confession of faith': M. Hunter (1981), p. 29. • 'It is their usual recreation': W. Charleton (1659), *The Immortality of the Human Soul, Demonstrated by the Light of Nature*, pp. 46–7. London. • 'He can hardly be a good scholar, who is not a general one': I. Barrow (1700), *The Works of the Learned Isaac Barrow*, Vol. II, Sermon XXII, 'Of Industry in our particular Calling, as Scholars', p. 220. J. Heptinstall, London. • 'It was proposed that some regular course': T. Birch (1754), *The History of the Royal Society of London*, Vol. I, p. 3. A. Millar, London. • 'a kind of Elizium': in M. Hunter (2009), p. 70. • 'Study of the Booke of Nature': ibid., p. 73. • 'make it very *easy* to solve': R. Boyle, in T. Birch (ed.) (1772), op. cit., Vol. 5, p. 301. • 'His greatest delight is chymistrey': in R. T. Gunther (1923), *Early Science in Oxford*, Vol. 1, p. 10. Oxford University Press, Oxford. • 'constantly open to the Curious': in S. Shapin (1988), p. 386. • 'more from men, real experiments': in W. E. Houghton (1941), p. 59. • 'two or three masons, and stone-cutters': ibid., p. 58. • 'mighty Design': in W. Eamon (1994), p. 334. • 'several of the other Lodgings, as their occasions do require': T. Sprat (1667/1959), p. 93. • 'three or four considerable': in M. Feingold, 'Robert Hooke: Gentleman of Science', in M. Cooper & M. Hunter (eds) (2006), p. 209. • 'Theory of Motion': R. Hooke, *Royal Society Classified Papers*, Vol. XX (Hooke papers), f.67. See Inwood (2002), p. 38. • 'the greatest Mechanick this day in the world': J. Aubrey (2000), p. 396. • 'to fill the vacancy of the ensuing page': R. Hooke (1676), 'A description of helioscopes and some other instruments', in R. T. Gunther (1931), *Early Science in Oxford* Vol. 8, pp. 151–2. Oxford University Press, Oxford. • 'a *decimate* of the *centesme*':

ibid. • 'A Way of Regulating all sorts of *Watches*': ibid. • 'The business & designe of the Royall Society': in M. Hunter (1995), p. 172. • 'An account of the Improvement of Optick Glasses': *Philosophical Transactions of the Royal Society* 1, 1, 6 March 1665. • 'they confin'd themselves to no order of subjects': T. Sprat (1667/1734), p. 115. • 'In the making of all kinds of Observations': R. Hooke, *A General Scheme or Idea of the Present State of Natural Philosophy*, in R. Waller (ed.) (1705/1969), op. cit., pp. 61–2. • 'science destroys the image of the familiar': C. Wilson (1995), p. 37. • 'Fearful of excluding anything': L. Daston & K. Park (1998), p. 316. • 'It is necessary that we *deceive our imagination*': N. Malebranche (1674–5), *De la recherche de la vérité*, 5.8, 6th edn, in *Oeuvres de Malebranche*, Vol. 2, ed. G. Rodis-Lewis, p. 207. J. Vrin, Paris, 1962–7. • 'too bizarre or singular to be classified': L. Daston & K. Park (1998), p. 231. • 'the greatest Curiosities': T. Sprat (1667/1959), p. 99. • 'An Account of *Four Suns*': *Philosophical Transactions of the Royal Society* 1, p. 219 (1665). • 'the *Society* approaches with as much circumspection': T. Sprat (1667/1959), p. 101. • 'It is not easy to know, what phaenomena may, and what cannot': R. Boyle, 'The Aerial Noctiluca', in T. Birch (ed.) (1772), op. cit., Vol. 4, pp. 393–4. • 'Whenever [the reader] finds that I have ventur'd': R. Hooke (1665/2007), p. 18. • 'Hypotheses are not to be regarded': I. Newton (1717), *Opticks*, 3rd edn (1717), Query 31, quoted in P. Dear, 'The Meanings of Experience', in K. Park & L. Daston (eds) (2006), p. 129. • 'a kind of violence': L. J. Daston, 'Classifications of Knowledge in the Age of Louis XIV', in D. L. Rubin (ed.) (1992), *Sun King: The Ascendancy of French Culture during the Reign of Louis XIV*, p. 210. Associated University Presses, Cranbury, NJ. • 'the Académie des Sciences has grasped': in R. J. W. Evans & A. Marr (eds) (2006), p. 46. • 'grainy with facts': L. Daston & K. Park (1998), p. 237. • 'Nature it self does not so exactly': R. Hooke, *A General Scheme or Idea of the Present State of Natural Philosophy*, in R. Waller (ed.) (1705/1969), op. cit. p. 38. • 'a completed natural science': A. C. Crombie (1990), p. 404. • 'We earnestly beg you to be so good': H. Oldenburg to M. Malpighi, 28 December 1667, in A. R. Hall & M. Boas Hall (eds) (1965–77), *The Correspondence of Henry Oldenburg*, Vol. 4, p. 92. Mansell, London. • 'the sovereign beneficence of God': in M. Boas Hall (ed.) (1970), p. 219. • 'we can rely on nothing with greater assurance': ibid., p. 220. • 'it is our principal intent to incite others': ibid., p. 221. • 'free correspondence with those several societies': ibid. • 'The principal and most useful occupation': ibid., p. 226. • 'all sorts of optical wonders': G. W. Leibniz (1675), 'An odd thought concerning a new sort of exhibition', in P. Wiener (ed.) (1951), *Leibniz: Selections*, pp. 585–94. Charles Scribner's Sons, New York. • 'All respectable people would want': ibid. • 'The history of science was formerly presented': C. Wilson (1995), p. 16. • 'The idea of harmony in nature': C. Webster (1982), p. 69. • 'lacking in imagination to make a virtue': M. Hunter (1981), pp. 17–18. • 'the Society's corporate policy': M. Hunter (2011), p. 103. • 'rewrote Restoration intellectual history': M. Winship (1996), *Seers of God: Puritan Providentialism in the Restoration and Early Enlightenment*, p. 131. Johns Hopkins University Press, Baltimore, MD. • 'had no place in Royal Society policy': M. Purver & E. J. Bowen (1960), p. 8. • 'lacking in imagination': M. Hunter (1981), pp. 17–18. • 'I shall not dare to think my self': in M. Hunter (1981), p. 91.

Chapter 6

'The variety of nature': M. Cavendish (1666/2001), p. 99. • *'Multi pertransibunt et augebitur scientia'*: in J. Henry (2002), p. 19. • 'In all my life I have seen nothing': in R. J. W. Evans & A. Marr (eds) (2006), p. 65. • 'There are trees of a thousand kinds': O. Dunn & J. E. Kelley (eds & transl.) (1989), *The Diario of Christopher Columbus's first Voyage to America, 1492– 1493*, pp. 106–7. University of Oklahoma Press, Norman. • 'there are stranger things to be seen': in S. Shapin (1996), p. 19. • 'because I have seen things as fantastic': in S. Greenblatt (1991), p. 22. • 'For myself, I find the unicorn much less improbable': M. Doran (1940), p. 163. • 'in taste . . . are altogether as good as our English peaze': T. Hariot (1588), p. 18. • 'The leaves thereof being dried': ibid., p. 22. • 'Most thinges they sawe with us': ibid., p. 39. • 'We [are] already overstocked with Books of Travels': J. Swift (1726/1985), p. 189. • 'a worthy precursor to Francis Bacon': P. Findlen, 'Natural history', in L. Daston & K. Park (eds) (2006), p. 451. • 'painters often make big mistakes': J. Faber (1651), in *Rerum medicarum novae hispaniae thesaurus*, p. 581. Vitale Mascardi, Rome. In D. Freedberg (2002), p. 357. • 'Painting is indeed deceptive': ibid. See Pliny, *Natural History*, Book 25, Ch. 4. • 'Nature was fashioned in such a way': Leonhart Fuchs (1542), *De historia stirpium commentarii insignes*, in S. Kusukawa (1997), 'Leonhart Fuchs on the importance of pictures', *Journal of the History of Ideas* 58, 411. • 'testify to his omnivorous recording': D. Freedberg (2002), p. 369. • 'doth, in a short Time, quite take away': in S. Inwood (2002), p. 397. • 'a Creature, bodied like an ox': in M. Hunter (2009), p. 236. • 'He relates that in the Persian Gulfe': R. Boyle, Work-diary 22, p. 50. See Boyle Papers 8, fol. 89v: http://www.bbk. ac.uk/boyle/workdiaries/WD22Orig.html. • 'often Entertain'd those who came to visite him': in M. Hunter (ed.) (1994), *Robert Boyle by Himself and His Friends*, p. xlii. William Pickering, London. • 'vain and vicious curiosity': in W. Williams, '"Out of the trying pan . . .": Curiosity, danger and the poetics of witness in the Renaissance traveller's tale', in R. J. W. Evans & A. Marr (eds) (2006), p. 29. • 'an honest curiosity to inquire': in R. J. W. Evans & A. Marr (eds) (2006), p. 27. • 'Nothing certain can be established':

in S. Bakewell (2010), *How To Live: A Life of Montaigne*, p. 130. Chatto & Windus, London. • 'Let this universal and natural reason': in B. Benedict (2001), p. 32. • 'a fragmented vision': P. Mauries (2002), p. 43. • 'for all manner of rare beasts': in W. E. Houghton (1942) (I), p. 69. • 'almost persuaded a Man might in one daye': in L. Jardine (1999), p. 255. • 'a natural Dragon, above two inches long': in M. Doran (1940), p. 160. • 'In the museum of Mr John Tradescant': The Ashmolean Museum, http://www.ashmolean.org/ash/amulets/tradescant/tradescant03.html • 'to gather up all raritye of flowers': ibid., http://www.ashmolean.org/ash/amulets/tradescant/tradescant02.html • 'without being forced to take paines': in M. Hunter (1995), p. 23. • 'the greatest virtuoso and curioso': in ibid., p. 37. • 'Dees and Kethes and many other Books': in L. Jardine (1999), p. 258. • 'the use and improvement of physic': in R. D. Altick (1978), *The Shows of London*, p. 229. Harvard University Press, Cambridge, MA. • 'shew'd them very freely to all lovers': in L. Jardine (1999), p. 264. • 'Chocolate is here us'd': ibid., p. 267. • 'so very curious, as to desire to carry part': ibid., p. 276. • 'a General Collection of all the Effects of Arts': http://royalsociety.org/Before-the-British-Museum-the-Repository-of-the-Royal-Society/. • 'to procure for the society a collection': in M. Hunter (1989), p. 126. • 'I am now making a collection': ibid., p. 127. • 'already drawn together into one Room': ibid., p. 37. • 'the secret parts of a woman': in S. Inwood (2002), p. 86. • 'tooth taken out of a testicle': in K. T. Hoppen (1976), p. 8. • 'not only Things strange and rare': N. Grew (1681), *Musaeum regalis societatis, or a Catalogue and Description of the Natural and Artificial Rarities Belonging to the Royal Society*, Part I, p. 151. W. Rawlins, London. • 'conick, with the *Turban*': in O. Impey & A. Macgregor (eds) (1985), p. 222. • 'Hardly a thing is to be recognized': in W. H. Quarrell & M. Mare (transl. & eds) (1934), *London in 1710; from the Travels of Z. C. von Uffenbach*, pp. 97–8. Faber & Faber, London. • 'there arises from a bad': F. Bacon (1606/1620/1944), p. 320. • 'we should, by learning the *Character*': in S. Clauss (1982), p. 539. • 'The central thesis of Wilkins' linguistic philosophy': ibid., pp. 545–6. • 'Do not hope ever to see such a language in use': ibid., p. 548. • 'Music is nothing else': in J. Godwin (1979), p. 66. • 'a book I am mighty glad of': S. Pepys, Diary, 22 February 1668, in R. Latham & W. Matthews (eds) (2000), *The Diary of Samuel Pepys 1668–9*, p. 85. University of California Press, Berkeley. • 'an appendix to the Curse of Babel': J. Wilkins (1668), *Essay Towards a Real Character and a Philosophical Language*, p. 13. London. • 'there is reason enough to expect the like Fate': ibid., epistolary dedication, no page. • 'it would not be too much to expect men': in S. Clauss (1982), p. 539. • 'it cluttered the mind with more': P. Findlen, 'Natural history', in L. Daston & K. Park (eds) (2006), p. 468. • 'Of this Engine, no Man': in M. B. Hesse (1966), p. 69. • 'the Discovery of the more internal Texture': ibid. • 'I did from an

Art of Invention': R. Hooke, 'A Description of Helioscopes', in R. T. Gunther (ed.) (1930), *Early Science in Oxford*, Vol. 8, pp. 146–7. Oxford University Press, Oxford. • 'A Philosophical Grammar': F. Bacon (1605/1620/1944), p. 306.

Chapter 7

'Before their eyes in sudden view appear': J. Milton (1667), *Paradise Lost*, Book V, 750–4. • 'Let us therefore not try to discover': J. Kepler (1599), letter to Hewart von Hohenburg, in Crombie (1969), p. 195. • 'The proposition that the sun is in the centre': Papal condemnation of Galileo, 22 June 1633, in, e.g., G. de Santillana (1955), *The Crime of Galileo*, pp. 306–10. University of Chicago Press, Chicago. • 'the study of the heavens was not undertaken': G. E. R. Lloyd (2002), p. 27. • 'Copernicus lived among scholars': J. D. North (1989), p. 17. • 'to reverse the entire science of astronomy': in A. Debus (1978), p. 98. • 'want of honesty and decency': ibid. • 'One single experiment or conclusive proof': in A. C. Crombie (1969), p. 148. • 'If an effect, which has succeeded': ibid., p. 147. • 'an isolated, and in fact quite untypical': A. Koestler (1988), *The Sleepwalkers*, p. 533. Penguin, London. • 'If there were a true demonstration': R. Bellarmine to P. A. Foscarini, 12 April 1615, in A. Favaro (ed.) (1890–1909), *Le Opere di Galileo Galilei*, Vol. XII, p. 172. Reprinted by Barbèra, Florence, 1968. • 'Scripture cannot err but its interpreters can': in W. R. Shea & M. Artigas (2003), p. 57. • 'with little prudence or self-control': P. Guicciardini to Cosimo II, 4 March 1616, in A. Favaro (ed.) (1890–1909), op. cit., Vol. XII, pp. 242–3. • 'What an admirable and angelic doctrine': in W. R. Shea & M. Artigas (2003), p. 142. • 'I have taken the Copernican side': ibid., p. 160. • 'I neither maintained nor defended': ibid., p. 186. • 'Not having seen it for so long': ibid., p. 190. • 'my error has been one of vainglorious': ibid. • 'I do not hold this opinion of Copernicus': ibid., p. 193. • 'a victim of bad luck and bad judgement': S. J. Gould (1988). • 'made no contribution to theoretical astronomy': A. Koestler (1988), op. cit., p. 358. • 'an immoral, rough, and quarrelsome soldier': in M. Caspar (1993), p. 34. • 'How intense was my pleasure': in S. Hawking (ed.) (2002), p. 630. • 'never in history has a book': O. Gingerich (1993), p. 309. • 'he has abundant wealth': in M. Caspar (1993), p. 87. • 'Therefore, one must take pains': ibid. • 'you will come not so much as guest': ibid., p. 100. • 'Now old age steals upon him': ibid., p. 102. • 'I quickly took advantage of the absence': in S. Hawking (ed.) (2002), p. 631. • 'In my case, Saturn and the Sun work together': J. Kepler (1599), letter to Hewart von Hohenburg, in J. M. Ross & M. M. McLaughlin (eds) (1985), *The Portable Renaissance Reader*, pp. 606–7. Penguin, Harmondsworth. • 'a childish or fateful desire': in M. Caspar (1993), p. 64. • 'so obscure that apparently': in A. C. Crombie (1959), p. 201. • 'I dare frankly to confess': in S. Hawking (ed.) (2002), p. 633. • 'God has established nothing': ibid., p. 654. • 'each thing, as far as is in its power': R. Descartes (1644/1991), *Principles of Philosophy*, transl. V. R. Miller, p. 59. Kluwer, Dordrecht. • 'upon that principle all the laws': in D. T. Whiteside (1991), 'The prehistory of the Principia from 1664 to 1686', *Notes and Records of the Royal Society of London* 45, 11–61, here p. 26. • 'so that others, trying and failing': ibid. • 'I never knew him to take any Recreation': H. Newton, letter to J. Conduitt, 17 January 1728, Keynes MS 135, King's College, Cambridge, pp. 2–3. Available at http://www.newtonproject.sussex.ac.uk/view/texts/normalized/THEM00033. • 'whilst he was musing in a garden': in R. S. Westfall (1980), p. 154. • 'is too well attested to be thrown out of court': ibid. • 'vulgarizes universal gravitation': ibid., p. 155. • 'certainly not mechanical': in A. R. Hall (1992), p. 222. • 'unphilosophical to seek for any other': ibid., p. 221. • 'This most elegant fabric of the Sun': ibid. • 'the most exact Care and Diligence': in L. Jardine (1999), p. 14. • 'They may as well ask why St Paul's': ibid., p. 171. • 'nothing has emerged from the Observatory': ibid. • 'scholastic occult quality': in K. Hutchison (1982), p. 250. • 'the pains have been chiefly bestowed': F. Bacon (1605/1620/1944), p. 86. • 'knowledge of stable universal principles': A. C. Crombie (1990), p. 165. • 'the mathematical sciences and arts': ibid., p. 170. • 'the crucial act in the scientific revolution': C. C. Gillispie (1960), *The Edge of Objectivity*, p. 54. Princeton University Press, Princeton. • 'the Copernican-cum-Newtonian world view': T. E. Huff (2011), p. 14. • 'a pioneer of a new natural philosophy': A. C. Crombie (1990), p. 352. • 'the uncritical curiosity of natural magic': ibid., p. 353. • 'In the book of nature things are written': in ibid., p. 356.

Chapter 8

'Perhaps a thousand other worlds that lie': J. Dryden (1692), *Eleanora*, II, 76–9, in M. Nicolson (1956), p. 41. • 'as soon as the art of Flying is Found out': J. Wilkins (1638/1684), p. 138. • 'Her spots thou seest/As clouds': J. Milton (1667), op. cit., Book VIII, 145–7. • 'To whatever region you direct your spyglass': in Galileo (1610/1989), p.62 • 'swarms of small stars placed exceedingly close': ibid. • 'the Great Star of France': in C. M. Coffin (1937), p. 117. • 'we have moreover an excellent and splendid argument': Galileo (1610/1989), p.84 • 'Men that inhere upon Nature only': in M. Nicolson (1935), p. 459. • 'And new Philosophy calls all in doubt': in D. Boorstin, p. 316. • 'as though heav'n suffered earthquakes': in M. Nicolson (1935), p. 456. • 'We have added to the world Virginia': in ibid., p. 452. • 'had tir'd out the *Sun*, and *Moon*, and *Stars*': T. Sprat (1667/1959), p. 416. • 'which may relieve their fellow-creatures': ibid. • 'the Moon, whose Orb/Through Optic

Glass': J. Milton (1667), op. cit., Book I, 287–91. • 'A broad and ample road': ibid., Book VII, 577–81. • 'All this visible world is but an imperceptible point': B. Pascal (1669), 'Man's disproportion', in *The Thoughts, Letters, and Opuscules of Blaise Pascal,* transl. O. W. Wright, p. 158. Houghton, Mifflin & Co., Boston, MA, 1893. • 'pendent World, in bigness as a star': J. Milton (1667), op. cit., Book II, 1052–3. • 'a trunk . . . no bigger than a flute-case': B. Jonson (1621), 'News from the New World Discovered in the Moon', in M. Nicolson (1935), p. 447. • 'Adam and Galileo are both motivated': P. Brantlinger (1972), p. 368. • 'gives no information, sometimes it gives': F. Bacon (1620/1626/1980), p. 24. • 'by reason either of the subtlety': ibid. • 'If anyone should look through': in B. J. Ford (1985), p. 18. • 'I have seen the secret use': in D. Freedberg (2002), p. 101. • 'in and out like a trombone': ibid., p. 102. • 'Giambattista della Porta wrote about': in S. J. Gould (1988). • 'I really am delighted': in D. Freedberg (2002), p. 105. • 'for the utility of this country alone': in D. J. Boorstin (1985), p. 314. • 'you are the first and almost the only person': in C. M. Coffin (1937), p. 117. • 'Oh telescope, instrument of much knowledge': in M. Caspar (1993), p. 201. • 'no one but Galileo has seen': in D. J. Boorstin (1985), p. 315. • 'The spyglass makes us see things': in W. R. Shea & M. Artigas (2003), p. 42. • 'Over a period of two years now': in D. J. Boorstin (1985), p. 316. • 'I discovered another very strange wonder': Galileo, in A. Favaro (ed.) (1890–1909), op. cit., Vol. X, pp. 409–10. Edizione Nazionale, Florence. In A. van Helden (1976), p. 105. • 'Now what is to be said about': in ibid., p. 107. • 'it was an illusion and a fraud': ibid. • 'I need not say anything definite': in S. J. Gould (1988). • 'For Levania seems to its inhabitants': in E. Rosen (ed.) (1967), p. 17. • 'Here is the thesis of the whole *Dream*': ibid., p. 82. • 'provide the fare for us who are being chased': ibid., p. xxi. • 'easily unite and reconcile the *Lunatique Church*': in M. Nicolson (1935), p. 455. • 'I suspect that the author of that impudent satire': in M. Nicolson (1940), p. 268. • 'There will certainly be no lack of human pioneers': in A. Koestler (1984), *The Watershed: A Biography of Johannes Kepler,* p. 195. University Press of America, Lanham, MD. See http://www.physics.emich.edu/aoakes/letter.html. • 'the new Embassador of the Gods': J. Wilkins (1638/1684), p. 59. • 'yet it seem'd such an uncouth Opinion': ibid., p. 13. • 'then why may not another of the Planets': ibid., p. 61. • 'we conjecture that none': ibid., p. 128. • 'by Men, and Beasts, and Plants': ibid., p. 127. • 'perhaps, they had some [sin] of their own': ibid. • 'of a quite Different Nature': ibid., p. 126. • 'tis not likely then that this Opinion': ibid., p. 2. • 'The Negative Authority of Scripture': ibid., p. 20. • 'those things which are obvious': ibid., p. 23. • 'If the Holy Ghost had intended': ibid., p. 21. • 'when Men look for the Grounds': ibid., p. 24. • ''tis not Ingratitude to speak against him': ibid., p. 18. • 'Now its our Advantage': ibid., pp. 31,

59–60. • ''Tis not unlikely but that there may be': R. Hooke (1665/2007), p. 21. • 'seems to be some very fruitful place': ibid., p. 414. • 'I do seriously, and upon good grounds': J. Wilkins (1638/1684), p. 159. • 'This Engine may be contrived': ibid. • 'Notwithstanding all these seeming impossibilities': ibid., pp. 159–60. • 'how did the incredulous World gaze at *Columbus*': ibid., p. 2. • 'by this means, 'tis easily conceivable': ibid., p. 160. • 'It surely stands to reason': in E. Rosen (1965), *Kepler's Conversation with Galileo's Sidereal Messenger*, p. 27. Johnson Reprint Corp., New York. • 'who are likely to bee of a wicked': F. Godwin (1638/2009), p. 113. • 'I was allow'd a Week to satisfy my Curiosity': S. Brunt (1727), p. 165. • 'the moon is a world like ours': C. de Bergerac (1657/1662/1976), p. 19. • 'For it would be as ridiculous': ibid., pp. 23–4. • 'I am convinced that the earth': ibid., pp. 24–5. • 'the majority of men, who only judge': ibid., p. 25. • 'there are perhaps a million things': ibid., pp. 47–8. • 'a single country where even the imagination': ibid., p. 55. • 'This Aristotle whose science': ibid., p. 64. • '"But", they all said to me': ibid., p. 69. • 'Such is what the priests deem': ibid., p. 70. • 'Unhappy country where the symbols': ibid., p. 97. • 'Who has been putting you to sleep': ibid., p. 103. • 'for if there is [no God]': ibid., p. 104. • 'possessed more understanding': ibid., p. 94. • 'is not fire, but the matter to which': ibid., p. 139. • 'here's a universe so large': B. de Fontenelle (1686/1990), p. 63. • 'More than the various tracts': in ibid., p. xxvii. • 'one would almost take him to be a Pagan': in ibid., p. xxviii. • 'What! and leave my Microscope?': in M. Nicolson (1956), p. 188. • 'It would seem': B. de Fontenelle (1686/1990), p. 15. • '"In that case", said the Marquise': ibid., p. 12. • 'Those dark places that are taken for seas': ibid., p. 38. • 'the lunatic asylum of the Universe': ibid., p. 49. • 'The art of flying has only just been born': ibid., p. 34. • 'Did the great seas seem to the Americans': ibid. • 'A Man that is of Copernicus's Opinion': C. Huygens (1698). This and the following extracts are taken from the unpaginated online version.

Chapter 9

'Reason is not to be much trusted': R. Boyle, 'Of Naturall Philosophie', in M. Hunter (2000), *Robert Boyle (1627–91): Scrupulosity and Science*, p. 31. Boydell, Woodbridge. • 'In my judgement the use of mechanical history': F. Bacon (1605/1620/1944), p. 49. • 'when by art and the hand of man': in S. Shapin (1996), p. 85. • 'the fashion to talk as if art': F. Bacon, *Descriptio globi intellectualis*, in J. Spedding, R. L. Ellis & D. D. Heath (eds) (1857–74), *The Works of Francis Bacon*, Vol. 5, p. 506. Longman, London. • 'things artificial differ from things natural': ibid. • 'Are there not enough [experiments]': T.

Hobbes (1661), 'Dialogus physicus', in S. Shapin & S. Schaffer (1985), p. 114.
• 'resistance of the vacuum': Galileo (1638/1991), p. 17. • 'belong to Artificiers
and Handy-Craft-Men': T. White (1665), *An Exclusion of Scepticks From all
Title to Dispute*, p. 73. John Williams, London. In R. F. Jones (1982), *Ancients
and Moderns: A Study of the Rise of the Scientific Movement in Seventeenth-Century
England*, p. 306. Courier Dover, New York. • 'droop and appear sick: R. Boyle
(1660), *New Experiments Physico-Mechanical*, Experiment 41. London. In J. B.
West (2005), p. 37. • 'whether or no, if a Man were rais'd': in ibid., p. 38. •
'a more gross and temperate air': in ibid. • 'take upon me to determine so
difficult': in S. Shapin & S. Schaffer (1985), p. 45. • 'the controversy about a
vacuum rather a metaphysical': ibid. • 'almost totally devoid of air': ibid.,
p. 46. • 'He read the Proposition': J. Aubrey (2000), p. 427. • 'He would now
and then sweare an Oathe': ibid. • 'Every man that hath spare money': in
L. Jardine (1999), p. 50. • 'It is laudable, I confess': in S. Shapin & S. Schaffer
(1985), p. 128. • 'solitary, poor, nasty, brutish': T. Hobbes (1651/1985), p. 186.
• 'There is a great affinity and consent': in J. Henry (2002), p. 121. • 'the
fundamental laws of nature': in ibid., p. 120. • 'to be revenged on an engine':
in S. Shapin & S. Schaffer (1985), p. 173. • 'from what the words signifie': in
ibid., p. 118. • 'the obvious acceptation of the word': in ibid., p. 181. •
'Whatever true cause I told you': in ibid., p. 140. • 'nothing is to be omitted':
R. Hooke (1665/2007), p. 19. • 'much *rigour* in admitting': ibid. • 'a *scrupulous*
choice': ibid., p. 17. • 'esteem the riches of our Philosophical treasure': ibid.
• 'Hopelessly continuing in mistakes': T. Browne (1672), Book, I, Ch. III.
• 'There is scarce any tradition': ibid., Book I, Ch. VIII. • 'we need a man
either very honest': M. Montaigne, 'Of cannibals', in D. M. Frame (transl.)
(1958), *The Complete Essays of Montaigne*, p. 152. Stanford University Press,
Stanford. • 'eyes of an idiot . . . a careful and practised': Galileo, 'Letter to
the Grand Duchess Christina', in S. Drake (transl.) (1957), *Discoveries and
Opinions of Galileo*, p. 196. Anchor Books, New York. • 'in the presence of
an illustrious assembly': R. Boyle (1660), *New Experiments Physico-Mechanical*,
in S. Shapin & S. Schaffer (1985), p. 58. • 'till the whole *Company*': T. Sprat
(1667/1959), p. 99. • 'which of the inventions and experiences': F. Bacon
(1620/1626/1980), p. 80. • 'For to be open, will be to expose every thing':
in M. Hunter & P. B. Wood (1986), 'Towards Solomon's House: rival strate-
gies for reforming the early Royal Society', *History of Science* 24, 49–108, p.
86. • 'I send you here enclosed a Chymical process': in L. Jardine (1999), p.
325. • 'though the *Society* entertains many men': in T. Sprat (1667/1959), p.
67. • 'The problem of what qualifies as knowledge': W. Eamon (1994), p.
351. • 'display new machines': in S. Shapin & S. Schaffer (1985), p. 112. •
'Any institutionalized method': ibid., p. 225.

Chapter 10

'All philosophy is based on two things only': B. de Fontenelle (1686/1990), p. 11. • 'So, naturalists observe, a flea': J. Swift, in J. Mitford (ed.) (1880), *The Poetical Works of Swift*, p. 176. Houghton, Mifflin & Co., Boston, MA. • 'great difficulty before we could come to find': S. Pepys, Diary, 14 August 1664. See http://www.pepysdiary.com/archive/1664/08/14/. • 'not so much as I expect': ibid. • 'It seems as big as a little prawn or shrimp': H. Power (1664/1966), *Experimental Philosophy*, pp. 1–2. Johnson Reprint, New York. • 'A most curious bauble it is': S. Pepys, Diary, 13 August 1664. See http://www.pepysdiary.com/archive/1664/08/13/. • 'the subtilty of the composition of Bodies': R. Hooke (1665/2007), p. 17. • 'the limits of Naturall knowledge': R. Hooke, *Royal Society Classified Papers* 1680–1740, Vol. XX (Hooke Papers), f.78. In S. Inwood (2002), p. 393. • 'from an incredible distance': in A. Crombie (1990), p. 199. • 'glasses and means to see': F. Bacon (1620/1626/1980), p. 76. • 'I have observed many tiny animals': in W. R. Shea & M. Artigas (2003), p. 121. • '"[It] has been made': in D. Freedberg (2002), p. 160. • 'lenticular Glasses of crystal': in M. Nicolson (1956), p. 161. • 'the most ingenious book that ever I read': S. Pepys, Diary, 21 January 1665. See http://www.pepys.info/1665/1665jan.html. • 'new Worlds and *Terra Incognita*'s': R. Hooke (1665/2007), p. 28. • 'commander of astronomy': in M. Kemp (2004), p. 52. • 'mischiefs, and imperfection, mankind has drawn': R. Hooke (1665/2007), p. 15. • 'rectif[y] the operations of the Sense': ibid. • 'the limits, to which our thoughts are confin'd': ibid., p. 16. • '*artificial Organs* to the *natural*': ibid., p. 17. • 'of avoiding *Dogmatizing*': ibid., p. 9. • 'a Land-scape of those things': ibid., p. 307. • 'And the Creature was so greedy': ibid., p. 369. • 'not unworthy of our more serious speculation': ibid., p. 222. • 'there may be as much curiosity': ibid., p. 314. • 'Even in those things': ibid., p. 180. • 'discover'd of what pitifull *bungling scribbles*': ibid., p. 48. • 'like a great splatch of *London* dirt': ibid., p. 49. • 'many thousands of times sharper': ibid., p. 46. • 'The Productions of art are such rude': ibid., p. 56. • 'the most curious works of Art': in L. Jardine (1999), p. 84. • 'be so sottish, as to think all': R. Hooke (1665/2007), p. 300. • 'the hidden and innermost recesses of them': R. Boyle, *The Christian Virtuoso*, in T. Birch (ed.) (1772), op. cit., Vol. 5, p. 516. • 'so incompetent and pitiful a cause': ibid., p. 514. • 'Nothing better corroborates': G. W. Leibniz, 'Reflections on the common concept of justice', in L. Loemker (ed.) (1969), *Leibniz: Philosophical Papers and Letters*, 2nd edn, p. 566. Reidel, Dordrecht. • 'I see more of God': T. Moffett (1634), *Theatre of Insects*, Preface, ed. & transl. T. de Mayherne. Printed in E. Topsell (1658), *The History of Four-Footed Beasts and Serpents*, Vol. 4. London. • 'God is greatest in the least of things': H. Power, 'In Commendation of the Microscope', in T. Bowles (1934), 'Dr Henry

Power's poem on the microscope', *Isis* 21, 71–80, here p. 73. • 'there may be more intelligence': P. Bayle (1702), 'Rorarius', in *Dictionnare historique et critique*, 2nd edn, p. 2604. Rotterdam. • 'Nature is not capricious': B. de Fontenelle, *La République des philosophes, ou Histoire des Ajaoiens* (1768), in L. Daston, 'Description by omission', in J. B. Bender & M. Marrinan (eds) (2005), *Regimes of Description in the Archive of the Eighteenth Century*, p. 24. Stanford University Press, Stanford. • 'Those effects of Bodies': R. Hooke (1665/2007), p. 41. • 'the magnetical effluviums of the loadstone': H. Power (1664), *Experimental Philosophy*, in M. Boas Hall (ed.) (1970), pp. 125–6. • 'we are always, in this new world, intruders': C. Wilson (1995), pp. 62–3. • 'In every *little particle* of matter': R. Hooke (1665/2007), p. 17. • 'which had stood but a few days': A. van Leeuwenhoek, 'Observations concerning Little Animals', in C. Dobell (1960), p. 117. • 'As the size of a full grown animalcule': ibid., p. 123. • 'two tiny limbs near the head': ibid., p. 110. • 'different colours, some being whitish': ibid., pp. 110–11. • 'no more pleasant sight': ibid., p. 133. • 'a person unlearned both in sciences': in ibid., p. 43. • 'I do not gladly suffer contradiction': ibid., p. 42. • 'by all who saw them they were verily believed': ibid., p. 260. • 'the bigger they were magnify'd': R. Hooke (1665/2007), p. 174. • 'for reasons best known to himself': in S. Inwood (2002), p. 261. • 'whether this man has not assisted his eyes': L. Bellini, letter to M. Malpighi, 8 October 1686, in H. Adelmann (1966), *Marcello Malpighi and the Evolution of Embryology*, Vol. 1, p. 506. Cornell University Press, Ithaca. • 'Our imagination loses itself in this thought': N. Andry de Boisregard (1701), *An Account of the Breeding of Worms in Human Bodies*, p. 189. Rhodes & Bell, London. • 'These animalcules were smaller': 'Observationes D. Anthonii Lewenhoeck, de Natis è semine genitali Animalculis', *Philosophical Transactions of the Royal Society* 12, 1040–3, here p. 1041. 1677–8. • 'agrees with the Figure of the *Foetus*': N. Andry de Boisregard (1701), op. cit., p. 178. • 'decay is the beginning of all rebirth': in J. Jacobi (ed.) (1979), *Paracelsus: Selected Writings*, p. 143. Princeton University Press, Princeton. • 'all those things that we suppose': in S. Inwood (2002), p. 75. • 'there might living Creatures be seen': D. Defoe (1722/1969), *A Journal of the Plague Year*, ed. L. Landa, p. 203. Oxford University Press, London. • 'Picture the universe, therefore': C. de Bergerac (1657/1662/1976), p. 82. • 'You mustn't think that we see': B. de Fontenelle (1686/1990), pp. 44–5. • 'Nature has distributed the animals': ibid., p. 45. • 'Your Majesty will find, That there are': N. Grew (1682), *The Anatomy of Plants*, in M. Nicolson (1956), p. 215. • 'the use of Mechanical helps': R. Hooke (1665/2007), p. 28. • 'I am not able to give a solid judgement': M. Cavendish (1666/2001), p. 50. • 'more profitable studies, than in useless': ibid., p. 106. • 'artificial glasses . . . can present the natural figure': ibid., p. 50. • 'appear'd so terrible to her sight': this and subsequent extracts from *The Description of a New World* come from the online version of M. Cavendish (1668). • 'Their skins appeared so coarse and uneven': J. Swift (1727/1985), p. 158. • 'I could distinctly see the limbs': ibid., p. 152. • 'This made me reflect upon the fair skins': ibid., p. 130. • 'Undoubtedly, philosophers are in the right': ibid., p. 125. • 'These, concealed/By the kind art of forming Heaven': in M. Nicolson (1956), p. 211. • 'those creatures, or parts of creatures': M. Cavendish (1666/2001), p. 60. • 'By which it is evident': ibid., p. 99. • 'But oh a stranger thing, this Dame': in A. Battigelli (1998), p. 112. • 'It is exceeding difficult in some Objects': R. Hooke (1665/2007), p. 41. • 'several Plants bound up': N. Grew (1682), op. cit., in M. Nicolson (1956), p. 215. • 'Chain of Ideas coyled up': in S. Inwood (2002), p. 329. • 'These studies seem to me to be gardens': in H. Adelmann (1966), *Marcello Malpighi and the Evolution of Embryology* Vol. 1, pp. 562–3. Cornell University Press, Ithaca. • 'Much the same has been the Fate of Microscopes': R. Hooke (1726), 'Discourse concerning Telescopes and Microscopes', in W. Derham (ed.) (1967), *Philosophical Experiments and Observations*, p. 261. Cass, London. • 'The Works of nature are contrived': J. Locke, 'Some Thoughts concerning Education', in J. L. Axtell (ed.) (1968), *The Educational Writings of John Locke*, p. 301. Cambridge University Press, Cambridge. • 'be in a quite different world': J. Locke (1689), *An Essay Concerning Human Understanding*, Book II, Ch. 23, Sec. 12, in K. P. Winkler (ed.) (1996), *An Essay Concerning Human Understanding* p. 122. Hackett, Indianapolis. • 'there's but little resemblance': J. Glanvill (1661/1970), *The Vanity of Dogmatizing*, p. 211. Georg Olms Verlag, Hildesheim.

Chapter 11

'There is a *Master* of *Arts*': in E. N. Harvey (2005), p. 135. • 'In America there are birds so luminous': B. de Fontenelle (1686/1990), p. 59. • 'into bed with me, and holding it': R. Boyle (1663), 'Short account', in T. Birch (ed.) (1772), op. cit., Vol. 1, p. 799. • 'The annals of seventeenth-century': L. Daston & K. Park (1998), p. 312. • 'Yesterday, when I was about to go to bed': R. Boyle (1672), 'Some observations about shining flesh', in T. Birch (ed.) (1772), op. cit., Vol. 3, pp. 651–5. • 'wonder and delight': ibid. • 'All colours that can be conceived': in J. Gage (1995), *Colour and Culture*, p. 141. Thames & Hudson, London. • 'Some things, indeed, are not seen': Aristotle, *De anima*, Book II, Ch. 7, Sec. 4, in E. N. Harvey (2005), p. 23. • 'the putrified and rotten wood': Pliny, *Natural History*, Book XI, Ch. 37, in E. N. Harvey (2005), p. 28. • 'rare and marvellous plants': C. Gesner (1555), *De Raris et admirandis herbis quae sive quod noctu lucenat . . .* Zurich. In E. N. Harvey (2005), p. 66. • 'it is not the property of fire alone': F. Bacon (1605/1620/1944), p. 131. • 'red, and resembling the common *Glowworms*': *Philosophical Transactions of the Royal*

Society 1, p. 204 (1666). • 'one brisk stroke of a bodkin': R. Boyle, in T. Birch (ed.) (1772), op. cit., Vol. 5, p. 29. • 'like a cannon bullet taken red hot': in J. Emsley (2000), p. 32. See R. E. W. Maddison (1969), *The Life of the Honourable Robert Boyle, FRS*, pp. 158–164. Taylor & Francis, London. • 'a mixture of strangeness, beauty': in M. Cooper & M. Hunter (eds) (2006), p. 112. • 'if a considerably big piece': *Philosophical Transactions of the Royal Society* 12, 867 (1677–8). • 'icicles, which are the element': in A. E. Waite (1992), *Hermetic and Alchemical Writings of Paracelsus the Great*, Part II, p. 19. Alchemical Press, Edmonds, WA. • 'somewhat that belonged to the body': in J. Emsley (2000), p. 34. • 'the resuscitation of plants': J. Ince (1858), 'Ambrose Godfrey Hanckwitz', *Pharmaceutical Journal* 18, 126–30, 157–62, 215–22, here p. 160. See L. Principe (1998), pp. 134–6. • 'a poor young beginner': J. Ince (1858), op. cit., p. 159. • 'were lodged . . . in my own bedchamber': ibid. • 'terrible bawling creature': ibid., p. 160. • 'This is the fruits of meddling': ibid. • 'most rigorously if he allows himself': C. Philipp (1680), letter to Leibniz, 27 March/6 April, in *Gottfried Wilhelm Leibniz Sämtliche Schriften und Briefe* (1923–), ed. Preussische Akademie der Wissenschaften, Ser. 2, Vol. 3, pp. 368–9. • 'he has found his dupes': C. Philipp (1680), letter to Leibniz, 10/20 April, in *Sämtliche Schriften*, op. cit., Ser. 2, Vol. 3, p. 386. • 'the late time of the night': in J. Emsley (2000), p. 40. • *considerable quantity of Man's Urine*': R. Boyle (1694), *Philosophical Transactions of the Royal Society* 17, 583–4, here p. 583. • 'faint blewish light': ibid. • 'to the pleasure of a delightful': in J. V. Golinski (1989), p. 25. • 'We went to the house': in W. H. Quarrell & M. Mare (transl. & eds) (1934), op. cit., pp. 148–9. • 'Chymistry was very *phantastick*': in J. V. Golinski (1989), p. 12. • 'Chymistry doth daily present those': in ibid., p. 25. • 'by their use, not their strangeness': in ibid., p. 24. • 'many *Experiments* are obnoxious to failing': in ibid., p. 34. • 'burned all his clothes': in J. Emsley (2000), p. 20. • 'I could not look upon without': in ibid., p. 42. • 'suspicion of being Prodigies': in J. V. Golinski (1989), p. 31. • 'If we had a mind': in ibid., p. 27. • 'Dining at Mr Pepys'': J. Evelyn (1685), Diary, 13 December, in W. Bray (ed.) (1850), *Diary and Correspondence of John Evelyn F.R.S.*, Vol. 2, p. 245. Henry Colburn, London. • 'There resides in this phosphorus': in J. V. Golinski (1989), p. 20. • 'rotten Wood, rotten Fish': R. Hooke (1665/2007), p. 120. • 'that all kind of *fiery burning Bodies*': ibid., p. 119. • 'That the *Bononian stone* shines': ibid. • 'the Rob of Urine': R. Hooke (1680), in R. Waller (ed.) (1705), *The Posthumous Works of Robert Hooke*, p. 112. Samuel Smith & Benjamin Walford, London. • 'dominated the science of optics': R. S. Westfall (1980), p. 640. • 'those are in error who say': in R. B. Burke (transl.) (1962), op. cit., Vol. 2, p. 609. • 'vague, speculative, and wearisome': C. B. Boyer (1959/1987), *The Rainbow: From Myth to Mathematics*, p. 173. Princeton University Press, Princeton. • 'is such a remarkable phenomenon of nature':

R. Descartes (1637/1965), *Discourse on Method. Optics, Geometry, and Meteorology*, transl. P. J. Olscamp, p. 332. Bobbs-Merrill, Indianapolis. • 'the matter that has given me': in R. L. Lee & A. B. Stewart (2001), p. 182. • 'two refractions and one reflection': R. Descartes (1637/1965), op. cit., p. 334. • 'differently refrangible': I. Newton, *Philosophical Transactions of the Royal Society* 80, 3075–87, here p. 3079 (1671–2). Available at http://www.newton-project.sussex.ac.uk/view/texts/normalized/NATP00006. • 'While Newton certainly did make': A. R. Hall (1992), p. 119. • 'what I shall tell concerning them': in ibid., p. 121. See I. Newton, 'Draft of "A Theory Concerning Lights and Colors"', Additional MS 3970.3, ff. 460–6, Cambridge University Library. Available at http://www.newtonproject.sussex.ac.uk/view/texts/normal-ized/NATP00003. • 'I am purposing . . . to be considered of': in D. Brewster (1855), *Memoirs of the Life, Writings, and Discoveries of Sir Isaac Newton*, Vol. 1, p. 71. Edinburgh. • 'not a little pleased': R. Hooke (1672), in *The History of the Royal Society* 3, 10–15, here p. 10 (1757). Available at http://www.newtonproject.sussex.ac.uk/view/texts/normalized/NATP00005. • 'For all the experiments': ibid., p. 11. • 'To challenge the perfectly uniform simplicity': A. R. Hall (1992), p. 102. • 'The theory of Mr Newton': in H. Turnbull (ed.) (1959), *The Correspondence of Isaac Newton*, Vol. 1, p. 135. Cambridge University Press, Cambridge. • 'knows well it is not for one man': ibid., p. 172. • 'already by that little use I have made': in R. S. Westfall (1980), p. 246. • 'I desire that you procure': in H. Turnbull (ed.)(1959), op. cit., Vol. 1, p. 262. • 'I do justly value': ibid., p. 412. • 'what passes between friends': ibid., p. 416. • 'Perhaps the whole frame of nature': in D. Brewster (1855), op. cit., Vol. 1, p. 392. • 'It is not impossible': I. Newton (1704), *Opticks*, Book II, Part 3, p. 64. Available at http://www.newtonproject.sussex.ac.uk/view/texts/normal-ized/NATP00037. • 'as many Curiosities in its Fabrick': R. Hooke (1665/2007), p. 110. • 'I found that up and down': ibid., p. 111. • 'whether from these causes': ibid., p. 112.

Chapter 12

"Tis below a virtuoso to trouble himself': T. Shadwell (1676/1966), p. 72. • 'for his excellence': S. Butler (c.1670s/1835), 'The Elephant in the Moon', p. 128. • 'I content myself with the speculative part': T. Shadwell (1676/1966), p. 47. • 'so much advanc'd in the art': ibid., pp. 44–5. • 'A sot that has spent two thousand pounds': ibid., p. 22. • 'Damned dogs': in S. Inwood (2002), p. 239. • 'People almost pointed': ibid. • 'subtile, Eagle-ey'd Schoolmen': J. Glanvill (1668), *Plus ultra*, in R. H. Syfret (1950), p. 32. • 'hot and restless head': in ibid., p. 37. • 'that would reduce all learning': in ibid., pp. 41–2. • 'anything more Absurd and Impertinent': in R. J. W. Evans & A. Marr (eds)

(2006), p. 93. • 'hunting after Novelties': in W. E. Houghton (1942) (I), p. 196. • 'what is unusual, far fetch'd': ibid. • 'Through worlds unnumber'd': A. Pope (1732–4/1763), *An Essay on Man*, Epistle I, 21–2, p. 4. A. Millar & J. & R. Tonson, London. • 'He, who through vast immensity': ibid., 23–8, p. 4. • 'The bliss of man': ibid., 189–96, 69–72; pp. 18, 7. • 'There are some men whose heads': T. Addison, *The Tatler*, 26 August 1710, in R. H. Syfret (1950), p. 49. • 'the mind of man . . . is capable': ibid., p. 50. • 'But could experimental philosophers': M. Cavendish (1666/2001), p. 52. • 'has Abandoned the Society of Men': M. Astell (1696), 'The Character of a Virtuoso', in *An Essay in Defense of the Female Sex*, pp. 96–7. London. • 'To what purpose is it': ibid., pp. 102–3. • 'the world more injury than benefit': M. Cavendish (1666/2001), p. 51. • 'The Skin of a Snake bred': T. Browne (1683), *Certain Miscellany Tracts*, Tract XIII: *Musæum clausum*, MS Sloane 1847. Available at http://penelope.uchicago.edu/misctracts/museum.html. • 'An exact account': ibid. • 'the *Royal Society*, by their great': in W. E. Houghton (1942) (I), p. 54. • 'has broken his brains about the nature': T. Shadwell (1676/1966), p. 22. • 'he knew his name so well': ibid., p. 71. • 'What does it concern a man': ibid., p. 69. • 'Choose your air': ibid., p. 103. • 'O yes, I have sent one': ibid., p. 104. • 'We virtuosos never find out': ibid., p. 119. • 'Most certainly the world is very foolish': ibid., p. 110. • 'The endeavours to improve': in R. H. Syfret (1950), p. 58. • 'they envied him': T. Shadwell (1676/1966), p. 43. • 'a thousand other ridiculous Volumes': A. Behn, *The Emperor of the Moon*, I, i, in M. Summers (ed.) (1913), *The Works of Aphra Behn*, Vol. 3. London. Quotes taken here from http://www.gutenberg.org/cache/epub/10039/pg10039.html. • 'lunatic we may call him': ibid. • 'my Master's conjuring for you': ibid. • 'Are not you then the Emperor': ibid., III, iii. • 'To the Danish Agent late': in D. Stimson (1932), 'Ballad of Gresham Colledge', *Isis* 18, 103–17. Available at http://en.wikisource.org/wiki/Ballad_of_Gresham_College. See also A-H. Maehle, 'Literary responses to animal experimentation in seventeenth and eighteenth-century Britain', *Medical History* 34, 27–51 (1990). • 'too much curiosity': A. Behn (1686), *The Lucky Chance*. In B. Benedict (2002), p. 122. • 'Maggots breed in Rotten Cheese': S. Butler (1663–78), *Hudibras*, Part II, Canto III, 318. • 'What is the nat'ral cause why fish': S. Butler (c.1670s/1835), 'A Satire', p. 157. • 'A learned society of late': S. Butler (c.1670s/1835), 'Elephant', 1–16, 24–6, pp. 123–4. • 'A wonder, more unparallel'd': ibid., 123–6, 131–2, pp. 126–7. • 'in the next Transaction': ibid., p. 142. • 'that they would never recant': ibid., 465–6, 469, p. 137. • 'A comet, and without a beard': S. Butler (1663–78), *Hudibras*, Part II, Canto III, 427–8, in M. Nicolson (1956), p. 38. • 'prodigious swarms of flies': S. Butler (c.1670s/1835), 'Elephant', 487–8, p. 137. • 'That those who greedily pursue': ibid., 509–20,

p. 138. • 'a maid in Holland': in T. Birch (1754), op. cit., Vol. 1, p. 53. • 'Some of the questionnaires they sent': R. H. Syfret (1950), p. 47. • 'Whether in the Island of Sambrero': in ibid., p. 48. See 'Travellers' Tales', *Chambers' Edinburgh Journal*, 18 March 1848, p. 179. • 'As there is no lie too great': T. Shadwell (1676/1966), p. 70. • 'by the light of her Beauty': this and subsequent extracts come from M. Cavendish (1668), *Blazing-World*, at http://digital. library.upenn.edu/women/newcastle/blazing/blazing.html. • 'by parading her own allegedly boundless will': A. Battigelli (1998), p. 105. • 'always so wrapped up in cognition': J. Swift (1726/1985), p. 201. • 'cones, cylinders, parallelograms': ibid., p. 203. • 'Although they are dexterous': ibid., p. 205. • 'contrive new rules and methods': ibid., pp. 221–2. • 'such *things* as were necessary': ibid., p. 230. • 'which evidently shows them': ibid., p. 213. • 'The first man I saw was of meagre': ibid., p. 223. • 'hands and clothes daubed': ibid., p 224. • 'His employment since his first coming': ibid. • 'mean and filthy things': in T. Shadwell (1676/1966), Introduction, p. xx. • 'from the highest to the lowest': in J. Swift (1727/1985), Introduction, p. 9. • 'We ought to have a great dread': in R. H. Syfret (1950), pp. 43–4. • 'a few Ignorant and Comical': in ibid., p. 46. • 'the sly insinuations': in ibid., p. 44. • 'Howe far this may deaden the industry': in C. Wilson (1995), p. 33. • '*Wits* and *Railleurs*': T. Sprat (1667/1959), p. 417. • 'of the most fertil subject': ibid. • 'To Whitehall, where in the Duke's chamber': S. Pepys, Diary, 1 February 1664. Available at http://www.pepysdiary.com/archive/1664/02/. In Inwood (2002), p. 59. • 'Odds fish, brother, you are in the right': I. D'Israeli, *Essay on the Royal Society*, Available at http://www.readbookonline.net/ readOnLine/42830/. In R. H. Syfret (1950), p. 45. • 'there are many such tales': related to the author by Jenny Uglow at a talk at the Royal Society, 'Charles II: "Patronus et fundator", 1 October 2011. See J. Uglow (2009). • 'I found myself better disposed': W. King (1699), *A Journey to London*, in M. Lister (1823), *An Account of Paris at the Close of the Seventeenth Century*, p. xiii. J. Rutter, Shaftesbury. • 'By giving footboys leave to interpose': S. Butler (*c*.1670s/1835), 'Elephant', p. 148. • 'those who 'ave purchased': ibid., p. 132. • 'For Truth is too reserved and nice', ibid., p. 131. • ''Tis strange, the miser should his cares': A. Pope (1731), 'Epistle to Burlington', 1–4, in W. Walker (ed.) (n.d.), *The Essential Poetry of Pope*, p. 28. Routledge & Sons, London. • 'in the late Transactions': W. King (1700), *The Transactioneer*, in R. H. Syfret (1950), p. 62. • 'it comes first to fluidity': T. Shadwell (1676/1966), p. 102. • 'From this, Lyndamore, we may learn': in R. H. Syfret (1950), p. 60. • 'When I have perfected it': T. Shadwell (1676/1966), p. 112. • 'By this princes may converse': ibid., p. 113. • 'the reluctance of late-humanist culture': C. Preston (2005), *Thomas Browne and the Writing of Early Modern Science*, p. 156. Cambridge University Press, Cambridge.

Chapter 13

'The Universe is so full of Wonders': H. Baker (1744), *The Microscope Made Easy*, in M. Nicolson (1956), p. 233. • 'For Art and Science': W. Blake, *Jerusalem*, 55:62. • 'looked like snot': M. Hopkin, 'Swimming in syrup is as easy as water', Nature News 20 September 2004, http://www.nature.com/ news/2004/040920/full/news040920-2.html. • 'the forces required to drag sheep': J. T. Harvey et al., *Applied Ergonomics* 33, 523–31 (2002). • 'licentious indulgence, as well as': B. Benedict (2002), p. 37. • 'Let us not too curiously examine': J. Hoefnagel (1592), *Archetypa studiaque patris Georgii Hoefnageli*, in M. Warner (2004), *Fantastic Metamorphoses*, p. 227. Oxford University Press, Oxford. • 'the point of invoking curiosity': N. Kenny (2004), p. 1. • 'a sudden surprise of the soul': R. Descartes (1642/1989), *The Passions of the Soul*, transl. S. H. Voss, p. 56. Hackett, Indianapolis. • 'Wonder was the reward': L. Daston & K. Park (1998), p. 323. • 'Nature is never so wondrous': B. de Fontenelle (1709), *Histoire du renouvellement de l'Académie Royale des Sciences en M.DC. XCIX et les éloges historiques*, p. 21. Pierre de Coup, Amsterdam. • 'The science only adds to the excitement': R. Feynman, interviewed on *Horizon*, BBC, 1981. • 'is a form of perception': M. B. Campbell (1999), p. 5. • 'a disreputable passion': L. Daston & K. Park (1998), pp. 14–15. • 'the Power, Wisdom and Goodness': in *Catholic Encyclopaedia*, http://www.newadvent.org/ cathen/02783b.htm. • 'elite relative': M. B. Campbell (1999), p. 5. • 'Do not all charms fly': Keats, 'Lamia'. E.g. *The Poetical Works of Coleridge, Shelley, and Keats*, Section III, p. 39. A. & W. Galignani, Paris, 1829. • 'I worked on true Baconian principles': in J. Henry (2002), p. 2. • 'the false, the unreal, the surreal': M. B. Campbell (1999), p. 100. • 'In order to be made useful': L. Daston & K. Park (1998), p. 354. • 'the world is not there to delight us': C. Wilson (1995), p. 17. • 'betwixt my eye and the bone': in R. S. Westfall (1980), p. 94. • 'both anxious doubts': R. Boyle, 'The Excellency of Theology, Compared with Natural Philosophy', in T. Birch (ed.) (1744), op. cit., Vol. 3, p. 428. • 'or some other *collection*': in ibid., p. 44. • 'The world was made to be inhabited': T. Browne (1643), *Religio medici*, in H. Fisch (1953), p. 255. • 'without any intention for ends': R. Cudworth (1678/1845), *The True Intellectual System of the Universe*, transl. J. Harrison, Vol. II, p. 612. London. • 'an hypocritical veil': in H. Fisch (1953), p. 262. • 'Not only fantastical philosophy': F. Bacon (1605/1620/1944), p. 328. • 'operose, solicitous and distractious': in H. Fisch (1953), p. 262. • 'artificial nature, which': R. Cudworth (1678/1845), op. cit., p. 594. • 'creative nature . . . pregnant automaton': in H. Fisch (1953), p. 263. • 'particularly those that make up': in ibid., p. 264. • 'the admirable windings of Providence': in H. Fisch (1953), p. 264. • 'Isolated facts soon become uninteresting': A. J. Desmond & J. Moore (1992), *Darwin*, p. 190. Penguin, London. • 'During the twentieth century biology': R.

Weinberg (2010), 'Hypotheses first', *Nature* 464, 678. • 'curiosity-driven approach seems increasingly': A. Zewail (2010), 'Curiouser and curiouser: managing discovery making', *Nature* 468, 347. • 'cannot dictate the products of science': L. Daston (1995), p. 404. • 'view of intelligence as neatly detached': ibid., p. 403. • 'The instant I saw the picture': J. Watson (1969), *The Double Helix*, p. 107. Mentor, New York. • 'the most concentrated solution of happiness': P. McEuen (2011), *Spiral*, p. 66. Dial Press, New York.

Bibliography

J. Aubrey (2000). *Brief Lives*, ed. J. Buchanan-Brown. Penguin, London.

F. Bacon (1605/1620/1944). *Advancement of Learning & Novum Organum*. Willey.

F. Bacon (1620/1626/1980). *The Great Instauration and New Atlantis*. Harlan Davidson, Arlington Heights, IL.

F. Bacon (1824). *The Works of Francis Bacon*, Vol. I. W. Baynes & Son, London

F. Bacon (1857). *The Works of Francis Bacon*, eds J. Spedding, R. L. Ellis & D. D. Heath, Vol. III. Longman, Green, Longman & Roberts, London.

F. Bacon (1858). *The Works of Francis Bacon*, eds J. Spedding, R. L. Ellis & D. D. Heath, Vol. IV. Longman, Green, Longman & Roberts, London.

R. Barbour & C. Preston (eds) (2008). *Sir Thomas Browne: The World Proposed*. Oxford University Press, Oxford.

A. Battigelli (1998). *Margaret Cavendish and the Exiles of the Mind*. University Press of Kentucky, Lexington.

A. Behn (1687). *The Emperor of the Moon*, in M. Summers (ed.) (1913), *The Works of Aphra Behn*, Vol. 3. London. Available at http://www.gutenberg.org/cache/epub/10039/pg10039.html.

M. Ben-Chaim (2004). *Experimental Philosophy and the Birth of Empirical Science: Boyle, Locke, and Newton*. Ashgate, Aldershot.

B. Benedict (2002). *Curiosity: A Cultural History of Early Modern Enquiry*. Chicago University Press, Chicago.

C. de Bergerac (1657/1662/1976). *Other Worlds: The Comical History of the States and Empires of the Moon and Sun*, transl. & ed. G. Strachan. New English Library, London.

M. Boas Hall (ed.) (1970). *Nature and Nature's Laws*. Macmillan, London.

D. J. Boorstin (1985). *The Discoverers*. Vintage, New York.

P. Brantlinger (1972). 'To see new worlds: curiosity in *Paradise Lost*', *Modern Language Quarterly* 33, 355–69.

T. Browne (1672). *Pseudodoxia epidemica*, 6th edn. Available at http://penelope.uchicago.edu/pseudodoxia/.

S. Brunt (1727). *A Voyage to Cacklogallinia*. J. Watson, London. Available at http://www.gutenberg.org/files/16202/16202-h/16202-h.htm.

P. Burke (2000). *A Social History of Knowledge: From Gutenberg to Diderot*. Polity Press, Cambridge.

S. Butler (1663–78). *Hudibras*. Available at http://www.gutenberg.org/cache/epub/4937/pg4937.html.

S. Butler (c.1670s/1835). 'A Satire Upon the Royal Society', in *The Poetical Works of Samuel Butler*, Vol. 2, p. 157. William Pickering, London.

S. Butler (c.1670s/1835). 'The Elephant in the Moon', in *The Poetical Works of Samuel Butler*, Vol. 2, pp. 123–38. William Pickering, London. (Note that this is followed by an alternative version in 'long verse', that is, iambic pentameter.)

M. B. Campbell (1999). *Wonder and Science*. Cornell University Press, Ithaca.

M. Caspar (1993). *Kepler*, transl. & ed. C. D. Hellman. Dover, Mineola, NY.

M. Cavendish (1666/2001). *Observations Upon Experimental Philosophy*, ed. E. O'Neill. Cambridge University Press, Cambridge.

M. Cavendish (1668). *The Description of a New World, Called the Blazing-World*. Maxwell, London. Available at http://digital.library.upenn.edu/women/newcastle/blazing/blazing.html.

S. Clauss (1982). 'John Wilkins' Essay Toward a Real Character: its place in the seventeenth-century episteme', *Journal of the History of Ideas* 43, 531–53.

C. M. Coffin (1937). *John Donne and the New Philosophy*. Columbia University Press, Morningside Heights.

M. Cooper & M. Hunter (eds) (2006). *Robert Hooke: Tercentennial Studies*. Ashgate, Aldershot.

N. Copernicus (1543). *De revolutionibus orbium coelestium*. Transl. in S. Hawking (ed.) (2002), *On the Shoulders of Giants*. Running Press, Philadelphia.

R. Crease (2003). *The Prism and the Pendulum*. Random House, New York.

A. C. Crombie (1969). *Augustine to Galileo Vol. 2: Science in the Later Middle Ages and Early Modern Times 13th-17th Century*. Penguin, Harmondsworth.

A. C. Crombie (1990). *Science, Optics and Music in Medieval and Early Modern Thought*. Hambledon Press, London.

L. Daston (1995). 'Curiosity and early modern science', *Word and Image* 11(4) 391–404.

L. Daston & K. Park (1998). *Wonders and the Order of Nature 1150–1750*. Zone Books, New York.

A. G. Debus (1978). *Man and Nature in the Renaissance*. Cambridge University Press, Cambridge.

A. G. Debus (2006). *The Chemical Promise*. Science History Publications, Sagamore Beach, MA.

G. B. Della Porta (1658/1957). *Natural Magick*. Basic Books, New York.

F. J. Dijksterhuis (2004). *Lenses and Waves: Christiaan Huygens and the Mathematical Science of Optics in the Seventeenth Century*. Kluwer, Dordrecht.

C. Dobell (1960). *Antony van Leeuwenhoek and His 'Little Animals'*. Dover, New York.

M. Doran (1940). 'On Elizabethan "credulity", *Journal of the History of Ideas* 1(2), 151–76.

W. Eamon (1994). *Science and the Secrets of Nature*. Princeton University Press, Princeton.

J. Elsner & R. Cardinall (eds) (1994). *The Cultures of Collecting*. Reaktion, London.

J. Emsley (2000). *The Shocking History of Phosphorus*. Macmillan, London.

R. J. W. Evans (1973). *Rudolf II and His World*. Clarendon Press, Oxford.

R. J. W. Evans & A. Marr (eds) (2006). *Curiosity and Wonder from the Renaissance to the Enlightenment*. Ashgate, Aldershot.

P. Findlen (ed.) (2004). *Athanasius Kircher: The Last Man Who Knew Everything*. Routledge, London.

B. le Bovier de Fontenelle (1686/1990). *Conversations on the Plurality of Worlds*, intr. N. R. Gelbart, transl. H. A. Hargreaves. University of California Press, Berkeley.

B. J. Ford (1985). *Single Lens*. Heinemann, London.

D. Freedberg (2002). *The Eye of the Lynx*. University of Chicago Press, Chicago.

H. Fisch (1953). 'The scientist as priest: a note on Robert Boyle's natural theology', *Isis* 44, 252–65.

Galileo Galilei (1610/1989) *Sidereus nuncius, or The Sidereal Messenger*, ed. & transl. A. van Helden. University of Chicago Press, Chicago.

Galileo Galilei (1633). *Dialogue Concerning the Two Chief World Systems: Ptolemaic and Copernican*. Transl. in S. Hawking (ed.) (2002), *On the Shoulders of Giants*. Running Press, Philadelphia.

Galileo Galilei (1638/1991). *Discourses Concerning Two New Sciences*, transl. H. Crew & A. de Salvio. Prometheus Books, Buffalo, NY.

S. Gaukroger (2001). *Francis Bacon and the Transformation of Early Modern Philosophy*. Cambridge University Press, Cambridge.

S. Gaukroger (2006). *The Emergence of a Scientific Culture*. Oxford University Press, Oxford.

B. Gettelfinger & E. L. Cussler (2004). 'Will humans swim faster or slower in syrup?', *American Institute of Chemical Engineers Journal* 50, 2646–7.

O. Gingerich (1993). *The Eye of Heaven*. American Institute of Physics, Melville, NY.

C. Ginzburg (1989). 'Clues: Roots of an evidential paradigm', in *Clues, Myths and the Historical Method*, transl. J. Tedeschi & A. Tedeschi, pp.96–125. Johns Hopkins University Press, Baltimore.

C. Ginzburg (1990). 'High and low: the theme of forbidden knowledge in the sixteenth and seventeenth centuries', in *Myths, Emblems, Clues*, transl. J. Tedeschi & A. Tedeschi. Hutchinson, London.

J. M. Glide (1970). 'Shadwell and the Royal Society', *Studies in English Literature 1500–1900* 10, 469–90.

F. Godwin (1638/2009). *The Man in the Moone*, ed. W. Poole. Broadview, Peterborough, Ontario.

J. Godwin (1979). *Athanasius Kircher*. Thames & Hudson, London.

J. V. Golinski (1989). 'A noble spectacle: phosphorus and the public culture of science in the early Royal Society', *Isis* 80, 11–39.

P. Gouk (1999). *Music, Science and Natural Magic in Seventeenth-Century England*. Yale University Press, New Haven.

S. J. Gould (1988). 'The sharp-eyed lynx, outfoxed by nature', *Natural History*, May.

A. Grafton (1991). *Defenders of the Text*. Harvard University Press, Cambridge, MA.

A. Grafton (1996). 'The new science and the traditions of humanism', in J. Kraye (ed.), *The Cambridge Companion to Renaissance Humanism*. Cambridge University Press, Cambridge.

E. Grant (1978). 'Aristotelianism and the longevity of the medieval world view,' *History of Science* 16, 93–106.

E. Grant (1996). *The Foundations of Modern Science in the Middle Ages*. Cambridge University Press, Cambridge.

S. Greenblatt (1991). *Marvellous Possessions: The Wonder of the New World*. University of Chicago Press, Chicago.

J. Gribbin (2005). *The Fellowship: The Story of a Revolution*. Penguin, London.

L. Guerrini (2008). 'The "Accademia dei Lincei" and the New World'. Preprint, Max-Planck Institute for the History of Science.

A. R. Hall (1992). *Isaac Newton: Adventurer in Thought*. Cambridge University Press, Cambridge.

T. Hariot (1588). *A Briefe and True Report of the New Found Land of Virginia*. London. Available in facsimile at digitalcommons.unl.edu/cgi/viewcontent.cgi?article=1020&context=etas.

J. Hart (2003). *Columbus, Shakespeare and the Interpretation of the New World*. Palgrave, New York.

E. N. Harvey (2005). *A History of Luminescence: From the Earliest Times to 1900*. Dover, Mineola.

S. Hawking (ed.) (2002). *On the Shoulders of Giants*. Running Press, Philadelphia.

J. Heilbron (2010). 'In retrospect: The celestial message', *Nature* 467, 398–9.

L. Hendrix (1995). 'Of hirsutes and insects: Joris Hoefnagel and the art of the wondrous', *Word and Image* 11(4), 373–90.

J. Henry (2002). *Knowledge is Power*. Icon, Duxford.

J. Henry (2008). 'The fragmentation of Renaissance occultism and the decline of magic', *History of Science* xlvi, 1–48.

M. B. Hesse (1966). 'Hooke's philosophical algebra', *Isis* 57, 67–83.

T. Hobbes (1651/1985). *Leviathan*, ed. C. B. Macpherson. Penguin, London.

R. Holmes (2008). *The Age of Wonder*. Harper Press, London.

R. Hooke (1665/2007). *Micrographia*. BiblioBazaar, Charleston, SC.

K. T. Hoppen (1976). 'The nature of the early Royal Society', *British Journal for the History of Science* 9, 1–24 & 243–273.

W. E. Houghton (1941). 'The History of Trades: its relation to seventeenth-century thought', *Journal of the History of Ideas* 2(1), 33– 60.

W. E. Houghton (1942). 'The English Virtuoso in the Seventeenth Century (I)', *Journal of the History of Ideas* 3, 51–73.

W. E. Houghton (1942). 'The English Virtuoso in the Seventeenth Century (II)', *Journal of the History of Ideas* 3, 190– 219.

T. E. Huff (2011). *Intellectual Curiosity and the Scientific Revolution: A Global Perspective*. Cambridge University Press, Cambridge.

W. H. Huffman (1988). *Robert Fludd and the End of the Renaissance*. Routledge, London.

D. Hume (1748/2008). *An Enquiry Concerning Human Understanding*. Oxford University Press, Oxford.

M. Hunter (1981). *Science and Society in Restoration England*. Cambridge University Press, Cambridge.

M. Hunter (1983). *Elias Ashmole 1617–1692: The Founder of the Ashmolean Museum and His World*. Ashmolean Museum, Oxford.

M. Hunter (1989). *Establishing the New Science*. Boydell & Brewer, Woodbridge, Suffolk.

M. Hunter (1995). *Science and the Shape of Orthodoxy*. Boydell & Brewer, Woodbridge, Suffolk.

M. Hunter (2009). *Boyle: Between God and Science*. Yale University Press, New Haven.

M. Hunter (2011). 'The Royal Society and the decline of magic', *Notes and Records of the Royal Society* 65, 103–119.

K. Hutchison (1982). 'What happened to occult qualities in the Scientific Revolution?', *Isis* 73, 233–253.

C. Huygens (1698). *Cosmotheoros*. Adriaan Moetjens, The Hague. Available at http://www.phys.uu.nl/~huygens/cosmotheoros_en.htm.

O. Impey & A. Macgregor (eds) (1985). *The Origins of Museums: The Cabinet of Curiosities in Sixteenth and Seventeenth Century Europe*. Clarendon Press, Oxford.

S. Inwood (2002). *The Man Who Knew Too Much: The Strange and Inventive Life of Robert Hooke 1635–1703*. Macmillan, London.

J. R. Jacob (1983). *Henry Stubbe, Radical Protestantism and the Early Enlightenment.* Cambridge University Press, Cambridge.

L. Jardine (1974). *Francis Bacon: Discovery and the Art of Discourse.* Cambridge University Press, Cambridge.

L. Jardine & A. Stewart (1998). *Hostage to Fortune: The Troubled Life of Francis Bacon 1561–1626.* Victor Gollancz, London.

L. Jardine (1999). *Ingenious Pursuits: Building the Scientific Revolution.* Little, Brown & Co., London.

F. R. Johnson (1940). 'Gresham College: precursor of the Royal Society', *Journal of the History of Ideas* 1(4), 413–438.

K. Jousten (ed.) (2008). *Handbook of Vacuum Technology.* Wiley-VCH, Weinheim.

T. DeCosta Kaufmann (1993). *The Mastery of Nature.* Princeton University Press, Princeton.

M. Kemp (2004). *Leonardo.* Oxford University Press, Oxford.

N. Kenny (1998). *Curiosity in Early Modern Europe: Word Histories.* Harrassowitz, Wiesbaden.

N. Kenny (2004). *The Uses of Curiosity in Early Modern France and Germany.* Oxford University Press, Oxford.

J. Kepler (1619). *Harmonice mundi.* Transl. in S. Hawking (ed.) (2002), *On the Shoulders of Giants.* Running Press, Philadelphia.

D. Knight (1986). 'Science fiction of the seventeenth century', *Seventeenth-Century* 1, 69–79.

R. L. Lee & A. B. Fraser (2001). *The Rainbow Bridge.* Penn State University Press, University Park.

C. Salaman (ed.) (1975–2010). *The Letters of Marsilio Ficino*, 8 vols. Shepheard-Walwyn, London.

C. F. Lloyd (1929). 'Shadwell and the Virtuosi', *Proceedings of the Modern Language Association* 44, 472–94.

G. E. R. Lloyd (2002). *The Ambitions of Curiosity.* Cambridge University Press, Cambridge.

J. Locke (1690/2008). *Essay Concerning Human Understanding*, ed. P. Phemister. Oxford University Press, Oxford.

P. Marshall (2006). *The Theatre of the World.* Harvill Secker, London.

P. Mauries (2002). *Cabinets of Curiosities.* Thames & Hudson, London.

T. More (1516/1965). *Utopia*, transl. P. Turner. Penguin, Harmondsworth.

W. R. Newman (2005). *Promethean Ambitions: Alchemy and the Quest to Perfect Nature.* University of Chicago Press, Chicago.

I. Newton (1687). *Philosophiae naturalis principia mathematica.* Transl. in S. Hawking (ed.) (2002), *On the Shoulders of Giants.* Running Press, Philadelphia.

C. Nicholls (1997). *The Chemical Theatre.* Akadine Press, Pleasantville, NY.

M. Nicolson (1935). 'The "new astronomy" and the English literary imagination', *Studies in Philology* 32, 428–62.

M. Nicolson (1940). 'Kepler, the Somnium, and John Donne', *Journal of the History of Ideas* 1(3), 259–80.

M. Nicolson (1956). *Science and Imagination*. Cornell University Press, Ithaca.

J. D. North (1989). *The Universal Frame*. Hambledon Press, London.

R. H. Nuttall (1988). 'Pepys and the microscope', *Notes and Records of the Royal Society of London* 42, 133–8.

K. Park & L. Daston (eds) (2006). *The Cambridge History of Science Volume 3: Early Modern Science*. Cambridge University Press, Cambridge.

G. Parry (2011). *The Arch-Conjuror of England: John Dee*. Yale University Press, New Haven.

Plato (4th century BC/1971). *Timaeus and Critias*, transl. D. Lee. Penguin, Harmondsworth.

W. Poole (2005). 'The origins of Francis Godwin's "The Man in the Moone", *Philological Quarterly* 84 (spring).

L. Principe (1998). *The Aspiring Adept: Robert Boyle and His Alchemical Quest*. Princeton University Press, Princeton.

M. Purver (1967). *The Royal Society: Concept and Creation*. Routledge & Kegan Paul, London.

M. Purver & E. J. Bowen (1960). *The Beginning of the Royal Society*. Clarendon Press, Oxford.

P. M. Rattansi (1968). 'The intellectual origins of the Royal Society', *Notes and Records of the Royal Society* 23, 129–43.

I. A. Richter (ed.) (2008). *Leonardo da Vinci: Notebooks*. Oxford University Press, New York.

E. Rosen (ed.) (1967). *Kepler's Somnium*. University of Wisconsin Press, Madison.

P. Rossi (1968). *Francis Bacon: From Magic To Science*, transl. S. Rabinovitch. Routledge & Kegan Paul, London.

T. Shadwell (1676/1966). *The Virtuoso*, eds M. H. Nicolson & D. Rodes. Edward Arnold, London.

S. Shapin (1988). 'The house of experiment in seventeenth-century England', *Isis* 79, 373.

S. Shapin (1996). *The Scientific Revolution*. University of Chicago Press, Chicago.

S. Shapin & S. Schaffer (1985). *Leviathan and the Air-Pump*. Princeton University Press, Princeton.

W. R. Shea & M. Artigas (2003). *Galileo in Rome*. Oxford University Press, Oxford.

W. R. Shea & M. Artigas (2006). *Galileo Observed*. Science History Publications, Sagamore Beach, MA.

P. H. Smith (1994). *The Business of Alchemy: Science and Culture in the Holy Roman Empire*. Princeton University Press, Princeton.

T. Sprat (1667/1959). *History of the Royal Society*, ed. J. I. Cope & H. W. Jones. Routledge & Kegan Paul, London.

T. Sprat (1667/1734). *The History of the Royal Society of London*, 4th edn. J. Knapton et al., London.

C. Swan (1995). 'Ad vivum, near het leven, from the life: defining a mode of representation', *Word and Image* 11(4), 353–72.

J. Swift (1726/1985). *Gulliver's Travels*, ed. P. Dixon & J. Chalker. Penguin, London.

R. H. Syfret (1947). 'The origins of the Royal Society', *Notes and Records of the Royal Society of London* 5, 75–136.

R. H. Syfret (1950). 'Some early critics of the Royal Society', *Notes and Records of the Royal Society of London* 8, 20–64.

L. Thorndike (1923). *History of Magic and Experimental Science*, Book II. Columbia University Press, New York.

L. Thorndike (1923). *History of Magic and Experimental Science*, Book IV. Columbia University Press, New York.

L. Thorndike (1941). *History of Magic and Experimental Science*, Book V. Columbia University Press, New York.

P. Turner (ed.) (1962). *Pliny's Natural History*. Centaur, London.

J. Uglow (2009). *A Gambling Man: Charles II and the Restoration*. Faber & Faber, London.

A. van Helden (1976). 'Saturn and his Anses', *Journal for the History of Astronomy* 5, 105–21.

C. Webster (1970). 'Macaria: Samuel Hartlib and the Great Reformation', *Acta Comeniana* 26, 147–64.

C. Webster (1982). *From Paracelsus to Newton*. Cambridge University Press, Cambridge.

J. B. West (2005). 'Robert Boyle's landmark book of 1660 with the first experiments on rarified air', *Journal of Applied Physiology* 98, 31–9.

R. S. Westfall (1977). *The Construction of Modern Science*. Cambridge University Press, Cambridge.

R. S. Westfall (1980). *Never at Rest: A Biography of Isaac Newton*. Cambridge University Press, Cambridge.

J. Wilkins (1638/1684). *The Discovery of a World in the Moone*, 5th edn. London. Available via European Cultural Heritage Online, http://echo.mpiwg-berlin.mpg.de/.

C. Wilson (1995). *The Invisible World: Early Modern Philosophy and the Invention of the Microscope*. Princeton University Press, Princeton.

B. Woolley (2001). *The Queen's Conjurer*. Henry Holt, New York.

F. Yates (1972). *The Rosicrucian Enlightenment*. Routledge & Kegan Paul, London.

F. Yates (2001). *The Occult Philosophy in the Elizabethan Age*. Routledge, London.

J. P. Zetterberg (1980). 'The mistaking of "the Mathematicks" for magic in Tudor and Stuart England', *Sixteenth Century Journal* 11(1), 83–98.

E. Zilsel (1941). 'The origins of William Gilbert's scientific method', *Journal of the History of Ideas* 2(1), 1– 32.

Image Credits

Index

www.vintage-books.co.uk